T5-DIE-854

21.50
MW
01

Advances in

OPTICAL *and* ELECTRON
MICROSCOPY

Volume 4

Advances in

OPTICAL *and* ELECTRON MICROSCOPY

Volume 4

WITHDRAWN
FAIRFIELD UNIVERSITY
LIBRARY

Edited by

R. BARER

Department of Human Biology and Anatomy,
University of Sheffield, England

AND

V. E. COSSLETT

Department of Physics,
Cavendish Laboratory, University of Cambridge,
England

ACADEMIC PRESS · 1971
LONDON AND NEW YORK

ACADEMIC PRESS INC. (LONDON) LTD
Berkeley Square House
Berkeley Square
London, W.1

U.S. Edition published by
ACADEMIC PRESS INC.
111 Fifth Avenue
New York, New York 10003

Copyright © 1971 by Academic Press Inc. (London) Ltd.

All Rights Reserved

NO PART OF THIS BOOK MAY BE REPRODUCED IN ANY FORM BY PHOTOSTAT, MICROFILM
OR ANY OTHER MEANS WITHOUT WRITTEN PERMISSION FROM THE PUBLISHER

Library of Congress Catalog Card Number: 65–25134
Standard Book Number: 12–029904–6

PRINTED IN GREAT BRITAIN AT
The Whitefriars Press Ltd., London and Tonbridge
England

Contributors

M. E. BARNETT, *Imperial College, London, England* (p. 249).

C. BEADLE, *Metals Research Ltd., Cambridge, England* (p. 361).

H. BETHGE, *Institut für Festkörperphysik und Elektronenmikroskopie der Deutschen, Akademie der Wissenschaften zu Berlin, Halle, Germany* (p. 237).

F. C. BILLINGSLEY, *Jet Propulsion Laboratory, California Institute of Technology, Pasadena, California, U.S.A.* (p. 127).

A. B. BOK, *Philips P.I.T.C.A.E.O., Eindhoven, Netherlands* (p. 161).

H. DE LANG, *Philips Research Laboratories, Waalre, Netherlands* (p. 233).

F. GABLER, *C. Reichert Optische Werke AG, Vienna, Austria* (p. 385).

KARL-JOSEPH HANSZEN, **Department of Physics, University of Arizona, Tucson, Arizona, U.S.A.* (p. 1).

J. HEYDENREICH, *Institut für Festkörperphysik und Elektronenmikroskopie der Deutschen, Akademie der Wissenschaften zu Berlin, Halle, Germany* (p. 237).

K. KROPP, *C. Reichert Optische Werke AG, Vienna, Austria* (p. 385).

J. B. LE POOLE, *Tech. Physische, Dienst, TNO-TH Delft, Netherlands* (p. 161).

A. J. F. METHERELL. *Cavendish Laboratory, Cambridge, England* (p. 263).

ROBERT NATHAN, *Jet Propulsion Laboratory, California Institute of Technology, Pasadena, California, U.S.A.* (p. 85).

J. ROOS, *Technische Physische Dienst, TNO-TH Delft, Netherlands* (p. 161).

* Present address the Physikalisch-Technische Bundesanstalt, Braunschweig, Germany.

114450

Preface

The present volume, by accidents of timing, is weighted strongly on the side of electron microscopy. The articles on this subject deal with aspects which have come rapidly to the fore in the past few years. Experimental as well as theoretical work on phase contrast effects in electron images has developed to the point where their importance for extracting the maximum of information from micrographs has become clear, especially in the attempt to achieve resolution of individual atoms.

In these developments a proper understanding of optical transfer theory as applied to the electron microscope is very necessary. A detailed exposition of it is provided here by Hanszen. The processing of images has been actively pursued in the astronomical field, particularly on photographs of the moon. An outline of how these techniques can be applied to electron micrographs is contributed by Nathan, whilst Billingsley gives a more detailed description of digital processing by computer.

New types of electron microscope are continually being devised and their potential applications explored. Two such instrumental developments are included among these articles: the mirror electron microscope, by Bok and others, and energy analysing electron microscopes by Metherell.

As regards light optical microscopy, two topics of great current interest are dealt with. Gabler and Kropp survey progress in the automation of photomicrography, a particularly timely article in view of the large number of automatic and semi-automatic instruments now appearing on the market. The concluding contribution, by Beadle, concerns the Quantimet computer microscope, which serves as a good example of the trend towards quantitative analysis of microscope images.

We shall continue to endeavour to keep abreast of new advances in the various areas of microscopy. The topics on which articles have been commissioned are listed below. Suggestions for, and offers of, articles on other subjects of current interest are invited.

Reflectance measurements on minerals.
Use of lasers in microscopy.
Remote control microscopy.

New techniques in optical lens design.
Fluorescence microspectrometry.
High resolution electron microscopy.
Display of microprobe analysis results.
Superconducting electron lenses.
Localization of enzymes by electron microscopy.
Lorentz microscopy of magnetic films.
Negative staining for electron microscopy.
Phase contrast electron microscopy.
Environmental cells for the electron microscope.
Scanning electron microscopy in biology.

March, 1971   V. E. COSSLETT
                                                         R. BARER

Contents

CONTRIBUTORS V

PREFACE vii

The Optical Transfer Theory of the Electron Microscope: fundamental principles and applications

KARL-JOSEPH HANSZEN

I. The Problem of Image Formation in the Electron Microscope 1
II. Symbols and Definitions; Numerical Data 3
III. The Illumination 7
IV. The Electron Microscopical Object 9
V. The Formation of the Optical Image 23
VI. Contrast by Means of Phase Shift: Contrast Transfer-Functions for Axial Illumination 46
VII. Contrast by Means of Phase Shift: Contrast Transfer-Functions for Oblique Illumination 65
VIII. The Problem of Point Resolution 67
IX. Aperture Contrast 71
X. Contrast Transfer for Partially Coherent and for Incoherent Illumination 75
XI. Test Objects 79
XII. Historical Remarks 80
Acknowledgements 82
References 82

Image Processing for Electron Microscopy: I. Enhancement Procedures

ROBERT NATHAN

I. Introduction 85
II. Problems of High Resolution............................... 86
III. Computer Image Processing 87
IV. Enhancement of Periodic Images 102
V. High Resolution by Computer Synthesis....................... 113
Acknowledgements 119
References 119
Appendix 120

Image Processing for Electron Microscopy: II. A Digital System

F. C. BILLINGSLEY

I. Introduction 127
II. General Considerations 129
III. Resolution 130
IV. Quantization 135
V. Electronic Cameras 145
VI. Data Recording 149
VII. Computer System 153
References 158

Mirror Electron Microscopy

A. B. BOK, J. B. LE POOLE, J. ROOS, H. DE LANG, H. BETHGE, J. HEYDENREICH, AND M. E. BARNETT

I. Introduction 161
II. Contrast Formation in a Mirror Electron Microscope with Focused Images 167
III. Description and Design of a Mirror Electron Microscope with Focused Images 207
IV. Results and Applications 218
V. Appendix 227
References 259

Energy Analysing and Energy Selecting Electron Microscopes

A. J. F. METHERELL

I. Introduction 263
II. Energy Analysing and Selecting Devices 266
III. Energy Analysing Electron Microscopes 318
IV. Energy Selecting Electron Microscopes 340
Acknowledgements 358
References 359

The Quantimet Image Analysing Computer and its Applications

C. BEADLE

I. Introduction 362
II. Desirable Design Features 364
III. Description of the Quantimet 368
IV. Accuracy of the Quantimet to other Problems 375
V. Application of the Quantimet to other Problems 378
   References 382

Photomicrography and its Automation

F. GABLER AND K. KROPP

I. Introduction 385
II. Some Details on Photomicrographic Technique 386
III. The Automation of Photomicrography 390
   References 413

AUTHOR INDEX 415

SUBJECT INDEX 419

CUMULATIVE LIST OF AUTHORS 423

CUMULATIVE LIST OF TITLES 423

The Optical Transfer Theory of the Electron Microscope: fundamental principles and applications†

KARL-JOSEPH HANSZEN‡

Department of Physics, University of Arizona, Tucson, Arizona, U.S.A.

I.	The Problem of Image Formation in the Electron Microscope							.		1
II.	Symbols and Definitions; Numerical Data.				.	.	.	.		3
III.	The Illumination	.	.	.	.	.	.	.	.	7
IV.	The Electron Microscopical Object	.	.	.	.	.	.		9
V.	The Formation of the Optical Image .	.	.	.		.	.		23
VI.	Contrast by Means of Phase Shift: Contrast Transfer-Functions for Axial Illumination				.	.	.	.	.	46
VII.	Contrast by Means of Phase Shift: Contrast Transfer-Functions for Oblique Illumination				.	.	.	.		65
VIII.	The Problem of Point Resolution		.	.	.	.	.		67
IX.	Aperture Contrast	.	.	.	.	.	.	.	.	71
X.	Contrast Transfer for Partially Coherent and for Incoherent Illumination				.	.	.	.	.	75
XI.	Test Objects	.	.	.	.	.	.	.	.	79
XII.	Historical Remarks .	.	.	.	.	.	.	.		80
	Acknowledgements	.	.	.	.	.	.	.		82
	References	.	.	.	.	.	.	.	.	82

I. THE PROBLEM OF IMAGE FORMATION IN THE ELECTRON MICROSCOPE

IN transmission electron microscopy, the object, the thickness of which is small compared with its diameter, is penetrated by a pencil of electrons that suffer various types of interactions with the atoms of the object. As a result,

(a) Electrons are scattered out of their initial path.

(b) The scattering occurs with or without energy losses.

(c) A total loss of kinetic energy is possible (*electron absorption*).

In the ideal case, an electron microscope would map each point§ of the object plane in a one-to-one correspondence onto an element of the image

---

† This work has been supported by Public Health Service Grant No. GM 11852–5.

‡ Present address: Physikalisch-Technische Bundesanstalt, Braunschweig, Germany.

§ For reasons of brevity we shall henceforth write "object points" instead of "elements of the object plane".

plane. Such an ideal mapping is impossible because of geometrical aberrations and the diffraction error.

In the object plane, an intrinsic diversity of interactions occurs; in the image plane, however, the recording of only one physical parameter is possible, that is the local distribution of the electron current density (e.g. as brightness distribution on a screen or as density distribution in a photographic emulsion). Even if we could eliminate all aberrations from the imaging system, it would be impossible to record the multiplicity of interactions mentioned above on a screen or on a plate. Therefore we could not retrieve all the information about the object structure that is carried by the ray pencil leaving the object. It is certainly possible to predict unequivocally the current distribution in the image if we know the atomic structure of the object and the imaging properties of the microscope. To determine the object structure from the image is impossible, as a rule.

Nevertheless the electron microscopist wants to determine an unknown or partially known object structure from the electron micrograph. This he can only do if the object and the imaging system fulfil certain conditions. The following article deals with these cases.

In order to obtain unambiguous images, the interaction parameters have to be reduced to one. Therefore, either (1) we restrict ourselves to sufficiently thin objects and to electron pencils of sufficiently high energy that the energy losses are of minor importance, or (2) we use a filter lens to remove the decelerated electrons.

Since in the first case there is no absorption, the only interaction is scattering without energy losses. Consequently the interaction can be treated wave-mechanically as Kirchhoff diffraction at the atomic potential distribution in the object, in which case the distribution behaves as a phase structure. Hence the object structure can be derived unambiguously from the image if the electron optical system permits an unequivocal imaging.

In the second case, all electrons eliminated by the filter lens can be treated as absorbed electrons in image formation (Boersch 1947, 1948). In addition to the information about the phase structure, the mononergetic electron pencil used for image formation now provides information about amplitude structure in the object. In order to obtain unambiguous information, the imaging of one of these structures must be prevented. Fortunately we can suppress the phase structure while improving the imaging of the amplitude structure by means of a sufficiently wide illumination aperture. Therefore, an essential requirement for an unequivocal image interpretation is also fulfilled.

Previous studies about electron microscopical resolution have mainly dealt with atomic interactions, while the question of image formation has been of secondary importance. In those studies, approximate distributions of particular model objects were used, and by so doing, the contrast properties and the possibility of resolving atomic distances on the basis of the point resolution theory were discussed. The conclusions drawn in this way are valid only for that particular object and that distance. The imaging properties of the microscope, however, depend strictly on the object under investigation. It is impossible, therefore, to explain the imaging properties and the resolution of the electron microscope merely from the result of those investigations. Being aware of this, we seek to discover objects the imaging properties of which are equal, in the sense that they depend only on the parameters of the imaging system. It will be shown later, that usually in electron microscopy these objects are the weak phase and weak amplitude objects. The imaging properties of these important objects can be described by very simple equations. The characteristic features of the image formation of other objects and the problems arising will also be discussed.

It is the aim of the writer to present the electron microscopical imaging theory in a form which is clear and complete within the limits indicated and comprehensible to the user of the instrument.

II. Symbols and Definitions; Numerical Data

A. *Symbols and Definitions (see Fig. 1)*

The coordinate along the optic axis will be denoted by z, the coordinates in the perpendicular directions by x and y, with the x-coordinate lying in the plane of the drawing. The z-coordinate of the effective electron source (entrance pupil) is denoted by z_Q, the coordinate of the object by z_O, the coordinates of the principal planes of the objective, z_H and $z_{H'}$, that of the exit pupil (briefly called "pupil") by z_P, of the Gaussian image plane by z_B and of the recording plane by z_S. The x and y coordinates in these planes are marked with the same index as the z-coordinate; in the object plane, the index o is omitted for simplicity.

We consider the cardinal elements to be independent of the magnification. Therefore, in a strict sense, our equations are only valid for Newtonian imaging fields. When, as commonly happens, the object is

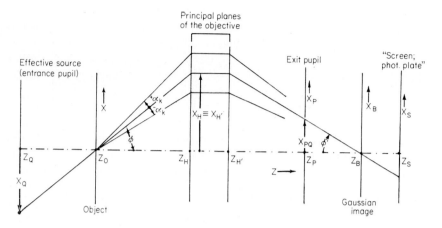

FIG. 1. Explanation of the symbols used. φ is considered to be so small, that the corresponding ray on the image side strikes the axis in the Gaussian image plane.

located within the field, the real cardinal elements are to be used (Glaser 1956, especially p. 194ff†).

We now examine a ray, emerging from a single point of the electron source located off the axis. This ray strikes the axis in the object plane at a small angle φ and in the image plane at a small angle φ', so that $\varphi, \varphi' \ll \dfrac{\pi}{2}$. For $\varphi \to 0$ we have

$$\varphi = \frac{x_Q}{z_Q - z_O} = \frac{x_H}{z_H - z_O}; \qquad \varphi' = \frac{x_{H'}}{z_{H'} - z_B} = \frac{x_H}{z_{H'} - z_B}. \qquad (1)$$

The signs are so determined that the z-coordinate increases downstream along the beam. The angles are positive if a ray, leaving its point of intersection with the axis, is entering in the first quadrant of the coordinate system and they are negative if it enters the fourth quadrant. By means of φ and φ', the magnification M' is also determined. If we denote the object size by x and the Gaussian image size by x_B, then we have‡

$$M' =_{df} \frac{x_B}{x} = \frac{\varphi}{\varphi'}. \qquad (2)$$

The magnification is negative, when the image is inverted.

---

† It must be pointed out, however, that in this case the position of the pupil cannot be obtained by constructing the asymptotes of the rays in the image space. Here, we are confronted with a similar problem to that discussed by Hanszen and Lauer (1965) (see especially Fig. 3 in that article).

‡ $=_{df}$ means "is defined to be equal to".

In order to avoid a repeated use of the magnification factor, we use the reduced image coordinate

$$x' =_{df} \frac{-x_B}{M'}.$$ (3)

The front and rear focal lengths of the objective, assumed to be a single lens, are denoted by f and f', and the coordinates of the focal planes by z_F and $z_{F'}$. Then we have:

$$f = -f'; \quad (f \text{ is negative}).$$ (4)

For the object distance $(z_O - z_H)$ and the image distance $(z_B - z_{H'})$ we have in the case of Gaussian imaging

$$z_O - z_H = f(1 - 1/M'); \qquad z_B - z_{H'} = f'(1 - M').$$ (5)

The image of the electron source located in the plane of the exit pupil has a magnification M'_Q. This quantity can be expressed by

$$\left.\begin{array}{l} M'_Q =_{df} \dfrac{z_P - z_{H'}}{z_Q - z_H}; \\[2mm] z_Q - z_H = f(1 - 1/M'_Q); \\[2mm] z_P - z_{H'} = f'(1 - M'_Q). \end{array}\right\}$$ (6)

For $z_Q \to -\infty$, z_P coincides with $z_{F'}$.

We consider the object as composed of sinusoidal structures having the period lengths ϵ_k. A ray falling upon the object will be diffracted into the angles $\pm a_k$. For small φ and a_k, we have

$$a_k = \pm \lambda/\epsilon_k; \qquad \lambda = \text{electron wavelength}.$$ (7)

In the pupil plane, we make use of the reduced coordinates

$$S =_{df} - \frac{x_P M'}{\lambda f(M' - M'_Q)} = + \frac{x_P M'}{\lambda f'(M' - M'_Q)}.$$ (8)

In that plane, the undiffracted ray has the natural coordinate x_{PQ} and the reduced coordinate

$$Q =_{df} - \frac{x_{PQ} M'}{\lambda f(M' - M'_Q)} = \frac{\varphi}{\lambda}.$$ (9)

In the same plane, the diffracted rays have the distances $(x_{Pk} - x_{PQ})$ with respect to the undiffracted beam. The corresponding reduced coordinates are

$$R_k =_{df} S_k - Q = - \frac{(x_{Pk} - x_{PQ})}{\lambda f} \cdot \frac{M'}{(M' - M'_Q)} = \frac{a_k}{\lambda}.$$ (10)

The physical meaning of the reduced coordinate R_k can be understood by comparing (10) and (7). $|R_k|$ is identical with $1/\epsilon_k$. Also the distance x_{PQ} between the undiffracted ray and the axis is measured on this scale. Likewise, (9) and (8) can be explained.

In the important case $(z_Q - z_H) \to - \infty$, i.e. $M'_Q = 0$, we have

$$
\left.
\begin{aligned}
S(M'_Q = 0) &= - \frac{x_P}{\lambda f}; \\[2mm]
Q(M'_Q = 0) &= - \frac{x_{PQ}}{\lambda f}; \\[2mm]
R_k(M'_Q = 0) &= - \frac{x_{Pk} - x_{PQ}}{\lambda f}.
\end{aligned}
\right\}
\tag{11}
$$

We need other reduced coordinates besides x', Q, R, and S. They will be introduced and discussed later (see Section V).

We also need notations for spherical aberration and defocusing. The spherical aberration can be expressed either by the radius δ_s of the aberration disk in the image plane (transverse aberration)

$$
\delta_s =_{df} M' C_s a^3 = M'^4 C_s a'^3;
\tag{12}
$$

if $C_s = C_s(M') =$ spherical aberration constant and

$$
a' = a/M',
\tag{13}
$$

or by the longitudinal aberration

$$
\Delta z_s =_{df} - M'^4 C_s a'^2 \approx M'^4 C_s \left(\frac{-x_P}{z_B - z_P} \right)^2.
\tag{14}
$$

Here a and a' are the half aperture angles of the ray pencils on the object and image side, respectively; their values are determined by the object aperture stop.

In the electron microscope, defocusing is produced by deviations ΔI of the objective current I and ΔU of the accelerating voltage U from the nominal values. The voltage deviation leads to the wavelength alteration (in non-relativistic approximation)

$$
\Delta \lambda \approx - \frac{\lambda}{2U} \Delta U.
\tag{15}
$$

It is possible to express both deviations by the chromatic aberration of the cardinal elements Δf, Δz_H and $\Delta z_{H'}$ in terms of the chromatic aberration constant C_{ch}; see e.g. Hanszen (1966a). In microscopical optics, i.e. in the case of $|M'| \gg 1$, the defocusing dependence of $\Delta z_{H'}$ can be neglected. The defocusing by ΔI and ΔU can be compensated by a displacement Δz of the object:

$$
\Delta z =_{df} \left(1 - \frac{2}{M'} \right) \Delta f + \Delta z_H =_{df} C_{ch} \left(\frac{\Delta U}{U} - 2 \frac{\Delta I}{I} \right).
\tag{16}
$$

We regard Δz as the characteristic defocusing quantity and call it simply "defocusing". The corresponding displacement Δz_d on the image side is

$$\Delta z_d = z_B - z_S =_{df} \Delta z M'^2. \tag{17}$$

All other quantities are explained, later in this article. New definitions are always denoted by the sign $=_{df}$.

B. *Numerical Values*

In order to be independent of the characteristics of a particular electron microscope, the important results are given in dimensionless form. For illustration and in order to help the user of the instrument, several numerical examples will be given. These are based on the following set of data:

electron wave length: $\lambda = 3.7 \times 10^{-2}$ Å (corresponding to 100 keV electrons);

objective focal length: $f' = -f = -2.7$ mm;

magnifications: $M' = -27$; $M'_Q = -\dfrac{1}{46}$; $\dfrac{M'}{M' - M'_Q} = -1.001$;

spherical aberration constant $C_s = 4$ mm;

chromatic aberration constant $C_{ch} = 2.1$ mm;

object field diameter: $2x_e = 1000$ Å;

distance between object and electron source: $(z_O - z_Q) = 100$ mm;

beam aperture: $2\beta = 2x_e/(z_O - z_Q) \approx 10^{-6}$.

We call an objective having the above characteristics a "normal objective".

III. THE ILLUMINATION

A. *Fundamentals*

The entrance pupil acts as the effective illumination source. It can be either the cross-over of the electron beam formed by the field of the electron gun, the image of this cross-over produced by the condenser or especially, if the cross-over is imaged onto the object plane, the cross-section of the beam in the plane of the condenser stop. Usually there is a finite distance $(z_Q - z_O)$ between the entrance pupil and the object.

The pupil cross-section may have an intricate shape such as a circular disc, centred or not centred with respect to the axis, an annulus, a zone plate, etc. In Fig. 1, the radiation emerging from a single point of the effective source in the direction of the centre of the object plane is characterized by the ray drawn. The total radiation, arriving at each object point, can be calculated by appropriate integration over the radiating area of the source.

We now assume that the electron source is located at $z_Q = -\infty$, and identify the ray mentioned in Fig. 1 with the direction of propagation of a plane wave, having an oblique incident angle φ with the axis. In this simplified description, a "finite size of the source" means that the rays entering the object point form a finite angular interval $\varphi - \Delta\varphi < \varphi < \varphi + \Delta\varphi$. The illumination conditions can be determined by integration over φ, and they are equal in each object point. This is not the case when the source is located at a finite distance from the object. The resulting differences, however, lead only to higher order terms (Born and Wolf 1964, esp. p. 522ff), and can thus be neglected. A detailed discussion of the relation between the position of the exit pupil and that of the source and related problems will be given in Section V. The investigation of the influence of the illumination aperture has recently been completed. In Section X.E. a few examples of a finite aperture are given. Otherwise, our considerations are concerned exclusively with the imaging properties at vanishingly small illumination aperture, i.e. with coherent illumination.

B. *Coherent, Partially Coherent and Incoherent Illumination*

The maximum size of the illumination aperture that can be used for *coherent* illumination is given by the spatial coherence condition:

$$\sin(\Delta\varphi) \approx \Delta\varphi \ll \lambda/4x_e, \tag{18}$$

in which $\Delta\varphi$ is the semi-angular illumination aperture and $2x_e$ the object field diameter.

Example: In high resolution electron microscopy, structures between 5 Å and 2 Å are of main interest. If we desire an image-field diameter of only 500 image points, corresponding to the value $2x_e = 1000$ Å stated in Section IIB, the coherence imposes the severe condition $\Delta\varphi \ll 10^{-5}$: which has rarely ever been fulfilled in practice. In fact, the condition (18) can be weakened. For details see the recent studies by Hanszen and Trepte (1971c) on the influence of a finite illumination aperture on the imaging and the summary in Section XA.

As long as the illumination aperture is small compared to the objective aperture, we call the illumination *partially coherent*. As is shown by light-optical transfer theory, we may call an illumination *incoherent* when the size of the illumination aperture is at least equal to that of the objective aperture.

C. *The Equation for the Incident Wave*

In the sense of wave mechanics, we can express the *wave function* ψ_Q of a wave incident on the object plane under the angle φ at a fixed time by

$$\psi_Q(x) = \Psi \exp\left(- 2\pi i \varphi x / \lambda\right) = \Psi \exp\left(- 2\pi i Q x\right); \quad \Psi = \text{real},$$

(19)

where Ψ is the amplitude of this wave. The exponent gives the phase of the wave at the moment of its entering the object plane. We denote the *intensity* of the incident wave by

$$I_Q(x) =_{df} \psi_Q(x)\, \psi_Q^*(x).$$

(20)

It is proportional to the current density in the object, see e.g. Glaser (1956) Eq. (50.14). If the wave proceeds in the direction of the axis ($\varphi = 0$), the phase is always constant throughout the object plane.

IV. THE ELECTRON MICROSCOPICAL OBJECT

A. *Definition of the Object Transparency*

The amplitude and the phase are locally changed by the interactions of the incoming wave with the object. As has been pointed out before, the wavelength is assumed to be constant. Our investigations are limited to objects causing changes in amplitude and phase which can be described by a factor

$$F(x, y) = A(x, y) \exp\left[i\Phi(x, y)\right]; \quad A < 1 \text{ (real)}; \quad \Phi \text{ real}$$

(21)

multiplying the incoming wave. The amplitude modification of the wave is described by A, the phase modification by Φ. We call $A(x, y)$ the amplitude distribution and $\Phi(x, y)$ the phase distribution in the object. The wave function $\psi(x, y)$ and the intensity $I(x, y)$ as the wave emerges from the object are

$$\psi(x, y) = \psi_Q(x) \cdot F(x, y);$$

(22)

$$I(x, y) = \psi(x, y)\, \psi^*(x, y) = A^2(x, y)\, I_Q(x) =_{df} F_I(x, y) \cdot I_Q(x).$$

(23)

F is called the *object transparency for the wave function*, and $F_I = |F|^2 = A^2$ the *object transparency for the intensity*.

B. *Amplitude Object and Phase Object*

When the object transparency for the wave function is real, i.e. when

$$F(x, y) =_{df} F_A(x, y) = A(x, y),$$

(24)

we call the object an *amplitude object*. When the absolute value of the transparency is constant, and only its phase varies, i.e.

$$F(x, y) =_{df} F_\Phi(x, y) = \exp\left[i\Phi(x, y)\right]$$

(25)

we call the object a *phase object*. We now explain these special cases in detail.

A pure phase object is an idealization. Proceeding from a thick specimen to a thin one, one observes that $A(x, y)$ approaches unity faster than $\Phi(x, y)$ approaches zero in the whole object field. The amplitude modulation $A(x, y)$ therefore becomes imperceptible when the phase modulation is still observable. For this reason, thin objects are, to a good approximation, pure phase objects. When the stricter condition $A(x, y) = 1; \Phi \ll 1$ is fulfilled, we are dealing with *weak phase objects*. Many objects do indeed behave as weak phase objects.

A pure amplitude object is also an idealization. Only thick, unsupported, opaque objects could be in a close approximation pure amplitude objects. In this case, the image would appear as a silhouette, i.e. $\langle A \rangle = \langle 0;1 \rangle$. Therefore, these objects would always be strong amplitude objects. Pure amplitude objects that are weak do not exist under natural conditions.

C. *Fourier Representation of the Object Transparency*

In order to simplify the notation, we limit our considerations to one-dimensional objects. It is known from light optics (see Hopkins 1962 especially p. 485, and Hauser 1962, especially p. 138), that this limitation is possible in transfer theory without the loss of any general features. Minor misinterpretations which may occur when extending our considerations to two-dimensional objects will be dealt with in Section VM.

There are two significant possibilities for dividing the object transparency into independent contributions:

(1) Decomposition of the transparency into "object points", using the delta function:

$$F_O(x) = \int\limits_{-\infty}^{+\infty} F_O(x_i)\, \delta(x - x_i)\, dx_i \qquad (26)$$

in which $F_O(x_i)$ is the weight-function for the δ-sources.

(2) Disintegration of the real and imaginary parts of the transparency into sinusoidal gratings with the period lengths (grating constants) $\epsilon = 1/R$, i.e. into a Fourier integral. This disintegration is appropriate for describing image formation by the wave concept.

The period lengths and therefore also their reciprocals R, which are called spatial frequencies, are positive definite quantities. Because of the axial symmetry of the electron lenses, it is appropriate to expand in terms of cosines.

First, the position of the elementary gratings with respect to the axis intersection $x = 0$ has to be established. This is done by means of the term $\xi(R)$—the so-called lateral phase—in the argument of the cosine, see eq. (28). The lateral phase is related to the coordinate Δx of the first maximum on the side with negative coordinates by the expression:

$$\Delta x(R) =_{df} - \xi(R)/2\pi R; \qquad 0 \leq \xi(R) < 2\pi \qquad (27)$$

Δx is the *lateral displacement*.

To begin with, we give the formal mathematics of the Fourier expansion. The *physical* interpretation of these results is given in Sections IVD to IVL.

The Fourier integral for the object transparency of the wave function is

$$F(x) =_{df} \int_0^\infty \{\tilde{a}_{re}(R) \cos [2\pi Rx + \xi_{re}(R)] +$$
$$+ i\tilde{a}_{im}(R) \cos [2\pi Rx + \xi_{im}(R)]\}dR. \qquad (28)$$

Here, the coefficients $\tilde{a}_{re}$ and $\tilde{a}_{im}$ of the expansion, as well as ξ_{re} and ξ_{im}, are real quantities. In order to normalize, we put:

$$\tilde{a}_{re}(R = 0) = 2 \times 2x_e; \text{ where } 2x_e = \text{object field, cf (42) and (85)};$$
$$\tilde{a}_{im}(R = 0) = 0; \qquad \xi_{re}(R = 0) = 0. \qquad (29)$$

It is convenient to write eq. (28) in complex notation:

$$F(x) = \int_0^\infty \left\{ \frac{\tilde{a}_{re}(R)}{2} (\exp \{i[2\pi Rx + \xi_{re}(R)]\} + \right.$$
$$+ \exp \{- i[2\pi Rx + \xi_{re}(R)]\}) +$$
$$+ i \frac{\tilde{a}_{im}(R)}{2} (\exp \{i[2\pi Rx + \xi_{im}(R)]\} +$$
$$\left. + \exp \{- i[2\pi Rx + \xi_{im}(R)]\}) \right\} dR. \qquad (30)$$

Formally admitting negative values of R, we obtain

$$F(x) = 1 + \int_{-\infty}^{+\infty} \frac{\tilde{a}_{re}(R \neq 0)}{2} \exp \{i[2\pi Rx + \xi_{re}]\} dR +$$
$$+ i \int_{-\infty}^{+\infty} \frac{\tilde{a}_{im}(R \neq 0)}{2} \exp \{i[2\pi Rx + \xi_{im}]\} dR. \qquad (31)$$

This is valid if

$$\tilde{a}_{re}(R) = \tilde{a}_{re}(-R); \qquad \tilde{a}_{im}(R) = \tilde{a}_{im}(-R);$$
$$\text{i.e. } \tilde{a}_{re}, \tilde{a}_{im} \text{ are even functions}; \qquad (32)$$

$$\xi_{re}(R) = - \xi_{re}(- R); \qquad \xi_{im}(R) = - \xi_{im}(- R);$$
$$\text{i.e. } \xi_{re}, \xi_{im} \text{ are odd functions}. \qquad (33)$$

When the notations of eq. (I.1) in Table I are used (31) becomes

$$F(x) = 1 + \int_{-\infty}^{+\infty} \tilde{F}(R \neq 0) \exp{(2\pi i R x)}\, dR. \tag{34}$$

In a similar manner, the Fourier integral for the transparency of the intensity, as defined in (23), can be written:

$$F_I(x) =_{df} 1 + \int_{0}^{\infty} \tilde{A}_I(R \neq 0) \cos{[2\pi R x + \xi_I(R)]}\, dR$$

$$= \int_{-\infty}^{+\infty} |\tilde{F}_I(R \neq 0)| \exp{[i\xi_I(R)]} \exp{(2\pi i R x)}\, dR. \tag{35}$$

Here, we have for the real functions $|\tilde{F}_I(R)|$ and $\xi_I(R)$:

$$|\tilde{F}_I(R)| = |\tilde{F}_I(-R)|; \qquad \xi_I(R) = -\xi_I(-R). \tag{36}$$

Equation (35) has this simple form, because $F_I(x)$ is real.

D. *Diffraction at the Object*

As diffraction theory shows (Lohmann and Wegener, 1955), the coordinates $\pm R$ in the Fourier space, introduced in (31) to (34), correspond to the angles $a = \pm \lambda R$ at which a plane wave is diffracted at a plane elementary grating having a spatial frequency R; see (7). Knowing this, the formal extension of R toward negative values becomes clear. Thus, if we interpret $R = a/\lambda$ as *reduced angular coordinate*, the diffraction spectrum behind the object is given by $\tilde{F}(R)$ in (34) for $-\infty < R < +\infty$. If, however, we consider R as a spatial frequency, the frequency representation of both the real part and the imaginary part of the transparency is given by (28) for $0 < R < +\infty$. Here the Fourier coefficients are subject to the symmetries (32) and (33).

Example: An amplitude object, the transparency of which is modulated only by the spatial frequency R_1.

Fourier representation according to (28):

$$F(x) = 1 + \tilde{a}_1 \cos{(2\pi R_1 x - \pi/2)} = 1 + \tilde{a}_1 \sin{2\pi R_1 x}. \tag{37}$$

Fourier representation according to (34)

$$F(x) = 1 - i\frac{\tilde{a}_1}{2} \exp(2\pi i R_1 x) + i\frac{\tilde{a}_1}{2} \exp(-2\pi i R_1 x) \tag{38}$$

Because of this, the wave field behind the object consists only of the following partial waves:

1. The undiffracted wave with the relative amplitude 1.

2. Two waves, diffracted into the angles $a = \pm R_1 \lambda$, having the relative amplitude $\tilde{a}_1/2$ and the phases $\pm \pi/2$ with respect to the undiffracted wave.

TABLE I

Relations between the Fourier-Coefficients

General relations between the complex Fourier coefficients in Equations (28)–(36):

(I.1)

$$\tilde{F}(R = 0) =_{df} |\tilde{F}| \exp(i\theta) =_{df} \frac{\tilde{a}_{re}}{2} \exp(i\xi_{re}) + i\,\frac{\tilde{a}_{im}}{2} \exp(i\xi_{im});$$

$$\tilde{F}(R = 0) =_{df} \tilde{a}_{re}(R = 0)/2 = 2x_e;$$

$$|\tilde{F}(R \neq 0)|^2 = \tfrac{1}{4}(\tilde{a}_{re}\cos\xi_{re} - \tilde{a}_{im}\sin\xi_{im})^2 + \tfrac{1}{4}(\tilde{a}_{re}\sin\xi_{re} + \tilde{a}_{im}\cos\xi_{im})^2;$$

$$\cos\theta\,(R \neq 0) = \tfrac{1}{2}(\tilde{a}_{re}\cos\xi_{re} - \tilde{a}_{im}\sin\xi_{im})/|\tilde{F}(R \neq 0)|$$

Both $\tilde{F}(R)$ and $\theta(R)$ have no symmetry with respect to $R = 0$

	Coefficients of the mathematical Fourier expansion (28):		Relations replacing (I.1) when *weak objects* are present; see (57)—(61):		Coefficients resulting from the empirical division of the object in phase and amplitude components; see (57)–(61):
(I.2)	$\tilde{a}_{re}(\pm R)$	=	$2|\tilde{F}_A(\pm R)|$	=	$\tilde{A}(+R)$
(I.3)	$\tilde{a}_{im}(\pm R)$	=	$2|\tilde{F}_\Phi(\pm R)|$	=	$\tilde{\Phi}(+R)$
(I.4)	$\xi_{re}(+R) = -\xi_{re}(-R)$	=	$\theta_A(+R) = -\theta_A(-R)$	=	$\xi_A(+R)$
(I.5)	$\xi_{im}(+R) = -\xi_{im}(-R)$	=	$\theta_\Phi(+R) = -\theta_\Phi(-R)$	=	$\xi_\Phi(+R)$
(I.6)			$\tilde{A}(R = 0) = 2 \times 2x_e;\ \tilde{\Phi}(R = 0) = 0;$ (Normalization see (29))		

We now know the Fourier description of the object transparency and its relation to the diffraction spectrum, but the applicability of the concept "object transparency" is still uncertain. Furthermore we do not know how the amplitude and phase components of the object can be read from the transparency. Information on these details will be given in the next sections.

E. *The Influence of the Object Thickness (a purely geometrical consideration)*

Equation (34) is equivalent to physical diffraction only for objects of vanishing thickness. When thick objects are present, their z-dimension can be divided into thickness-elements Δz, and the diffraction of each partial wave at each consecutive element must be considered, for details and further references see Jeschke and Niedrig (1970). Within the wave-field emerging from the object, there will be partial waves in each direction which originate from different elements Δz. These waves may weaken one another by interference.

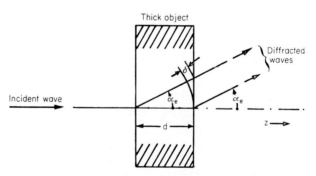

Fig. 2. Optical path difference δ between two diffracted waves, originating from front and rear of a thick object.

We call an object thin, if the phase differences of the waves diffracted at the front and at the rear of the object by the highest spatial frequency are much smaller than $\lambda/2$, so that there is no extinction for any of the diffracted waves. If d is the object thickness, the optical path-difference between the two waves is, according to Fig. 2:

$$\delta = d\alpha_e^2/2 \tag{39}$$

According to (7) and (10), the resulting thickness limitation is

$$d \ll \frac{1}{\lambda R_e^2} = \frac{\epsilon_e^2}{\lambda} \tag{40}$$

Example: If we are interested in image structures of the size of 2 Å, with the numerical values given in Section IIB, we obtain the thickness limitation $d \ll 100$ Å.

F. *Inadequacy of the Concept "transparency" for Thick Objects*

In (21), the transparency was introduced as a function dependent on position (x, y) only, and in particular, independent of the direction of the incident wave. If the illumination is at an angle to the axis, the wave field behind the object is also inclined without other change in the diffracted waves (Morgenstern, 1965), as is shown in Fig. 3a and b. This is undoubtedly true for diffraction in weak objects. Nor is there any objection to applying (21) to thick, amorphous objects, if the permissible thickness as a function of the angle of incidence is carefully considered according to (40).

Thick crystalline objects raise serious difficulties, however, because diffracted waves appear only at discrete incident angles (Bragg angles); see Fig. 3c and d. Thus the transparency should be described by a function of both the object coordinate and the angular position of the lattice planes, with respect to the incident wave. In order to avoid complications of this type, we restrict our considerations to objects, the thickness d of which is small compared to the extinction thickness d_{ex}:

$$d_{ex} \approx \lambda \frac{U}{\Delta U_h} \; ; \tag{41}$$

$U =$ acceleration voltage;

$\Delta U_h =$ Fourier coefficient of the crystal potential in the direction of the penetrating beam;

$\lambda = \lambda(U).$

Example: According to Komoda (1964), the extinction thickness, valid for diffraction at the (111)-planes of a gold crystal is 161 Å. According to transfer theory, therefore, the thickness restrictions for crystalline objects do not seem to be more severe than those for amorphous ones.

G. *The Object Modulation*

The undiffracted wave $R = 0$ contains the information about the average intensity (background). The diffracted waves $R \neq 0$ carry the essential information about the object structure, i.e. about the amplitude and the phase components of each spatial frequency, and the lateral phases of these frequencies. Some new concepts will now be introduced, enabling us to describe the optical transfer of the amplitude and phase components quantitatively in terms of the spatial frequency.

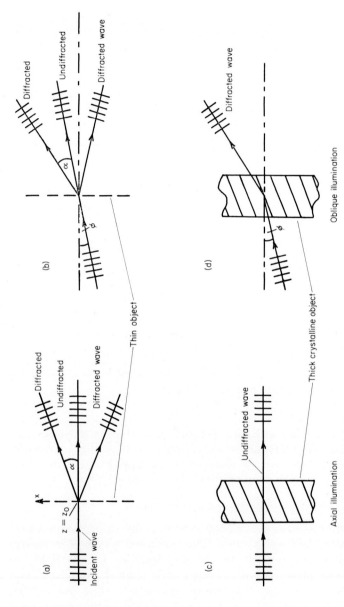

Fig. 3 Electron diffraction at a crystalline object, with incident waves of different inclination to the axis. (a) and (b): thin object; (c) and (d): thick object. (Morgenstern, 1965).

In the object, both an amplitude distribution $A(x)$ and a phase distribution $\Phi(x)$ are present, as a rule. It is now interesting to find out which spatial frequencies are present in the amplitude and which ones in the phase distribution. In other words, we are looking for the Fourier integrals

$$A(x) = \frac{\tilde{A}(R = 0)}{2 \times 2x_e} + \int_0^\infty \tilde{A}(R \neq 0) \cos [2\pi Rx + \xi_A(R)]dR;$$

$$\tilde{A}(R) = \text{real}; \; \tilde{A}(R = 0) = 2\int_{-\infty}^{+\infty} A(x)dx \qquad (42)$$

$$\Phi(x) = \int_0^\infty \tilde{\Phi}(R \neq 0) \cos [2\pi Rx + \xi_\Phi(R)]dR; \qquad \tilde{\Phi}(R) = \text{real}. \qquad (43)$$

For completeness we give a similar expression for the intensity transparency:

$$F_I(x) = A^2(x) = \frac{\tilde{A}_I(R = 0)}{2 \times 2x_e} + \int_0^\infty \tilde{A}_I(R \neq 0) \cos [2\pi Rx + \xi_I(R)]dR;$$

$$\tilde{A}_I(R) = \text{real}; \; \tilde{A}_I(R = 0) \; 2\int_{-\infty}^{+\infty} F_I(x)dx. \qquad (44)$$

Because the x-independent term has a special meaning, it has been taken outside the integral. In the phase expansion, (43), it was put equal to zero. The strength of each spatial frequency $R > 0$ with respect to the background and the existing lateral phases will be described by the following quantities:

Modulation of the absolute value of the wave function in the object:

$$K_A(R) =_{df} \frac{\tilde{A}(R)}{\tilde{A}(R = 0)} \exp [i\xi_A(R)]; \qquad (45)$$

Phase modulation in the object:

$$K_\Phi(R) =_{df} \frac{\tilde{\Phi}(R)}{1} \exp [i\xi_\Phi(R)]; \qquad (46)$$

Intensity modulation in the object:

$$K_I(R) =_{df} \frac{\tilde{A}_I(R)}{\tilde{A}_I(R = 0)} \exp [i\xi_I(R)]. \qquad (47)$$

The phase distribution is given in an angular scale. Since there is no constant term in (43), the reference quantity in (46) can be prescribed arbitrarily. We use the radian as our angular unit. For simplicity we normalize:

$$\frac{\tilde{A}(R = 0)}{2 \times 2x_e} = 1; \qquad \frac{\tilde{A}_I(R = 0)}{2 \times 2x_e} = 1. \qquad (48)$$

It is possible to describe all objects by two of the above given equations, i.e. by (45) and (46), or by (47) and (46). For describing a pure amplitude object we need (45) or (47) only, and for a pure phase object, only (46) is necessary.

Equations (45) to (48) give a description of the object based on the empirical separation into amplitude and phase components, while the description in equations (28) to (36) is actually related to the symmetry conditions in the diffraction pattern. The relationship between these two descriptions will be given in the next section.

H. *The Strong Phase Object*

The separation of the object transparency into amplitude and phase components according to (45) to (46) was a purely phenomenological one, and is not directly related to the object diffraction, described by (34). In the special case of a pure amplitude object, (42) is identical with the real part of (28). In the case of pure phase objects, however, there are differences between (43) and the imaginary part of (28), as explained by the following example†:

Example: A pure phase object, the phase of which is modulated by one spatial frequency only (see Hanszen et al., 1963). The transparency for the wave function of this object is given by

$$F(x) = \exp(i\tilde{a} \cos 2\pi R_1 x); \text{ with } \tilde{a} = 2\pi \varDelta n d/\lambda, \tag{49}$$

where (in non-relativistic approximation)

$$\varDelta n = \frac{1}{2} \frac{\varDelta U_m}{U} \tag{50}$$

is the deviation of the refractive index (due to the "mean inner potential" $\varDelta U_m$) and d the object thickness. The Fourier representation of $F(x)$ is

$$F(x) = \sum_{k=-\infty}^{+\infty} \tilde{F}_k \exp(2\pi i k R_1 x), \tag{51}$$

where the Fourier coefficients

$$\tilde{F}_k = \exp(ik\pi/2) \cdot J_k(\tilde{a}) \tag{52}$$

contain the Bessel-functions $J_k(\tilde{a})$. Although there is only one spatial frequency in the object, there is a sequence of diffraction orders k behind the object on each side.

The unequivocal correlation between *one* spatial frequency and *two* diffracted waves which was seen in the case of an amplitude object (see (37) and (38)) is therefore, not generally valid for phase objects. Yet this correlation is decisive for the possibility of establishing a transfer theory. Therefore, it is impossible to cover all phase objects by the

† Similar conditions exist in the Fourier expansion of the object intensity $I(x)$. Each spatial frequency in the intensity transparency has a spectrum with numerous diffraction orders of the wave function $|\psi| = \sqrt{I(x)}$. In the case of a coherently illuminated *amplitude object*, it is thus not expedient to expand the transparency of the intensity according to Fourier, although it is expedient for other reasons to use $I(x)$, which can be directly observed.

theory. On the contrary, we have to restrict our theory to objects whose higher diffraction orders are negligible. That holds true, indeed, for *weak objects*: For $\tilde{a} \to 0$, the first terms in the expansion of the Bessel functions are†:

$$\left.\begin{array}{l} J_o(\tilde{a}) = 1 - \dfrac{\tilde{a}^2}{4} + \dots ; \\[2ex] J_1(\tilde{a}) = \dfrac{\tilde{a}}{2} - \dots ; \\[2ex] J_2(\tilde{a}) = \dfrac{\tilde{a}^2}{8} - \dots ; \\[2ex] J_3(\tilde{a}) = 0 \dots . \end{array}\right\} \qquad (53)$$

Example: An object may be called weak, if

$$J_2(\tilde{a})/J_1(\tilde{a}) \leqq 0 \cdot 2; \text{ i.e. } \tilde{a} \leqq 0 \cdot 8. \qquad (54)$$

According to (49) and (50) the object thickness is restricted to

$$d \leqq \frac{0 \cdot 8 \lambda(U) \cdot U}{\pi \Delta U_m}; \, \Delta U_m = \text{mean inner potential.} \qquad (55)$$

Except for the slightly different meaning of ΔU, this value differs from that given by (41) only by a factor $\approx 0 \cdot 25$. The limitations of thickness as presented in the previous section are therefore correct as to the order of magnitude. For example, the maximum permitted thickness of carbon specimens ($\Delta U_m \approx 6\text{V}$) is 150 Å.

It must be pointed out that the above estimate is based on the numerical value of the inner potential, i.e. the average of the local potential distribution in matter. In the imaging of atoms the real potential, which may differ locally much more from the anode potential than the inner potential, must be taken into account. For such cases the above theory may not be sufficient. For treating problems of this kind see Hauser (1962).

I. *The Weak Phase Object*

A weak phase object is defined by $\Phi(x, y) \ll 1$. Then, (25) can be evaluated as a series. Thus we obtain for the one dimensional object

$$F_\Phi(x) = 1 + i\Phi(x) - \tfrac{1}{2}\Phi^2(x) + \dots \qquad (56)$$

We truncate this series after the linear term and expand $F_\Phi(x)$ according to (34) with $\tilde{F}(R) \equiv i\tilde{F}_\Phi(R) = i|\tilde{F}_\Phi(R)| \exp [i\theta_\Phi(R)]$; where $\tilde{F}(R)$ is defined in Table I, eqn. (I.1), and obtain

$$F_\Phi(x) = 1 + i \int_{-\infty}^{+\infty} |\tilde{F}_\Phi(R \neq 0)| \exp \{i[2\pi Rx + \theta_\Phi(R)]\} \, dR; \qquad (57)$$

† Details about the spectra of strong phase objects can be found in the article of Nath (1939), esp. in Fig. 10.

or, we expand $\Phi(x)$ in (56) according to (43) and obtain

$$F_\Phi(x) = 1 + i \int_0^{+\infty} \tilde{\Phi}(R \neq 0) \cos\left[2\pi Rx + \xi_\Phi(R)\right] dR. \tag{58}$$

Since the integral is real, $\tilde{F}_\Phi(R) = \tilde{F}_\Phi^*(-R)$. Comparing the coefficients of both equations leads to eqns. (I.3, I.5) in Table I, which show the connection between the coefficients $\tilde{F}_\Phi$; $\tilde{\Phi}$ and θ_Φ; ξ_Φ.

In this special case the integral (58) is identical with the imaginary part of (28) if the coefficients are identified as shown in (29) and (I.5, Table I). This means that the phase modulation *of a weak phase object* affects only the imaginary part of the transparency for the wave function.

These facts are illustrated in Fig. 4. In Fig. 4a, a strong phase object can be seen. Both the imaginary part (containing the odd Bessel functions) and the real part (containing the even Bessel functions) are modulated. In this case, the general Fourier transform (28) is to be used. Figure 4b shows a weak phase object. The cylindrical surface, containing the whole set of the $F_\Phi(x)$ values, can be approximated by the tangential plane. This means that (28) contracts to (58).

J. *The Strong and the Weak Amplitude Object*

The expansion of (24) according to (34) and (42); (48), i.e.

$$F_A(x) = 1 + \int_{-\infty}^{+\infty} |\tilde{F}_A(R \neq 0)| \exp\left\{i[2\pi Rx + \theta_A(R)]\right\} dR;$$

with $\tilde{F}(R) \equiv \tilde{F}_A(R) = |\tilde{F}_A(R)| \exp\left[i\theta_A(R)\right] = \tilde{F}_A^*(-R)$, $\tag{59}$

$$F_A(x) = 1 + \int_0^{+\infty} \tilde{A}(R \neq 0) \cos\left[2\pi Rx + \xi_A(R)\right] dR \tag{60}$$

does not depend on the object strength. Comparing both equations, we obtain a connection between the coefficients $\tilde{F}_A$; $\tilde{A}$ and ξ_A; θ_A, given in eqns. I.2 and I.4 of Table I. In this case, (60) is identical with the real part of (28), if the coefficients are identified as in (29) and (I.4, Table I). Therefore the amplitude modulation of any pure amplitude object has a real transparency.

Because strong amplitude objects present other difficulties, as shown in Section VD, we restrict our further considerations to weak amplitude objects. They are defined by $\tilde{A}(R \neq 0) \ll 1$. Examples of strong and weak amplitude objects are given in Fig. 4c and 4d.

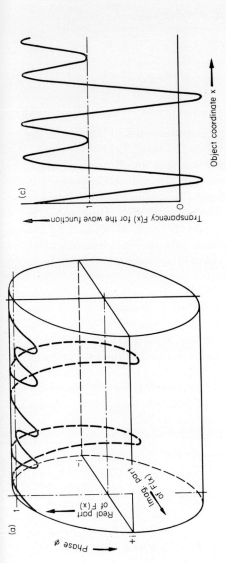

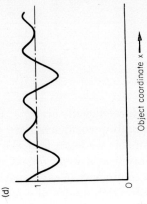

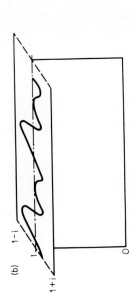

FIG. 4. The object transparency $F(x)$ for the wave function. The object has only two spatial frequencies $\neq 0$. (a) strong, (b) weak phase object; (c) strong, (d) weak amplitude object. (Hanszen and Morgenstern,

K. *The Weak Object with Amplitude and Phase Components*

For an object having both a weak amplitude and a weak phase modulation, we can write

$$F(x) = F_A(x)F_\Phi(x) \approx 1 + \int_{-\infty}^{+\infty} \{|\tilde{F}_A(R \neq 0)| \exp(i[2\pi Rx + \theta_A]) +$$
$$+ i|\tilde{F}_\Phi(R \neq 0)| \exp(i[2\pi Rx + \theta_\Phi])\} dR$$
$$\approx 1 + \int_0^\infty \{\tilde{A}(R \neq 0) \cos[2\pi Rx + \xi_A] +$$
$$+ i\tilde{\Phi}(R \neq 0) \cos[2\pi Rx + \xi_\Phi]\} dR, \qquad (61)$$

where all coefficients are $\ll 1$. Due to (29), (61) is equal to (28), when the corresponding coefficients are compared as shown in Table I.

From this we learn that the information about amplitude and phase modulation in a weak object is contained in the diffraction pattern in such a manner that each spatial frequency is unequivocally related to two diffraction angles of equal width but opposite sign. Unfortunately only $|\tilde{F}|$ and θ, but not the interesting quantities $\tilde{A}$, $\tilde{\Phi}$, ξ_A and ξ_Φ can be directly extracted from the diffraction pattern. In order to obtain these quantities, $|\tilde{F}|$ and θ must be divided into even and odd functions of R.

We are only interested in knowing whether it is possible in principle to solve this problem. To avoid complicated formulae, however, our further considerations are limited to simple examples of practical importance. For a more general treatize see Menzel (1958; 1960) and Hauser (1962). Equation (61) together with Table I is in fact the basis for the transfer theory of weak objects.

In order to develop the theory, an expression for the intensity transparency of a weak object is needed. From (44) and (61) we obtain the required approximation:

$$F_I(x) = \{1 + \int_0^{+\infty} \tilde{A}(R \neq 0) \cos[2\pi Rx + \xi_A] dR\}^2.$$
$$\approx 1 + 2 \int_0^{+\infty} \tilde{A}(R \neq 0) \cos[2\pi Rx + \xi_A] dR. \qquad (62)$$

In contrast to the remark in the footnote of Section IVH, each spatial frequency in the intensity transparency F_I of a weak object now has only one corresponding frequency in the transparency of the wave function F. The Fourier coefficients of F_I are twice as large as those of F.

L. *A Particular Relationship between Modulation and Diffraction of a Weak Object*

In (45) and (46) the concepts "amplitude and phase modulation" were introduced *ad hoc*. Knowing (61), we can now give the following relation between the transparency for the wave function $F(x)$ and the modulations $K_A(R)$ and $K_\Phi(R)$ in the case of a weak object, and only in that case:

$$F(x) = 1 + \tfrac{1}{2} \left\{ \int_{-\infty}^{\infty} [K_A(R \neq 0) + iK_\Phi(R \neq 0)] \exp(2\pi i Rx)\, dR; \right.$$

$$\text{with } K_A(-R) =_{df} K_A^*(+R); \quad K_\Phi(-R) =_{df} K_\Phi^*(+R). \tag{63}$$

Similarly, for the transparency of the intensity, we obtain:

$$F_I(x) = 1 + \int_{-\infty}^{\infty} K_A(R \neq 0) \exp(2\pi i Rx)\, dR. \tag{64}$$

Comparing (64) with (44) and (47), we find the following relation between the modulation of the amplitude of the wave function and the modulation of the intensity

$$K_I(R) = 2K_A(R). \tag{65}$$

In weak objects, the amplitude or phase modulation of a spatial frequency and the amplitude of the waves diffracted by these modulations, are strictly proportional. Therefore, the concept "weak object" has not only the meaning "the modulation of all spatial frequencies are small compared with 1", but also "the amplitude of the wave functions diffracted in any direction is small compared with the amplitude of the undiffracted wave". Thus, the transparency of a weak object can be obtained from the diffraction pattern.

The electron optical lens system alters the diffraction pattern in an undesired manner. Therefore, the image transparency is not identical with the object transparency. With the knowledge obtained in this section it is possible to evaluate the electron optical transfer properties directly from the difference between the so-called object-related and image-related diffraction patterns.

V. The Formation of the Optical Image

A. *Fundamental Problems*

In every plane which is traversed by the wave field behind the object, all the information about the interaction between the beam electrons and the atoms in the object is present. The information, however, is generally not separated according to the interactions at the different object points. In principle, the imaging system should decode this information in such a way that the current density in each image point gives information about the interaction in exactly one object point. Some reasons why the electron microscope is not capable of doing so have already been mentioned in the introduction.

Having now a complete knowledge about the weak object in (63)—(65), we can be sure that, even for such objects, it is generally impossible to determine the structure of the object from the image unequivocally. This is so because it is impossible, in principle, to draw unequivocal conclusions from the intensity modulation in a single micrograph about the intensity modulation of the amplitude *and* phase structure in the object. Therefore, we cannot profitably treat weak objects which have phase and amplitude components at the same time. It is better to limit our studies to pure phase or pure amplitude objects.

As previously stated, objects having a thickness of about 100 Å or less behave in high-voltage electron microscopy as weak phase objects without an amplitude component; for an experimental proof see Section VIE.

The Perfect Image. It is easy to indicate the conditions for perfectly imaging a pure amplitude object. In an ideal case, all electrons emerging from an object point should be focused into the corresponding image point (aberration-free image). Then in a reduced scale (3), the wave function $\psi'(x')$ in the image plane would be equal to the wave function $\psi(x)$ in the object plane; the same would be true for the related intensities $I'(x')$ and $I(x)$.

Under these conditions, however, the image of a phase object would be devoid of information, because with $A(x) = \text{const}$, (23) leads to $I(x)$ and $I'(x') = \text{const}$. In order to map the phase distribution $\Phi(x)$ onto the intensity distribution $I'(x')$, special measures are needed (see Section V J to V L).

The Real Image. A true point-to-point imaging is impossible. Diffraction at the pupil boundaries and the geometrical aberrations make the images of object points spread into aberration disks which overlap in a very complex manner. The relations between $\psi(x)$ and $\psi'(x')$ as well as between $I(x)$ and $I'(x')$ can be expressed by convolution integrals.

In this section, we only explain the processes taking place under coherent illumination†. Starting with an expression for the aberration

† For fundamentals see Born and Wolf (1964), p. 480 ff; for descriptions especially written from an electron microscopical point of view, see Hanszen, Morgenstern and Rosenbruch (1963) and Lenz (1965b), p. 274f.

disk of an object point (the so-called point-image), we present the imaging process in terms of spatial frequencies. Figure 5 illustrates the meaning of the following equations.

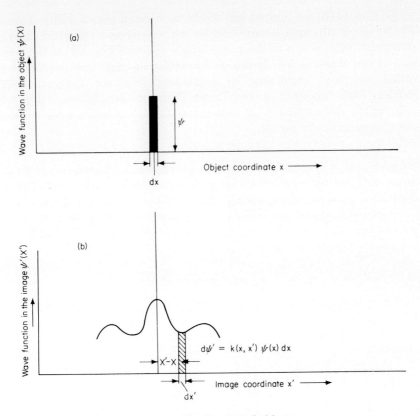

FIG. 5. Image formation of coherently illuminated objects.
(a) a radiating object point; (b) corresponding aberration disk in the image (the fact that ψ' is complex cannot be shown in this simplified figure).

The connection between the wave function $\psi(x)$ at a place x in the object plane and the wave function $\psi'(x')$ at a place x' in the image plane, produced by an imaging system having aberrations can be expressed by

$$d\psi'(x') = k(x, x')\, \psi(x)\, dx. \tag{66}$$

Here, $k(x, x')$ is called the *spread function* or *point image*. The contributions of all wave functions emerging from individual object points superimpose linearly at the image point x' in question. Therefore, the wave function in the image of an extended object is the integral

$$\psi'(x') = \int\limits_{-\infty}^{+\infty} k(x, x')\, \psi(x)\, dx. \tag{67}$$

In Section IIIB, we limited our study to a small paraxial object area. The image of this area is impaired by defocusing and spherical aberration only†. The shape of the point image does not depend on the position of the object point. An image formation having these properties is called *invariant, stationary*, or *isoplanatic* (e.g. see Linfoot, 1964; Röhler, 1967, especially p. 16). Under the above conditions, k depends only on the distance $(x' - x)$ between the image point in question and the Gaussian image point:

$$k(x, x') = k(x' - x). \tag{68}$$

By means of (67)–(68) the wave function is known at each image point. Since $\psi(x)$ is a function which, according to (22)–(26), depends only on the direction of the illuminating wave and the properties of the object and since the spread function $k(x' - x)$ is introduced as an instrument parameter, we are on the way to a solution of the imaging problem in the manner described in Section I.

B. *Introduction of Transfer Functions*

We replace the convolution (67)–(68) by a simpler but mathematically equivalent relation. For this purpose, we present not only the object and the image wave function, but also the spread function as a Fourier integral:

$$\psi(x) = \int\limits_{-\infty}^{+\infty} \tilde{\psi}(R) \exp\left(2\pi i R x\right) dR; \tag{69a}$$

$$\psi'(x') = \int\limits_{-\infty}^{+\infty} \tilde{\psi}'(R) \exp\left(2\pi i R x'\right) dR; \tag{69b}$$

$$k(x' - x) = \int\limits_{-\infty}^{+\infty} \tilde{k}(R) \exp\left[2\pi i R(x' - x)\right] dR. \tag{69c}$$

Using the Fourier inverse

$$(\tilde{\psi}R) = \int\limits_{-\infty}^{+\infty} \psi(x) \exp\left(- 2\pi i R x\right) dx \tag{70}$$

and those of (69b) and (69c) we obtain

$$\tilde{\psi}'(R) = \tilde{k}(R)\, \tilde{\psi}(R); \qquad \tilde{k}(R) = \text{complex}. \tag{71}$$

† If no mention is made, the axial astigmatism is considered to be corrected.

For reasons explained in Section VF, (71) is called the *filter-equation*, and $\tilde{k}(R)$ the *transfer-function for the wave function*. The intricate convolution (67); (68) of the wave function in position space can be replaced by a simple multiplication of its Fourier transform with the transfer-function $\tilde{k}(R)$.

When eq. (19) describing the illumination wave, and (22) and (34) describing the object transparency and its Fourier representation, and the Fourier inverse of (34) are introduced into (70), we obtain

$$\tilde{\psi}(R) = \Psi\tilde{F}(Q + R) \tag{72}$$

where $\tilde{\psi}(R)\,dR$ is the wave function, diffracted by the object in the direction $a = \lambda R$. Therefore, the filter equation takes the form

$$\tilde{\psi}'(R) = \Psi\tilde{F}(Q + R)\tilde{k}(R). \tag{73}$$

Here, $\tilde{\psi}'(R)\,dR$ is the wave function in the image, entering from the direction $a' = \lambda R/M'$. We call a' the *image-side diffraction-angle*. If we define the "image transparency for the wave function" by

$$F'(x') =_{df} \int\limits_{-\infty}^{+\infty} \tilde{F}'(Q + R)\exp\left[2\pi i(Q + R)x\right] d(Q + R), \tag{74}$$

and if we use its Fourier inverse and similar equations as (19) and (22) for the image side, then, with respect to (10), we have

$$\tilde{F}'(Q + R) = \tilde{k}(R)\,\tilde{F}(Q + R); \text{ or } \tilde{F}'(S) = \tilde{k}(R)\,\tilde{F}(S), \tag{75}$$

respectively.

By this result, the filter equation is related to the diffraction already known. We learn from (75) that the influence of the imaging system on the diffracted wave function can be described by a complex factor depending only on instrument parameters and the diffraction angle.

C. *Linear Transfer*

It was possible to describe the imaging process in a simple form (71)–(75), because the wave functions superpose linearly. Image formation which obeys these equations is called linear imaging, and it has the following properties. When a wave function $\psi_3(x)$ in the object is composed of the wave function $\psi_1(x)$ and $\psi_2(x)$ of two other objects according to

$$\psi_3(x) = C_1\psi_1(x) + C_2\psi_2(x), \tag{76}$$

the corresponding wave functions $\psi_1'(x')$; $\psi_2'(x')$, $\psi_3'(x')$ in the image are related by

$$\psi_3'(\mathbf{x}') = C_1'\psi_1'(x') + C_2'\psi_2'(x'). \tag{77}$$

Therefore, in the case of linear transfer, we can determine the optical transfer properties as follows. First we determine the transfer properties of objects having only one spatial frequency R_k. After having done this for objects with every possible R_k, we also know with certainty the transfer properties for all possible objects since, according to (76) and (77) the object and image properties of any compound object may be obtained by linear superposition of the corresponding properties of the elementary objects. In other words, linear transfer is characterized by the fact that the only object parameter present in the transfer function is the direction R of diffraction.

D. *The Imaging of Amplitude and Phase of the Object wave Function and of the Object Intensity into the Image Intensity*

Instead of the wave function $\psi'(x')$, the intensity

$$I'(x') = \psi'(x')\,\psi'^*(x') \tag{78}$$

is usually recorded in the image plane. Therefore, at the end of each imaging process, there is generally a quadratic relation which destroys the entire system of linear transfer for coherent illumination. Another quadratic relation, that is (23), is involved if one refers, as is usual, to the intensity distribution in the object instead of to the wave-function distribution. Also in the case of amplitude objects, to avoid difficulties, we must now restrict ourselves to *weak* objects. It has been shown in proceeding from (61) to (62), that the relation between the object transparencies for the wave function on the one hand and for the intensity on the other can be approximated by a linear equation. The same is true for the image quantities. Therefore, the imaging process of weak objects can be described to a good approximation by a sequence of linear equations.

The transfer function $\tilde{k}(R)$ indicates how the image-side wave function of the diffraction pattern arises from the corresponding object-side wave function. As will be shown later on, $\tilde{k}(R)$ has in general no symmetry properties with respect to $R = 0$. Therefore, $\tilde{\psi}'(R)$; $\tilde{F}'(R)$ and $\tilde{F}'_I(R)$ can be expressed by sets of coefficients, which are related in a complicated way to the set of coefficients for $\tilde{F}(R)$ and $\tilde{\psi}(R)$ respectively, given in Table I. These coefficients, however, carry the desired information about the modulation properties of the object and the image.

Since only a few functions $\tilde{k}(R)$ are of practical importance, we do not intend to evaluate general equations which are valid for any $\tilde{k}(R)$. Instead of this, we try in the next Section to study the physics of the imaging process in order to become acquainted with the physical significance of $\tilde{k}(R)$ and to learn which of the functions are of practical interest. The influence of these functions on the image properties will then be studied by means of simple *model objects*.

The limitation to the theory on weak objects has the effect that, contrary to the discussion in Section VC, only weak objects are usable as model objects, i.e. objects with a weak modulation (45)–(47). Such a model object is, for example:

$$F(x) = 1 + \tilde{A}_k \cos\left[2\pi R_k x + \xi_{A_k}\right] + i\tilde{\Phi}_k \cos\left[2\pi R_k x + \xi_{\Phi_k}\right];$$
$$\tilde{A}_k; \ \tilde{\Phi}_k \ll 1 \qquad (79)$$

$$F_I(x) = 1 + 2\tilde{A}_k \cos\left[2\pi R_k x + \xi_{A_k}\right] + \dots \qquad (80)$$

This object contains *two* frequencies $R = 0$ and $R = R_k$.

We try to determine the imaging properties of objects like (79)–(80) containing any possible R_k. Then, because of the linearity, we not only know the imaging properties of the electron microscope for all model objects, but also for all possible weak objects.

E. *The Physics of the Imaging Process*

The physics of the imaging process is shown in Fig. 6. The object within a field of radius x_e has only the spatial frequencies $R = 0$, R_1, and R_2. Therefore, diffracted beams appear only at the angles $a_1 = \pm \lambda R_1$ and $a_2 = \pm \lambda R_2$. The undiffracted beam is focused in the back focal point $(z_{F'}; \ x = 0)$, the diffracted beams in the points $(z_{F'}; \ \pm x_{P_1})$ and $(z_{F'}; \ \pm x_{P_2})$.

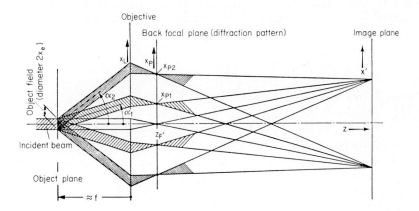

Fig. 6. Image formation in the electron microscope. The object is illuminated by a beam parallel to the axis. Each beam, diffracted by the spatial frequencies of the object is focused on a certain point in the pupil plane. Each image point receives waves from all radiating points of the pupil plane.

All points of the $z_{F'}$ = plane (plane of the exit pupil or diffraction plane) can be understood as point sources radiating coherently but with different phases. They radiate coherently because they are illuminated by a coherent beam, and they have different phases because the paths of the diffracted beams are not equal in length.

In the image plane, each beam interferes with the others. Therefore, every point in the image plane receives waves from every point in the diffraction plane. The intensity in each area of the image plane can be obtained by squaring the sum of the wave functions. The physical reason for neglecting the higher terms in (62) in the case of a weak object is the following. Only the interferences of the weak diffracted beams with the strong undiffracted beam are taken into account, but not the mutual interferences of the diffracted beams with one another.

We shall now study the physical conditions in the diffraction plane. As we have seen in Fig. 6, the information about the diffraction angles resides in the diffraction plane in such a way, that to each point of this plane there is a specific diffraction angle. Moreover, in the case of a weak object, *one* spatial frequency of the object belongs uniquely to *two* points of the diffraction plane at equal distance from but on opposite sides of the axis. Amplitude and phase at both points carry uniquely the information about amplitude and phase modulation of this frequency and the lateral phases of these components.

Accordingly we can say that the set of spatial frequencies, contained in the object, is imaged on the set of points in the diffraction plane in such a way that for each spatial frequency there are two spectral points. Correspondingly, the diffraction pattern is, in other words, the spatial frequency spectrum of the object. Strictly speaking the electron microscope up to the diffraction plane is a spectroscopic instrument for spatial frequencies.

We now abandon plane wave illumination and move the electron source to a finite distance from the object. The source should be small enough to fulfill the coherence condition. The position of the diffraction plane, which is no longer identical with the back focal plane, is now given by the last eq. of (6). Equation (10) denotes the position x_{Pk} of the spectral point, produced by the spatial frequency R_k on the side with positive coordinates. From this equation we learn further that the reduced pupil coordinate

$$R = -\frac{x_P - x_Q}{\lambda f} \frac{M'}{M' - M'_Q} \tag{81}$$

equals the spatial frequency when the intersection x_Q of the undiffracted beam with this plane (the zero order of the spectrum) is the zero point.

Expressions for the reduced coordinate Q of the undiffracted wave and the reduced coordinate $S = Q + R$, measured from the axis, have already been given in (8)–(9).

F. *Interpretation of the Filter Equation*

Having seen that R can be identified with the distance of the spectral points from the zero order, we realize that $\tilde{\psi}(R)\,dR$ in (69) is the wave function in the diffraction plane, not yet disturbed by aberrations in the imaging system. We call $\tilde{\psi}(R)$ the *object related wave function in the diffraction plane*.

In the diffraction plane, masking of different kinds can be introduced by inserting an aperture diaphragm, or by means of a phase plate causing a phase shift $\tilde{\Phi}_M(S) = \tilde{\Phi}_M(Q + R)$ and having a thin hole for the zero order, or by using zone plates, etc. Masking in the diffraction plane is especially advantageous, since each point of the mask has an influence on one single spatial frequency only. Because of this, the masking can be described by a complex factor $\tilde{F}_M(Q + R)$, depending only on the single object parameter R and on instrumental parameters:

$$\tilde{F}_M(Q + R) = |\tilde{F}_M(Q + R)| \exp[i\tilde{\Phi}_M(Q + R)]. \qquad (82)$$

In order to obtain the image-related wave function $\tilde{\psi}'(R)$ in the diffraction plane, the object-related wave-function $\tilde{\psi}(R)$ in the same plane must be multiplied with $\tilde{F}_M(Q + R)$:

$$\tilde{\psi}'(R) = \tilde{F}_M(Q + R)\tilde{\psi}(R);$$

or, as deriving of (75) from (71), we obtain:

$$\tilde{F}'(Q + R) = \tilde{F}_M(Q + R)\tilde{F}(Q + R); \qquad \tilde{F}'(S) = \tilde{F}_M(S)\tilde{F}(S). \qquad (83)$$

We call $\tilde{F}(S)$ the *transparency of the object-related*, and $\tilde{F}'(S)$ the *transparency of the image-related wave-function in the diffraction plane*. Equation (83) is a particular case of (71)–(75). As an example, we take axial illumination; i.e. $Q = 0$ and an aperture hole with the radius R_e. Then we have $\tilde{F}_M(R) = 1$ for $R \leq R_e$ and 0 for $R > R_e$ and we see that frequencies $> R_e$ are filtered out. The same is true for a zone plate. All frequencies, the spectral points of which are covered by the opaque areas of the plate are filtered out. This is why we speak about a filter equation (see Röhler, 1967, especially pp. 22 and 32). The influence of phase plates on the imaging process can be described by the phase $\tilde{\Phi}_M$ in (82).

In addition to the effects due to masking, those caused by the lens aberrations must also be considered, as will be done in Section VII. The transition from $\tilde{\psi}'(R)$ to $\psi'(x')$, i.e. from the image-related diffraction-pattern to the image, is the inverse to that from $\psi(x)$ to $\tilde{\psi}(R)$. This means that $\psi'(x')$ is obtained from $\tilde{\psi}'(R)$ by the inverse Fourier transform and is thus determined by (69b).

G. *Use of the Sampling Theorem*

In the Fourier expansion we now consider the finite limits $\pm x_e$ of the object field. The lowest spatial frequency contained in the object is $R_e = 1/2x_e$. All higher frequencies $R_k = k/2x_e$, with $k = 2,3 \ldots$ are multiples of this basic frequency. For this reason, a discrete spectrum of spatial frequencies exists instead of a continuous one. That the object field is not periodically repeated outside its borders is expressed by the fact that the distribution of the wave function in each spectral point has the shape of a $\sin(2\pi Rx)/2\pi Rx$-function. Therefore, the Fourier expansion of $\tilde{\psi}(R)$ can be written in the form

$$\tilde{\psi}(R) = \sum_{k=-\infty}^{+\infty} \tilde{\psi}\left(\frac{k}{2x_e}\right) \frac{\sin\left[\pi(R2x_e - k)\right]}{\pi(R2x_e - k)}. \tag{84}$$

This is the sampling theorem, given by Shannon (1949); see the summary by O'Neill (1963) and Röhler (1967). The spectral points with the coordinate values

$$R_s =_{df} k/2x_e; \qquad k = 0; \qquad \pm 1; \qquad \pm 2; \ldots \tag{85}$$

are called *sampling points*.

$$\Delta R_S = 1/2x_e; \qquad \Delta x_{PS} =_{df} \frac{\lambda f}{2x_e} \frac{(M' - M'_Q)}{M'} \tag{86}$$

is the distance between two sampling points in reduced and natural coordinates of the pupil plane. At each sampling point, the wave function has only one spatial frequency, because the contributions of all other frequencies vanish there.

As an illustration, the distance between two sampling points, when using the data of Section IIB, is $\Delta x_{PS} \approx 0.1 \ \mu m$. As a result of the sampling theorem, the wave function within the object field is determined by a countable number of spectral points, i.e. by the wave function at a countable number of points in the diffraction plane.

When the coherence condition was described in (18), information about the object field diameter was needed in order to determine the permissible size of the illumination aperture. Using (86), eqn. (18) becomes now

$$\Delta\varphi \ll \frac{\lambda}{2} \Delta R_S. \tag{87}$$

Accordingly, the illumination aperture should be small enough not to blur the zero-points in the distribution of each sampling point.

H. *Influence of Lens Aberrations in the Geometric-optical Treatment*

Aberration Disks in the Diffraction Spectrum. Each sampling point represents an image of the electron source, broadened by diffraction at the borders of the object field. All these images are further broadened by the lens aberrations. Because of this blurring, restrictions similar to those used in the last section for the illumination aperture should be made. First we study the broadening of the spectral points by spherical aberration of the objective. The border of the object field acts as an aperture stop for the strongly demagnified imaging of the electron source into the spectral points. On the whole, the ray paths are the inverse of those used in the highly magnified imaging of the object. Therefore, we expect to obtain aberration disks of the spectral points with very small radii δ_P. Detailed calculations, taking into account (5)–(6) and $C_s(M_Q') \approx C_s(M')/M_Q'^4$, (see Archard, 1958; Hawkes, 1968) lead to the result:

$$\delta_P = \left[\frac{x_e}{f} \frac{M'}{M_Q' - M'}\right]^3 C_s(M'). \tag{88}$$

The absolute value of δ_P should be small compared to the absolute value of $\Delta x_{PS}/2$. Assuming this, we obtain the relation:

$$x_e \ll f \frac{M' - M_Q'}{M'} \sqrt[4]{\frac{\lambda}{4C_s(M')}}. \tag{89}$$

Example: If we insert the instrumental data of Section IIB in the right-hand side of the equation, we find $x_e \ll 10 \ \mu m$.

The next question is, how far may the masking plane deviate from the plane of diffraction? As an answer we find that the absolute value of the aberration disk δ_D due to defocusing,

$$\delta_D = \Delta z_P \frac{x_e}{f} \frac{M'}{M_Q' - M'}, \tag{90}$$

should be small compared to the absolute value of $\Delta x_{PS}/2$. This leads to

$$x_e \ll f \frac{M' - M_Q'}{M'} \sqrt{\frac{\lambda}{4\Delta z_P}}. \tag{91}$$

Example: If we insert the instrumental data of Section IIB and if we allow a mechanical deviation $\Delta z_P = 0 \cdot 3$ mm of the mask from the diffraction plane, the allowable object field is limited by $2x_e \ll 2500$ Å.

Distortion of the Diffraction Pattern. All other geometric-optical third-order aberrations are omitted, because they do not contribute to an understanding of the main characteristics of optical transfer. We will discuss only the influence of distortion on the diffraction pattern as a particularly illustrative example.

The relative position of the spectral points is shifted by distortion. Therefore in the image-related wave-function of the diffraction plane, we do not find the set of exact whole numbers k given in (85). For this reason, the higher harmonics in the image are out of tune. Because each spatial frequency R_n in the image is built up by two sampling points $k = \pm (n + \nu)$ with $\nu \ll 1$, the local modulation of the spatial frequencies is correct in the paraxial region. Only at a larger distance from the axis does it become out of phase. Thus the resulting error leads to a limitation of the image field already known (isoplanatism patch, see Section VA). It is easy to show that a ray, diffracted at an angle a_k and having the transverse aberration (12) in the image plane due to spherical aberration, undergoes in the pupil plane a transverse aberration $\varDelta x_P$ which is smaller than (12) by a factor $\approx 1/(M' - M'_Q)$. From (10), the relative distortion of the diffraction pattern is

$$\frac{\varDelta x_P}{x_{Pk} - x_{PQ}} = \frac{\varDelta R_k}{R_k} = \frac{C_s(M')}{f} R^2 \lambda^2 \left(\frac{M'}{M' - M'_Q}\right)^2. \tag{92}$$

Since $C_s[M'/(M' - M'_Q)]^2/f \approx 1$, diffraction by atomic structures $(1/\mathrm{R} \approx 50\lambda)$ is affected by a relative distortion of less than 10^{-3}. Therefore in the image of an object-field diameter of 1000 Å, as was supposed in Section IIB, distortion is of minor importance.

Object and Exit Pupil in the Objective Field. Until now the object and the exit pupil have been assumed to be in field-free space. In reality, they are well within the magnetic field of the objective. Thus the rays are deflected by this field before reaching the object plane and will be deflected again after leaving the pupil plane. Both these effects can be expressed by a scaling factor. Since we can disregard scaling problems by using reduced coordinates, these effects are of little or no importance for the systematic theory. However, in designing zone plates for the pupil plane, these effects are of some practical importance. This problem has been solved in principle by Lenz (1964). According to his calculations, zone plates in the back focal plane of the Glaser-field with a lens power $k^2 = 0.6$ have radii that are 4% smaller than those calculated by using the equations above.

I. *The Wave Aberration*

The decisive influence of the lens aberrations on the imaging process can only be determined by wave optics. In order to understand it, we need the concept "wave aberration".

In the image space, which is considered field free, we construct reference spheres around each image point. These spheres intersect the axis in the pupil plane. For perfect imaging, the reference spheres would simultaneously be wave-fronts converging into the above image points. The aberrations of the lens system cause distortions of these wave-fronts. The deviation from spheres as a function of the distance

x_P from the axis in the pupil plane is the wave aberration $W(x_P)$. It is a function of the upper limit of the integral*

$$W(x_P) = -\int_0^{x_P} \frac{\Delta z_B x}{(z_B - z_P)^2}\, dx. \quad x = \text{current pupil coordinate.} \quad (93)$$

Here, Δz_B is the total longitudinal aberration of a ray intersecting the pupil plane at x_P. Equation (93) relates the ray and the wave concepts. Δz_B should be considered as a function of x_P, which is related to the spatial frequencies as shown in (10). Therefore, we know in principle that the lens aberrations influence the transfer of the spatial frequencies by producing phase shifts. For a more detailed study, $W(x_P)$ must be divided into the individual components of aberration. Since the object field is very small, only spherical aberration and defocusing are important. Therefore, Δz_B can be written as

$$\Delta z_B = \Delta z_s + \Delta z_d. \quad (94)$$

Here, Δz_s and Δz_d are the components of longitudinal aberration caused by spherical aberration and defocusing as defined in (14)–(17). Inserting these quantities in (94), expressing $(z_B - z_P)$ by f and M' according to (4)–(5), and considering that the spherical aberration constant C_s and defocusing are independent of x_P, we integrate (93) and obtain

$$W(x_P) =_{df} W_s(x_P) + W_d(x_P) = \frac{C_s(M')x_P^4}{4f^4}\left(\frac{M'}{M' - M_Q'}\right)^4 +$$
$$+ \frac{\Delta z x_P^2}{2f^2}\left(\frac{M'}{M' - M_Q'}\right)^2. \quad (95)$$

The same result, expressed in reduced coordinates according to (8) is

$$W(S) = \frac{C_s(M')}{4}\,\lambda^4 S^4 + \frac{\Delta z}{2}\,\lambda^2 S^2 \quad (96)$$

Since the objective has rotational symmetry, this function is even.

Generalized Coordinates. In (96) the wave aberration depends on the instrumental parameters f and $C_s(M')$, on the magnifications M' and M_Q', and on the defocusing Δz. Each electron microscope, therefore, has its own family of wave aberration curves which characterize its imaging properties. It is possible to condense such families into a single curve by introducing the following *generalized coordinates*†:

generalized wave aberration: $\quad \mathscr{W} =_{df} \dfrac{W}{\lambda};$ \quad (97)

---

* For the derivation of the differential form of this equation see e.g. O'Neill (1963), especially p. 50 eq. (4–7; 8).

† Corresponding equations given by Hanszen (1966a) contain an error. Due to an oversight $M'/(M' - 1)$ was given instead of $M'/(M' - M_Q')$. The latter expression approaches 1 as $M_Q' \to 0$, as it should.

generalized pupil coordinate:

$$\mathcal{X}_P =_{df} - \frac{x_P M'}{f(M' - M'_Q)} \cdot \sqrt[4]{\frac{C_s(M')}{\lambda}} = + S \sqrt[4]{C_s(M')\lambda^3}$$
$$= (R + Q) \sqrt[4]{C_s(M')\lambda^3} =_{df} \mathcal{R} + \mathcal{Q}; \qquad (98)$$

generalized defocusing:

$$\Delta =_{df} - \frac{\Delta z}{\sqrt{C_s(M')\lambda}}. \qquad (99)$$

Then, the generalized equation for the wave aberration has the form

$$\mathcal{W}(\mathcal{X}_P) = \frac{\mathcal{X}_P^4}{4} - \frac{\Delta}{2}\,\mathcal{X}_P^2. \qquad (100)$$

This family of curves (Δ as parameter) is valid for image formation using electrons of arbitrary wavelength, objective lenses of arbitrary focal length and spherical aberration, operating at a high, but not necessarily at an infinite magnification†, and with arbitrary defocusing of moderate value. Moreover, it is valid for light-optical models of the electron microscope; therefore, it is also helpful in describing holographic reconstruction of electron micrographs (see Section VIJ). The family of generalized wave-aberration curves is drawn in Fig. 7.

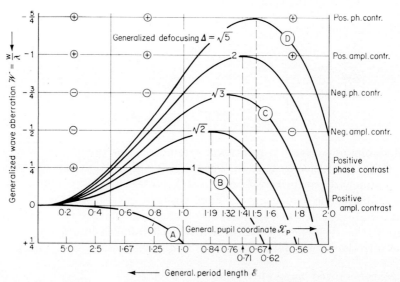

FIG. 7. Generalized wave aberrations. The contrast transfer functions of Figs. 11 and 12, belonging to these wave aberrations, are marked with the same letter. (Hanszen 1966a).

† It is assumed that the function $C_s(M')$ is known. For calculations see Albert (1966), for experiments Kunath and Riecke (1965).

Since $\mathscr{W}(\mathscr{X}_P)$ are even functions, only the positive values are shown. It should be pointed out that the maxima of these curves are of special importance, especially if they have the values $\mathscr{W}(\mathscr{X}_P) = -\dfrac{n}{4}; n = 1, 2 \ldots$.

In order to complete the system of generalized coordinates, we introduce the following quantities† in the object and the image plane:

generalized object and image coordinate:

$$\mathscr{X} = \frac{x}{\sqrt[4]{C_s \lambda^3}}; \quad \mathscr{X}' = \frac{x'}{\sqrt[4]{C_s \lambda^3}}; \tag{101}$$

generalized period length:

$$\mathscr{E}_k = \frac{1}{\mathscr{R}_k} = \frac{\epsilon_k}{\sqrt[4]{C_s \lambda^3}} = \frac{1}{R_k \sqrt[4]{C_s \lambda^3}} = \frac{1}{(S_k - Q)\sqrt[4]{C_s \lambda^3}}. \tag{102}$$

For the purpose of demonstration the generalized coordinates are compared in Fig. 8 with the natural ones that are based on the numerical data given in Section IIB. Rewriting Fig. 7 in natural coordinates as in Fig. 8, we see that under practical conditions the wave aberrations may considerably exceed $\lambda/4$.

The Significance of the Wave Aberration for the Imaging Process. The meaning of (96) is the following. The diffracted wave which intersects the diffraction plane at the point S acquires, along its whole path length from the object to the image, a path-length difference $W(S)$ due to lens aberrations. Therefore, the wave arrives at the image plane with a phase difference $\tilde{\Phi}_L = 2\pi W(S)/\lambda$. The effect is the same as if the phase shift occurred in the pupil plane. Thus, the influence of the wave aberration on the imaging process can be described by the factor

$$\tilde{F}_L(S) = \exp\left[i\tilde{\Phi}_L(S)\right] = \exp\left[i 2\pi W(S)/\lambda\right], \tag{103}$$

by which the object-related wave function $\tilde{\psi}(R)$ in the diffraction plane is to be multiplied. We combine this factor with the factor $\tilde{F}_M(Q + R)$ describing the masking. Thus, we define the *pupil function* $\tilde{F}_P(Q + R)$ as the product of both and obtain, with reference to (10),

$$\tilde{F}_P(Q + R) = |\tilde{F}_M(Q + R)| \exp\left\{i[\tilde{\Phi}_L(Q + R) + \tilde{\Phi}_M(Q + R)]\right\}$$
$$= |\tilde{F}_M(Q + R)| \exp\left\{i[2\pi W(Q + R)/\lambda + \tilde{\Phi}_M(Q + R)]\right\}. \tag{104}$$

Instead of (83) we can write

$$\tilde{\psi}'(R) = \tilde{F}_P(Q + R)\tilde{\psi}(R); \quad \tilde{F}'(Q + R) = \tilde{F}_P(Q + R)\tilde{F}(Q + R). \tag{105}$$

Comparing this with (71) we see that the pupil function $\tilde{F}_P(Q + R)$ is identical with the transfer function for the wave function $\tilde{k}(R)$:

$$\tilde{k}(R) = \tilde{F}_P(Q + R). \tag{106}$$

† These notations are not identical with those used by Hanszen (1966a).

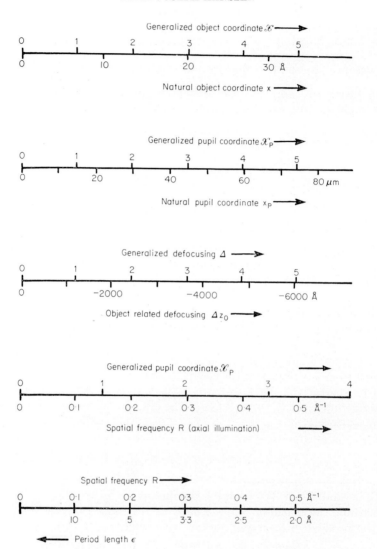

Fig. 8. Comparison of the generalized coordinates with the natural coordinates valid for the normal objective.

The equation has an imperfection, however. The diffraction pattern is symmetrical to the undiffracted beam, which does not coincide with the axis when oblique illumination is used. $\tilde{F}_L$ is symmetrical to the axis whereas $\tilde{F}_M$ can be of any shape. Therefore, we cannot expect that $\tilde{k}(R)$ has any symmetry with respect to the point $R = 0$. Because of this there are complications, as described in Section VD, in a mathematical formulation of the imaging process. In the interest of brevity, only the following special cases will be treated:

a. Axial illumination, i.e. $S = R$. Semitransparent masks are not used. The borders of the aperture are located symmetrically about the axis, i.e. $\tilde{F}_M(+R) = \tilde{F}_M(-R)$; $|\tilde{F}_M| = 0$ or 1. For the phase plates, we have $\tilde{\Phi}_M(+R) = \tilde{\Phi}_M(-R)$.

b. Oblique illumination, i.e. $S = Q + R$. Either no masking is used, i.e. $|\tilde{F}_M| \equiv 1$, or one side of the diffraction pattern is eliminated, i.e.

$$|\tilde{F}_M| \equiv 0 \quad \text{for} \; -\infty < R \leq 0$$

and

$$|\tilde{F}_M| = 1 \quad \text{for} \; 0 < R < +\infty;$$

phase shifts are not used, i.e. $\tilde{\Phi}_M = 0$.

Even and Odd Components of the Wave Aberration. For convenience, we separate the wave aberration W into an odd component W_u and an even component W_v with respect to $W(R = 0)$ (see Fig. 9):

$$W_u(Q + R) = -W_u(Q - R) =_{df} [W(Q + R) - W(Q - R)]/2 =_{df} \xi(R)\lambda/2\pi;$$

with

$$\xi/2\pi = \mathscr{R}^3\mathscr{Q} - \varDelta\mathscr{R}\mathscr{Q} + \mathscr{R}\mathscr{Q}^3, \tag{107}$$

$$W_v(Q + R) = W_v(Q - R) =_{df} [W(Q + R) + W(Q - R)]/2 - W(Q) =_{df} \zeta(R)\lambda/2\pi;$$

with

$$\zeta/2\pi = \mathscr{R}^4/4 - \varDelta\mathscr{R}^2/2 + 3\mathscr{R}^2\mathscr{Q}^2/2. \tag{108}$$

We call ξ the odd, and ζ the even phase shift of the wave aberration.

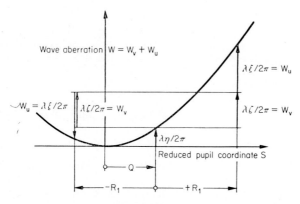

FIG. 9. Separation of the wave aberration W into an even component W_v and an odd component W_u. (Hanszen and Morgenstern, 1965); $\lambda\eta/2\pi \equiv W(Q)$ in eqn. (108).

J. *Phase Shifts and Masking Effects in the Diffraction Plane*

The results of the last two sections are:

(a) Both phase and amplitude modulation of the object can be derived from the object-related wave function in the diffraction plane according to (61) and (63);

(b) The image-related wave function $\tilde{\psi}'(R)$ in the diffraction plane can be obtained from the object related function $\tilde{\psi}(R)$ by multiplying the latter with the pupil function $\tilde{F}_P(R + Q)$. The whole image information is already contained in $\tilde{\psi}'(R)$.

(c) The essential pupil effects are shifts in phase and in the size of the aperture.

How these effects influence the optical transfer of phase and amplitude components of a compound object (79) into the image intensity is shown in Table II, lines 1 through 4 on p. 49. The results will be discussed in Section VL.

K. *The Amplitude-Contrast and Phase-Contrast Transfer-Functions of the Electron Microscope*

Image Modulation. As was done in eqn. (45) through (47) for the object, we can define the amplitude and phase modulation of the wave function as well as the intensity modulation *for the image*. As we have seen, only the intensity modulation is important in contrast transfer theory. We define this quantity by:

$$K'_I(R) = {}_{df} \frac{\tilde{A}'_I(R)}{\tilde{A}'_I(R = 0)} \exp[i\xi'_I(R)] \text{ with } \frac{\tilde{A}'_I(R = 0)}{2 \times 2x_e} = 1. \quad (109)$$

Examples for pure amplitude and phase objects are given in Table II, line 6a; b.

The Transfer of the Different Object Modulations into the Image. Within the scope of transfer theory, it is necessary and sufficient to know the following:

(a) How the modulation (47), (64) of the intensity of a pure amplitude object is transferred into the image intensity;

(b) How the phase modulation (46) of a pure phase object is transferred into the image intensity.

The information can be obtained from the following quantities:

Amplitude-contrast transfer-function (defined for pure, weak amplitude objects only):

$$D_I(R) = {}_{df} \frac{K'_I(R)}{K_I(R)} = \frac{\tilde{A}'_I(R)}{\tilde{A}_I(R)} \exp\left[i(\xi'_I - \xi_I)\right]; \quad (110)$$

TABLE II

Examples of masking-effects and contrast transfer-functions

(11.1) Object

$$F'(x) = 1 + \bar{A}_k \cos[2\pi R_k x + \xi_{Ak}] + i\bar{\Phi}_k \cos[2\pi R_k x + \xi_{\Phi k}] \quad \text{with } \bar{A}_k;\ \bar{\Phi}_k \ll 1;$$
$$\text{i.e.: } K_A(R_k) = \bar{A}_k \exp(i\xi_{Ak});\qquad K_\Phi(R_k) = \bar{\Phi}_k \exp(i\xi_{\Phi k})$$

(11.2) Object related diffraction spectrum

$$F(R=0) = 1;\qquad F(\pm R_k) = \frac{\bar{A}_k}{2}\exp(\pm i\xi_{Ak}) + i\frac{\bar{\Phi}_k}{2}\exp(\pm i\xi_{\Phi k})$$

		(A) even phase shift	(B) odd phase shift	(C) even and odd phase shift	(D) elimination of one side phase shift on the other side*	(E) dark field
(11.3)	Masking	$\mathcal{F}_P(R=0)=1;$ $\mathcal{F}_P(\pm R_k)=\exp(i\check{\xi})$	$\mathcal{F}_P(R=0)=1;$ $\mathcal{F}_P(\pm R_k)=\exp(\pm i\xi)$	$\mathcal{F}_P(R=0)=1;$ $\mathcal{F}_P(\pm R_k)=\exp[i(\zeta\pm\xi)]$	$\mathcal{F}_P(R=0)=1;$ $\mathcal{F}_P(+R_k)=0;$ $\mathcal{F}_P(-R_k)=\exp(i\chi)$	$\mathcal{F}_P(R=0)=0;$ $\mathcal{F}_P(\pm R_k)=1$
(11.4)	Image related diffraction spectrum	$F'(R=0)=1;$ $F'(\pm R_k)=\frac{\bar{A}_k}{2}\exp(\pm i\xi_{Ak})+i\frac{\bar{\Phi}_k}{2}\exp(\pm i\xi_{\Phi k})$	$F'(R=0)=1;$ $F'(\pm R_k)=\frac{\bar{A}_k}{2}\exp[\pm i(\xi_{Ak}+\xi)]+i\frac{\bar{\Phi}_k}{2}\exp[\pm i(\xi_{\Phi k}+\xi)]$	$F'(R=0)=1$ $F'(\pm R_k)=\exp(i\zeta)\left[\frac{\bar{A}_k}{2}\exp[\pm i(\xi_{Ak}+\xi)]+i\frac{\bar{\Phi}_k}{2}\exp[\pm i(\xi_{\Phi k}+\xi)]\right]$	$F'(R=0)=1;$ $F'(-R_k)=0$ $F'(+R_k)=\exp(i\chi)\left(\frac{\bar{A}_k}{2}\exp(i\xi_{Ak})+i\frac{\bar{\Phi}_k}{2}\exp(i\xi_{\Phi k})\right)$	$F'(R=0)=0$ $F'(\pm R_k)=\frac{\bar{A}_k}{2}\exp(\pm i\xi_{Ak})+i\frac{\bar{\Phi}_k}{2}\exp(\pm i\xi_{\Phi k})$
(11.5)	Transparency of the image intensity	$F'(x')=1+2\bar{A}_k\cos\zeta\cos(2\pi R_k x'+\xi_{Ak})-2\bar{\Phi}_k\sin\zeta\cos(2\pi R_k x'+\xi_{\Phi k})$	$F'(x')=1+2\bar{A}_k\cos(2\pi R_k x'+\xi_{Ak}+\xi)$	$F'(x')=1+2\bar{A}_k\cos\zeta\cos(2\pi R_k x'+\xi_{Ak}+\xi)-2\bar{\Phi}_k\sin\zeta\cos(2\pi R_k x'+\xi_{\Phi k}+\xi)$	$F'(x')=1+\bar{A}_k\cos(2\pi R_k x'+\xi_{Ak}+\chi)-\bar{\Phi}_k\sin(2\pi R_k x'+\xi_{\Phi k}+\chi)$	$F'(x')=\frac{\bar{A}_k^2}{2}[1+\cos[2\pi 2R_k x'+2\xi_{Ak}]]+\frac{\bar{\Phi}_k^2}{2}[1+\cos[2\pi 2R_k x'+2\xi_{\Phi k}]]$
(11.6a)	Intensity modulation in the image of a pure amplitude object i.e. $\bar{\Phi}_k=0$	$K_I^*(R_k)=2\bar{A}_k\cos\zeta\exp(i\xi_{Ak})$	$K_I^*(R_k)=2\bar{A}_k\exp[i(\xi_{Ak}+\xi)]$	$K_I^*(R_k)=2\bar{A}_k\cos\zeta\exp[i(\xi_{Ak}+\xi)]$	$K_I^*(R_k)=\bar{A}_k\exp[i(\xi_{Ak}+\chi)]$	$K_I^*(2R_k)=\exp(i2\xi_{Ak})$
(11.6b)	Intensity modulation in the image of a pure phase object i.e. $\bar{A}_k=0$	$K_I^*(R_k)=-2\bar{\Phi}_k\sin\zeta\exp(i\xi_{\Phi k})$	$K_I^*(R)=0$	$K_I^*(R_k)=-2\bar{\Phi}_k\sin\zeta\exp[i(\xi_{\Phi k}+\xi)]$	$K_I^*(R_k)=-\bar{\Phi}_k\exp[i(\xi_{\Phi k}+\chi-\pi/2)]$	$K_I^*(2R_k)=\exp(i2\xi_{\Phi k})$
(11.7a)	Amplitude-contrast transfer-function	$D_I(R)=\cos\zeta(R)$	$D_I(R)=\exp[i\check{\xi}(R)]$	$D_I(R)=\exp[i\check{\xi}(R)]\cos\zeta(R)$	$D_I(R)=(1/2)\exp[i\chi(R)]$	
(11.7b)	Phase-contrast transfer-function	$B(R)=-2\sin\zeta(R)$	$B(R)=0$	$B(R)=-2\exp[i\check{\xi}(R)]\sin\zeta(R)$	$B(R)=-\exp\{i[\chi(R)-\pi/2]\}=\exp\{i[\chi(R)+\pi/2]\}$	

The masking $\mathcal{F}_P(R=0)=1$: $\mathcal{F}_P(-R_k)=e^{i\chi}$; $\mathcal{F}_P(+R_k)=0$ would lead to: $D_I(R)=(\frac{1}{2})\exp[-i\chi(R)]$; $B(R)=-\exp\{-i[\chi(R)-\pi/2]\}=\exp\{-i[\chi(R)+\pi/2]\}$.

Phase-contrast transfer-function (defined for pure, weak phase objects only):

$$B(R) =_{df} \frac{K'_I(R)}{K_\Phi(R)} = \frac{\tilde{A}'_I(R)}{\tilde{\Phi}(R)} \exp[i(\xi'_I - \xi_\Phi)]. \qquad (111)$$

The value of these functions for a fixed spatial frequency $R = R_k$ is called the *amplitude-contrast or phase-contrast transfer-factor* for this frequency. Knowing these factors for all frequencies, we know the contrast transfer functions for all weak amplitude and phase objects.

The contrast transfer functions belonging to the pupil functions reviewed in Table II are presented on line 7a; b of this table (for particulars see Hanszen and Morgenstern, 1965).

L. *Discussion of Table II*

Even Phase Shift ζ (Column A). Both the amplitude and the phase structure are transferred into the image. In this case, the transfer is described by the contrast transfer functions given in (Table II.7). The functions also show that there is no phase contrast for $\zeta = n\pi$ and no amplitude contrast for $\zeta = (2n + 1)\pi/2$ where $n = 0; \pm1; \pm2 \ldots$

It must be pointed out that an even phase shift is capable of producing both a phase and an amplitude contrast. Therefore, the concept "contrast by means of a phase shift" is not identical with the concept "phase contrast".

Odd Phase Shift ξ (Column B). An odd phase shift cannot produce phase contrast. The strength of the amplitude contrast produced is independent of R. Unfortunately, lateral phase shifts ξ occur, the strength of which depends on R.

Even and Odd Phase Shift (Column C). Both the amplitude and the phase contrast are present. Both the amount and the lateral phase of D_I and B depend on R.

Elimination of the Diffraction Pattern on one Side; Phase Shift χ on the Other Side of the Pattern (Column D). By means of a mask for accomplishing the above, amplitude and phase contrast which are independent of R are produced simultaneously. There are lateral phases in both cases, however.

Elimination of the Zero Order, i.e. Darkfield (Column E). We examine the image of the object (Table II.1) and find the frequency $2R_k$ instead of the object frequency R_k. This is true, because we can no longer overlook the interference intensities between the diffracted beams; for details see Hanszen (1969a). If the object contains several spatial frequencies, the interferences of each diffracted wave with all the others must be taken into account. Thus the spatial frequencies are not transferred independently of each other; hence no transfer functions can be given.

Relations between D_I, B and $\tilde{k}$. There are, of course, relations between $D_I(R)$; $B(R)$ and $\tilde{k}(R)$. In the special case of an even phase shift only, i.e. $\tilde{k}(R) = \exp \{i\zeta(R)\}$ we can rewrite (Table II, 7)

$$D_I(R) = Re\ \tilde{k}(R); \qquad B(R) = -\ 2Im\ \tilde{k}(R). \qquad (112a)$$

According to Hauser (1962), we obtain in the general case:

$$D_I(R) = \frac{1}{2|\tilde{k}(Q)|^2} [\tilde{k}^*(Q)\ \tilde{k}(Q + R) + \tilde{k}(Q)\ \tilde{k}^*(Q - R)]; \qquad (112b)$$

$$B(R) = \frac{i}{|\tilde{k}(Q)|^2} [\tilde{k}^*(Q)\ \tilde{k}(Q + R) - \tilde{k}(Q)\ \tilde{k}^*(Q - R)]. \qquad (112c)$$

M. *Supplementary Comments on the Imaging of Two-Dimensional Objects*

As already pointed out, spatial frequencies masked by an aperture stop, especially by the opaque rings of zone plates, are irretrievably lost in the image, since both the diffraction spectrum and the image intensity of the objects correspond uniquely.

It was questioned (Langer and Hoppe, 1966) whether this statement, valid for one-dimensional objects, would still be valid for two-dimensional ones. To find the answer we study the radiation leaving the object point (x_k, y_k) and entering the image in the point (x'_k, y'_k). According to Fig. 6, this radiation extends over the entire pupil plane. The reduced coordinates in this plane are denoted by R_x and R_y. The intensity in every image point is generated by radiation coming from all points of the pupil plane. Therefore, masking at any point of this plane leads, as a rule, to intensity changes in each image point.

We now study the image points along the line $(y' = 0; x')$ of the image plane. This subset of image points is not only formed by the wave functions of the subset $(R_y = 0; R_x)$ of the pupil points, but also by the wave functions of the complete set of pupil points. It is possible to produce a given intensity distribution (e.g.

$$F'_I(x', y' = 0) = 1 + 2\tilde{A}_k \cos 2\pi R_{xk}x' \quad \text{with } \tilde{A}_k \ll 1)$$

in different ways (in the example by

$$\tilde{F}'(R = 0) = 1; \tilde{F}'(\pm R_k) = \pm \sqrt{R_{xk}^2 + R_y^2} = \tilde{A}_k/2$$

with an arbitrary value for R_y). Each of these spectra, however, produces a different intensity distribution in the residue set $(x', y' \neq 0)$ outside the line considered. If the points $R = \pm R_{xk}$ of the pupil plane are covered in this case, the period length $1/R_{xk}$ appears even then in the image along the line $(x'; y' = 0)$. However, one should not overlook the loss of information caused by masking, since there may be noticeable disturbances in other parts of the image plane.

In Fig. 10, an illustrative example is given. At the top we see a particular intensity distribution in the pupil and at the bottom the corresponding image intensity. If the line $(R_y = 0; R_x)$ in the diffraction figure is covered (above right), the intensity distribution in the line $(y' = 0; x')$ of the image at the bottom right is altered only imperceptibly (statement by Langer and Hoppe, 1966). Note, however, the spikes starting from the image points into the image field $(y' \neq 0; x')$. The spikes have to be regarded as "spurious structures"; see Section IXC. Our statement that the relation between diffraction spectrum and Gaussian image is unique is not valid for two arbitrarily selected sub-sets of spectral and image points but only for the complete sets. The information stored in each spectral point does not depend on the information in other points. Therefore, information that is lost by masking cannot be replaced by information stored in the remaining spectral points. It is another matter, whether the information eliminated by the mask is important or not. For example, the information that the electrons are blocked outside the object is unimportant. From this example we learn that information limitations by masking are less serious, if prior information is available.

N. *Summary and Conclusions*

We have succeeded in establishing the imaging properties of the electron microscope from the diffraction spectrum by presenting the contrast transfer-functions (Table II. 7a, b). In the following discussion, therefore, we do not need to study Gaussian image formation by means of interference of the wave functions $\check{\psi}'(R)$. In particular, we do not want to deal with the difficult problem of point image formation and "point resolution". On the contrary, we determine the image quality from the properties of the spatial frequency spectrum in the pupil plane. When the contrast transfer functions (Table II.7a, b) of an electron microscope are known (e.g. by knowing the lens aberrations and pupil conditions) it is easy to predict the properties of a micrograph of a known object, and it is possible to determine the structures of an unknown object from its micrograph apart from the structures which fall within a transfer gap. When the transfer-function is unknown, a micrograph of an object containing all spatial frequencies of interest gives much information about the shape of this function; see Section VIE.

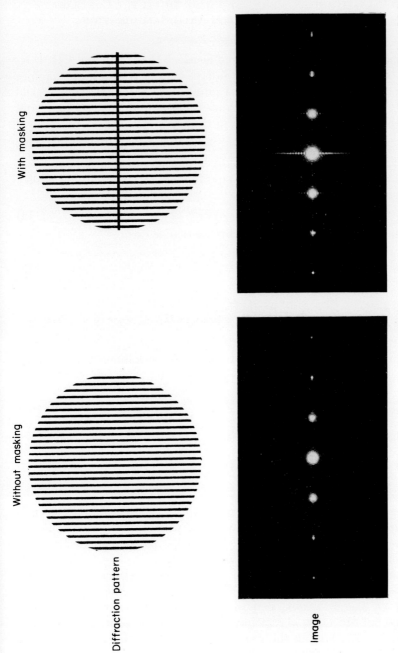

Fig. 10. Effect of masking in the pupil plane on the two dimensional image (demonstration-experiment with a light optical model for the electron microscope; see Hanszen 1968). Due to masking, each image point degenerates into a diffraction spectrum of the opaque strip which serves as a mask.

VI. Contrast by Means of Phase Shift: the Contrast Transfer-Functions for Axial Illumination

A. *The Pupil Function*

We can now combine the results of Sections IIA, IIIC, and VH, I, K. In the case of axial illumination, we have $Q = 0$, $R = S$ and therefore $\mathcal{Q} = 0$, $\mathcal{R} = \mathcal{X}_P$ for the generalized pupil coordinates. In this special case, $\mathcal{X}_P$ is identical with the generalized spatial frequency $\mathcal{R}$. The pupil function (104) can now be written in the form

$$\tilde{F}_P(R) = |\tilde{F}_M(R)| \exp\{i[2\pi W(R)/\lambda + \tilde{\Phi}_M(R)]\}. \tag{113}$$

Here $W(R)$ is the wave aberration given in (96) and $|\tilde{F}_M(R)| \exp(i\tilde{\Phi}_M)$ the masking. Using generalized coordinates and considering (96ff), (113) can be rewritten in the form

$$\tilde{F}_P(\mathcal{X}_P) = |\tilde{F}_M(\mathcal{X}_P)| \exp\{i[2\pi\mathcal{W}(\mathcal{X}_P) + \tilde{\Phi}_M(\mathcal{X}_P)]\}. \tag{114}$$

According to (102), we obtain in this particular case

$$\mathcal{X}_P = \mathcal{R} = \frac{1}{\epsilon} \sqrt[4]{C_s\lambda^3}. \tag{115}$$

B. *The Contrast Transfer-Functions in the Absence of a Mask*

In this case, we have $|\tilde{F}_M(\mathcal{X}_P)| \exp[i\tilde{\Phi}_M(\mathcal{X}_P)] \equiv 1$. Therefore, considering (100) and eqn. (II. 7a, b) of Table II, column A, we obtain:

Amplitude-contrast transfer-function

$$\mathcal{D}_I(\mathcal{R}=\mathcal{X}_P) = \cos\left[2\pi\mathcal{W}(\mathcal{X}_P)\right] = \cos\left[2\pi\left(\frac{\mathcal{X}_P^4}{4} - \Delta\frac{\mathcal{X}_P^2}{2}\right)\right]; \tag{116a}$$

Phase-contrast transfer-function

$$\mathcal{B}(\mathcal{R}=\mathcal{X}_P) = -2\sin\left[2\pi\mathcal{W}(\mathcal{X}_P)\right] = -2\sin\left[2\pi\left(\frac{\mathcal{X}_P^4}{4} - \Delta\frac{\mathcal{X}_P^2}{2}\right)\right]. \tag{116b}$$

Only one example for (116a) is given in Fig. 11. Otherwise, we restrict our considerations to (116b), i.e. we deal with transfer conditions for pure and weak phase objects only. For further particulars see Hanszen and Morgenstern (1965) and Hanszen (1966a).

Three families of phase contrast transfer-functions of practical importance are given in Fig. 12. The generalized defocusing Δ is the parameter. The first thing that we recognize is the contrast formation by means of objective aberrations only, i.e. phase contrast formation without any masking in the pupil. Unfortunately, this contrast varies with the spatial frequency. The contrast transfer functions oscillate

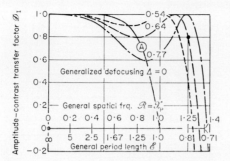

Fig. 11. Generalized amplitude-contrast transfer-functions for generalized defocusings $0 \leqq \varDelta < 1$. Transfer-function A of this figure belongs to the wave aberration A in Fig. 7. (Hanszen 1966a).

between $\pm$ 2. Therefore, one part of the spatial frequency spectrum is imaged positive while the other part shows a negative contrast. We call a contrast positive if object details of advancing phase, i.e. of more positive inner potential, appear bright in the image.

The contrast transfer-functions have zero points. Therefore, there are spatial frequencies which are not transferred to the image; i.e. the image has *frequency-gaps*.

Dependent on the sign of the transfer-function between two successive zero points, the corresponding frequency interval is called the *transfer interval* with positive or negative phase contrast. The position and magnitude of these intervals depend on $\varDelta$. The maximum values $|\mathscr{B}_{max}| = 2$ of $\mathscr{B}$ are reached at generalized frequencies $\mathscr{R}_{ex}$ for which

$$\mathscr{W} = n/4; \qquad n = \pm\, 1; \qquad \pm\, 3; \qquad \pm\, 5.\ldots \qquad (117)$$

Therefore, the transfer properties can also be easily determined from the wave aberration curves in Fig. 7. If the microscope is in focus, i.e. $\varDelta = 0$, the transfer intervals are very narrow because the curve of the related wave aberration is steep. If the defocusing counteracts spherical aberration, i.e. $\varDelta > 0$, the intervals are broadened. The above is true for a slightly underexcited lens.

Highly favourable transfer functions are obtained when the wave aberration curves have an extremum, the height of which is given by one of the values quoted in (117). The transfer interval, including such an extremum, is especially broad. We call it the *main transfer interval*. The extrema in the main transfer intervals are characterized by the following values of $\mathscr{W}$, $\mathscr{X}_{P}$, and $\varDelta$

$$\mathscr{W}_{ex} = -\,\varDelta_n^2/4; \qquad \mathscr{X}_{Pex} = \sqrt{\varDelta_n}; \qquad \varDelta_n = \sqrt{-\,n};$$
$$n = -\,1;\, -\,3;\, -\,5.\ldots \qquad (118)$$

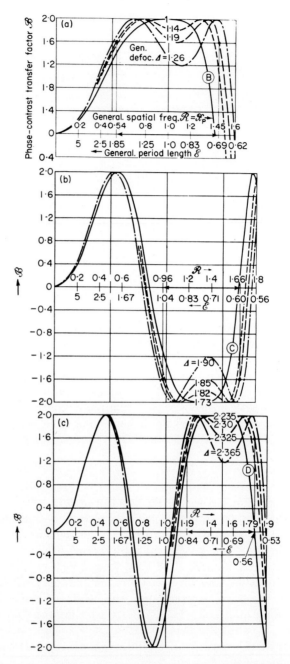

FIG. 12. Generalized phase-contrast transfer-functions in the vicinity of $\Delta = +1$; $+\sqrt{3}$ and $+\sqrt{5}$. The effective limits of the main transfer intervals are marked by arrows on the abscissa. The wave aberrations belonging to the transfer-functions B, C, and D are marked by the corresponding letter in Fig. 7 (Hanszen, 1966a).

The main transfer interval appears at higher spatial frequencies as defocusing is increased. It is impossible, however, to move this interval continuously along the frequency-coordinate by varying Δ.

The three graphs in Fig. 12 show the phase-contrast transfer-functions producing main transfer intervals with $\Delta = 1$; $\sqrt{3}$; $\sqrt{5}$. The useful width of the main transfer interval is marked by arrows; for details see Hanszen (1966a). Micrographs showing point distances below 4 Å are usually obtained by applying a defocusing $\Delta \approx \sqrt{5}$. In this case, these structures are transferred by the main transfer interval shown in Fig. 12c.

The essential information range of a micrograph is determined by the limits of the main transfer interval. According to (98), *both* limits depend on spherical aberration. If this aberration decreases, the main transfer interval moves towards higher spatial frequencies. However, in the scale of period lengths, it gets seriously narrower (see the numerical values given in Table III). The conventional assumption that the "resolution" of an electron microscope is improved by lowering the spherical aberration is, therefore, very dubious.

The defocusing steps available on an electron microscope should be small enough to give a wave aberration necessary for obtaining a main transfer interval. Figure 13, curve b, demonstrates the effect of a small deviation from the nominal defocusing value. In Table III, the values for the desired minimum steps are given. Compare these values

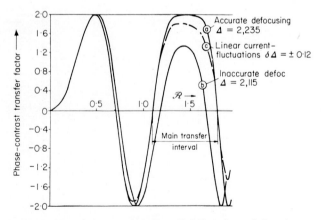

FIG. 13. Influence of an inaccurately adjusted defocusing, of electric current fluctuations and of a finite illumination aperture on the phase contrast transfer function with $\Delta = \sqrt{5}$.

(a): accurate adjustment; (b): "underfocus" with $\Delta I/I = 3.5 \cdot 10^{-6}$ in the case of the normal objective; (c): like (a) but with linear current fluctuations of maximal value $\Delta I/I = \pm 3.5 \cdot 10^{-6}$; (Hanszen 1966b). $\delta\Delta$ in the figure equals δ_l in eq. (121).

TABLE III

Electron microscopical data for high resolution microscopy at 100kV

		A	B
Positive phase contrast according to Fig. 12a	Period lengths, transferred with a contrast higher than 80%	12,4—4,6 Å	7,4—2,9 Å
	Radii in the pupil plane, limiting the main transfer interval	≈ 8,1—21,6 μm	≈ 6,0—16,2 μm
	Defocusing $\Delta I/I$	— 3,5 · 10^-5	— 3,6 · 10^-5
	Smallest defocusing steps required	1,5 · 10^-6	1,5 · 10^-6
Negative phase contrast according to Fig. 12b	Period lengths, transferred with a contrast higher than 80%	7,0—4,0 Å	4,2—2,4 Å
	Radii in the pupil plane, limiting the main transfer interval	≈ 14,3—24,7 μm	≈ 10,7—18,5 μm
	Defocusing $\Delta I/I$	— 5,4 · 10^-5	— 5,6 · 10^-5
	Smallest defocusing steps required	0,9 · 10^-6	0,9 · 10^-6
Positive phase contrast according to Fig. 12c	Period lengths, transferred with a contrast higher than 80%	5,6—3,8 Å	3,4—2,2 Å
	Radii in the pupil plane, limiting the main transfer interval	≈ 17,7—26,7 μm	≈ 13,3—19,9 μm
	Defocusing $\Delta I/I$	— 6,8 · 10^-5	— 7,1 · 10^-5
	Smallest defocusing steps required	0,75 · 10^-6	0,75 · 10^-6

Column A: Using a "normal" objective, the data of which are given in Section IIB.
Column B: Using a high-efficiency objective, $f' = 1\cdot2$ mm; $C_s = 0\cdot5$ mm; $C_{ch} = -0\cdot7$ mm.

to the relative energy width of 1×10^{-5} or greater which is usually present in the electron beams because of the thermal velocity distribution and Boersch-effect; for details see Section VID.

The results of the present section can be summarized as follows: The portion of the electron microscope from the electron source to the exit pupil can be considered as a spectroscopic instrument for spatial frequencies. The lens aberrations act as a filter for those frequencies. Some frequencies are completely missing. The modulation of others has decreased or even changed sign. The filter properties depend on the amount of defocusing.

C. *The Effect of Axial Astigmatism*

When axial astigmatism is present, defocusing depends on the azimuth $\theta = \arctan(R_y/R_x)$, if the axes of astigmatism are located in the R_x and R_y directions of the diffraction plane and $R = \sqrt{R_x^2 + R_y^2}$ is the spatial frequency in question. We can write

$$\Delta(\theta) = \Delta_m - \Delta_a \cos 2\theta \tag{119}$$

where $2\Delta_a$ is the defocusing difference due to astigmatism and Δ_m the mean defocusing. Experiments and calculations on the effect of astigmatism on the frequency spectrum of the image have been published by

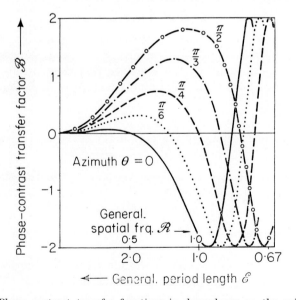

FIG. 14. Phase contrast transfer functions in dependence on the azimuth θ; the astigmatic defocusing difference is assumed to be $2\Delta a = 0.72$; ($\Delta_m = +0.48$).

Thon (1968b). Transfer theory characterizes this aberration by means of a family of transfer functions with θ as a parameter. An example is given in Fig. 14.

D. *Current and Voltage Fluctuations and the Energy Width of the Illuminating Beam*

According to what has already been explained about transfer theory, the electron microscope image has frequency gaps, and the transfer intervals following the main transfer intervals become smaller and smaller. The number of these intervals is not limited. The maximum contrast $|\mathscr{B}_{max}| = 2$ should be reached at a certain frequency within each interval.

The situation is different if we consider fluctuations of voltage or current in the instrument. According to (16) they cause variable defocusing resulting in temporal fluctuations of the contrast transfer functions. The effective transfer function is the mean value of the function during the time of recording. In order to obtain this mean value, we have to know the temporal variation of the accelerating voltage U and lens current I. We give here some results only and refer for details to Hanszen and Trepte (1970; 1971a).

When the fluctuations are symmetrical with respect to the mean value, the effective phase contrast transfer function is given by

$$\overline{\mathscr{B}} = G \cdot \mathscr{B} = -2G \sin (2\pi \mathscr{W}), \qquad (120)$$

where $G = G(\delta, \mathscr{R})$ is the envelope of the transfer function. G depends only on the generalized spatial frequency $\mathscr{R}$ and on a quantity δ which characterizes the fluctuation. For linear current fluctuations (e.g. sawtooth fluctuations) with the maximum generalized deviation $\pm \delta_l$, we have, for instance,

$$G_l = \sin (\pi |\delta_l| \mathscr{R}^2)/\pi |\delta_l| \mathscr{R}^2. \qquad (121)$$

For sinusoidal fluctuations with a maximum deviation $|\delta_s|$, we obtain

$$G_s = J_0(\pi |\delta_s| \mathscr{R}^2), \qquad (122)$$

where J_0 is the zero order Bessel-function; and for fluctuations obeying the *Gaussian error function* we obtain

$$G_g = \exp [-(\pi \delta_g/4)^2 \mathscr{R}^4/\ln 2] \qquad (123)$$

where δ_g characterizes the half-width of the fluctuations in generalized coordinates.

In each case, the zero points of the ideal transfer functions (Table II.7) remain unchanged. An example of the influence of (121) on the phase contrast is shown in Fig. 13, curve c. Comparing this curve with curve b, which characterizes an imperfect defocusing, we see that fluctuations between the defocusing values $\pm \delta_l$ have less influence on contrast than a defocusing difference δ_l. This is easy to understand, because the fluctuations are characterized by a *mean* value and the imperfect defocusing by an *extreme* value.

Looking back at the special case given in (123), a family of envelopes for the contrast transfer function, with δ_g as parameter, are drawn in Fig. 15. When G_g drops below a certain value, no contrast is produced

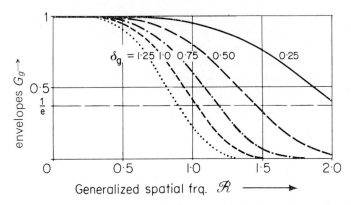

FIG. 15. Envelopes G_g of the contrast transfer functions when current or voltage fluctuations, having a Gaussian distribution, occur. The half width of this distribution is expressed by the generalized defocusing δ_g.

in practice. If this value is $1/e$, the $\mathscr{R}$-value for which $G_g = 1/e$, i.e.

$$\mathscr{R}_e = 2\sqrt[4]{ln2/(\pi\delta_g)^2} \qquad (124)$$

gives the practical limit of the contrast transfer function. In terms of the smallest transferred period length ϵ_e we obtain the following equation using eqns. (16) and (99); (102) with a mean current fluctuation $\Delta I/I$:

$$\epsilon_e = \sqrt{\frac{\pi}{2(ln2)^{\frac{1}{2}}} \lambda |C_{ch}| \left| \frac{\Delta I}{I} \right|}. \qquad (125)$$

From this we see the importance of chromatic aberration as it affects the transfer limit.

Example: Using the above mentioned data for the normal objective and assuming current fluctuation of the order 1×10^{-5}, we obtain $\epsilon_e = 3{\cdot}8$ Å. This value is too large for resolving atomic distances.

Since the energy distribution in the illuminating beam is asymmetric, its influence on the optical transfer is more difficult to calculate. It can, however, in most cases be described approximately by (123) using adapted values for δ_g; see Hanszen and Trepte (1971a).

E. Experimental Confirmation of the Transfer Theory

Thon's experiments (1965–1968) can be regarded as proof of contrast theory. He used carbon foils 100 Å thick as objects. As known from electron diffraction these foils have a continuous frequency spectrum, the intensity of which, although decreasing with $1/\mathscr{R}^4$, is still clearly noticeable for $\mathscr{R} = 2$. It is a necessary but not sufficient criterion for perfect imaging that the spectrum of a micrograph is the same as that of the object; the differences between object spectrum and micrograph spectrum give evidence about optical transfer.

Since electron micrographs are highly magnified, the structures on the photographic plate are large enough to be studied by diffraction of light waves. For this purpose a small aperture is put into a laser beam, in order to spread the beam by diffraction and illuminate a sufficiently large area of the micrograph. The diffraction pattern of this area— called a diffractogram—appears around the image of the aperture produced by a lens and can be photographed.

We assume that the transparency for the wave function of the object and the transparency for the image intensity are

$$F(x) = 1 + i \int_0^\infty \tilde{\Phi}(R) \cos\left[2\pi Rx + \xi_\Phi(R)\right] dR; \tag{126}$$

$$F'_i(x') = 1 + \int_0^\infty B(R)\tilde{\Phi}(R) \cos\left[2\pi Rx' + \xi_\Phi(R)\right] dR$$

$$= 1 - 2 \int_0^\infty \tilde{\Phi}(R) \sin\left[2\pi\mathscr{W}(R)\right] \cos\left[2\pi Rx' + \xi_\Phi(R)\right] dR \tag{127}$$

$$= 1 - \int_{-\infty}^{+\infty} \tilde{\Phi}(R) \sin\left[2\pi\mathscr{W}(R)\right] \exp\left\{i[2\pi Rx' + \xi_\Phi(R)]\right\} dR,$$

since $2\pi\mathscr{W}$ is even in R.

We assume further that the density on the photographic plate is roughly given by the same expression. Local variations in thickness occurring in photographic emulsions can be compensated by using an appropriate immersion liquid. Then, the Fourier transform of the light-optical transparency for the wave function in the plate is

$$\tilde{\tilde{F}}(R = 0) = 2x_e; \qquad \tilde{\tilde{F}}(R \neq 0) = \tilde{\Phi}(R \neq 0) \sin\left[2\pi\mathscr{W}(R)\right] \exp\left[i\xi_\Phi(R)\right] \tag{128}$$

and the intensity-transparency of the diffractogram is

$$\left(\frac{\tilde{\bar{F}}(R=0)}{2x_e}\right)^2 = 1; \qquad \left(\frac{\tilde{\bar{F}}(R \neq 0)}{2x_e}\right)^2 = \tilde{\varPhi}^2(R \neq 0)\sin^2\left[2\pi\mathscr{W}(R)\right]/4x_e^2$$

$$= \tilde{\varPhi}^2(R \neq 0)B^2(R)/16x_e^2. \quad (129)$$

In other words, the diffraction pattern shows the product of the object spectrum and the square of the phase contrast transfer function. In Fig. 16 this behavior is demonstrated for a particular object (126) under

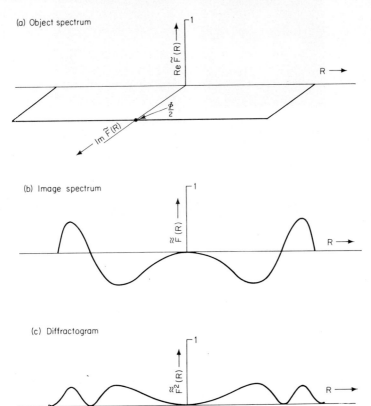

FIG. 16. The diffractometer method. Object: $F(x) = 1 + i\tilde{\varPhi}\int\cos(2\pi Rx)dx$.
(a) Fourier representation $\tilde{F}(R)$ of the object, identical with the object related diffraction transparency; (b) Fourier representation $\tilde{\bar{F}}(R)$ of the micrograph, identical with the transparency of the wave function occurring in the diffraction plane of the diffractometer; (c) Transparency $\tilde{\bar{F}}^2(R)$ of the intensity in the recording plane of the diffractometer, similar to the density-distribution in the diffractogram.

the assumption $\tilde{\varPhi} = \text{const}$; $\xi_\varPhi = 0$. Figure 16c represents a photometric curve recorded diagonally through the diffractogram. If the objective has axial astigmatism, these curves depend on the azimuth θ.

This means that the diffractogram method can also be used to determine astigmatism; compare Fig. 17 with Fig. 14. Figure 18 shows the different appearance of the same object area at different degrees of defocusing. Some characteristic diffractograms are also given. A detailed discussion

a)   b)

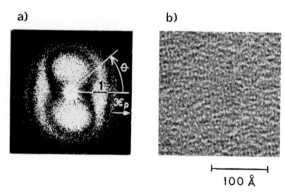

100 Å

FIG. 17. Diffractogram (a) of a micrograph (b) having astigmatism. The diffractogram describes the same facts as Fig. 14. (Thon 1968b, Fig. 18.)

of the structures appearing in different micrographs is given by Hanszen (1967). There is no "sharp" image within the micrographs of a focal series. There is no way of knowing how such an image could be distinguished from the others (Thon, 1967).

In order to evaluate the diffractograms of a focal series, the generalized defocusing is related to an arbitrary zero point; as a function of this parameter, diameter and width of the diffraction rings are drawn in a diagram. By displacing this diagram along the Δ — axis of another diagram showing the theoretical curves, the experimental data can be fitted to the theoretical functions, as is done in Fig. 19. In this manner, the absolute defocusing can be determined from the entire focal series. The experimental values fit only to the theoretical curves characterizing phase contrast, but not to the curves characterizing amplitude contrast. Therefore in Fig. 19, we have an experimental proof that a thin carbon foil is in a good approximation a pure phase object.

Obviously, the diffractograms do not give the sign of the contrast. The fact that the neighbouring rings belong to transfer intervals of opposite sign is an additional piece of theoretical information.

Contrast inversions dependent on defocusing can also be directly recognized in the micrographs. Especially when there is a considerable amount of defocusing, i.e. $(\Delta/2)\mathcal{X}_P^2 \gg \mathcal{X}_P^4$, the phase contrast transfer functions are odd functions of Δ. Changing the sign of Δ, therefore,

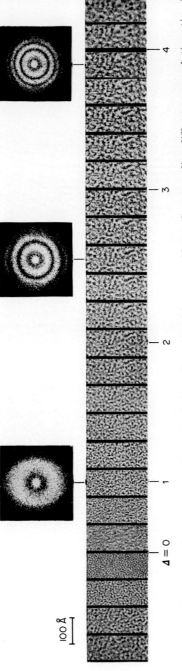

FIG. 18. Below: Focal series of a carbon foil, showing the same object area. Above: Corresponding diffractograms. Astigmatism is completely corrected (Thon 1966a).

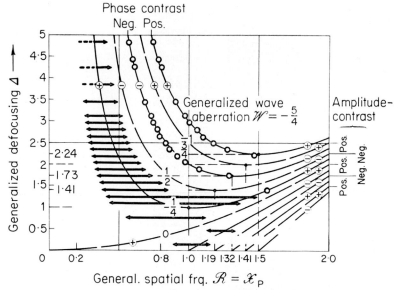

FIG. 19. Maximum contrast (○) and width (⟷) of the first transfer interval, data plotted in a $\Delta(\mathscr{R})$—diagram; experimental values given by Thon (1968b).

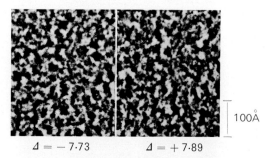

FIG. 20. Total change of phase contrast by inverting the sign of a strong defocusing (Thon, 1967).

results in micrographs that are negatives of the previous ones, although each micrograph is composed of transfer intervals with alternating sign (see Fig. 20). Since amplitude-contrast transfer functions are even, Fig. 20 gives a further proof to the fact that a thin carbon foil is a pure phase object.

F. *Limitation of the Objective Aperture*

If an aperture with the generalized diameter $\mathscr{X}_{pe} = \mathscr{R}_e$ is used, we have:

$$|\tilde{F}(\mathscr{X}_P)| = 1; \quad \tilde{\Phi}_M(\mathscr{X}_P) = 0 \qquad \text{for } |\mathscr{X}_P| \leq |\mathscr{X}_{Pe}|,$$

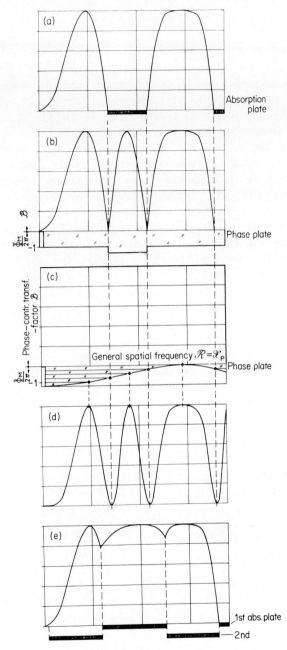

FIG. 21. Methods for contrast improvement, demonstrated for the case of curve D in Fig. 12c. (a) eliminating the transfer intervals with negative sign (Lenz, 1963); (b) inversion of the sign in these intervals (Hanszen and Morgenstern, 1965); (c) correction of the wave aberration in order to obtain the value $(n - 1)\, \lambda/4$ by a phase plate having locally variable thickness (Hanszen and Morgenstern, 1965); (d) light optical reconstruction (Gabor, 1949); (e) double exposure of a micrograph using proper defocusing and complementary zone plates (Hanszen, 1966b).

and

$$\tilde{F}(\mathscr{X}_P) = 0 \qquad \text{for } |\mathscr{X}_P| > |\mathscr{X}_{Pe}|.$$

Taking into consideration (104), we obtain the same contrast-transfer functions (Table II.7) in the interval $|\mathscr{X}_P| \leq |\mathscr{X}_{Pe}|$ as would be obtained without using an aperture. However, for $|\mathscr{X}_P| > |\mathscr{X}_{Pe}|$, the contrast transfer functions vanish. Unlike the case of incoherent illumination (see Section XB) the aperture stop now merely acts as a low-pass filter. It has, however, no influence on the contrast transfer of the remaining spatial frequencies and consequently does not raise the contrast of weak objects. This is true because higher terms in the ($R = 0$) part of (62) could be omitted. Therefore, in the image of a weak object, amplitude or phase contrast cannot be produced by using an aperture stop. On the other hand, in the image of strong objects, aperture contrast can be produced. In this case, however, the linear transfer theory fails, so that no transfer function can be defined. In weak objects, the aperture can eliminate only transfer intervals with the wrong sign.

G. *Absorption Plates*†

The improvement of transfer when using zone plates for amplitude objects (Hoppe, 1961; 1963) and phase objects (Lenz, 1963)‡ can be described by transfer theory as follows. If the transfer intervals with negative sign are eliminated by opaque zone rings, a contrast transfer function is obtained which has only positive values. There are, however, wide frequency gaps (see Fig. 21a). These theoretical predictions have been confirmed by diffractometer experiments of Möllenstedt *et al.* (1968) (see Fig. 22). It is not possible to replace the lost information by

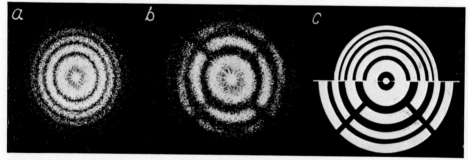

Fig. 22. Diffractograms of a carbon foil by Thon (1968) demonstrating the action of a Lenz-plate: (a) using no zone plate, (b) using a zone plate, (c) schematic diagram explaining the filter effect; see Hoppe *et al.* (1969), Fig. 12.

† Compare with Section VI G–J, also Tsujiuchi (1963).
‡ For a correction of the radii given, see Hanszen, Morgenstern and Rosenbruch (1963), especially p. 485.

other information stored in the same micrograph, as discussed in Section VE. If objective apertures, especially zone plates, are used, the image points deteriorate into diffraction patterns of the aperture. Maskings of this kind are therefore advantageous only if the resulting disks are smaller than the original disks due to lens aberrations.

H. *Weakening of the Undiffracted Beam*

If the zero order is totally removed, the phenomena described in Table II.E occur. In this case, there is no linear transfer. Thus, the contrast depends on the object and no contrast transfer functions can be defined; for further investigations see Hanszen (1969a).

In order to increase the modulation (i.e. the "contrast") in the image intensity, the wave function of the zero diffraction order can be weakened, as long as the wave functions of all diffraction orders are small compared to the zero order. There is an experimental complication, however, because every weakening is combined with phase delays. A suggestion by Le Poole, for performing the weakening without any perturbations in the image field due to diffraction in the absorption plate, is mentioned by Hanszen (1969a, p. 126).

I. *Phase Plates*

The possibilities of correcting micrographs by means of phase plates are discussed by Hanszen and Morgenstern (1965) on the basis of transfer theory. From (113)–(114) we see that the pupil function can be made unity, if one introduces in the pupil plane a non-absorbing, phase shifting foil with a hole for the zero diffraction order. The thickness of this foil should vary locally in such a manner, that the wave aberration of the objective is compensated up to a remainder $2\pi n$; this means that

$$2\pi \mathscr{W}(\mathscr{X}_P) + \tilde{\varPhi}_M(\mathscr{X}_P) = 2\pi n; \qquad n = 0; \qquad \pm 1; \qquad \pm 2 \ldots . \quad (130)$$

A foil of this kind would produce an amplitude-contrast transfer-function $\mathscr{D}_I(\mathscr{R}) \equiv 1$. In this case, the image of an amplitude object would be perfect; i.e. all spatial frequencies would be transferred with an equal and optimal positive amplitude contrast factor.

In order to obtain equal and optimal positive contrast for phase objects, i.e. $\mathscr{B}(\mathscr{R}) \equiv 2$, the phase shift $\tilde{\varPhi}_M$ of the foil should be given by

$$2\pi \mathscr{W}(\mathscr{X}_P) + \tilde{\varPhi}_M(\mathscr{X}_P) = \begin{cases} 2\pi(n - 1/4); & \text{for } \mathscr{X}_P \neq 0 \\ 2\pi n; & \text{for } \mathscr{X}_P = 0 \end{cases} \quad (131)$$

Since non-absorbing phase plates are realizable in a sufficient approximation, the manufacturing of plates according to (131) should be possible in spite of technical difficulties. The periodic potential of the atoms does not noticeably disturb the image field as long as the decisive period lengths are small compared with the diameter of the sampling points; for in this case the beams diffracted by the foil fall outside the image field.

Until zone plates of this kind are produced we have to be satisfied with a discontinuous zone plate. According to Fig. 21b such plates should have the same radii as the absorption plate in Fig. 21a. The transparent zones of the absorption plate should be replaced by zones which shift the phase by $2\pi n$, the opaque zones by zones which shift the phase by $2\pi \ (n + 1/2)$. Plates of this kind do not suppress the frequency intervals of negative contrast, but rather invert the contrast in these intervals. According to Fig. 21b, however, considerable transfer gaps still remain. Although manufacturing of such plates is not more difficult than that of absorption plates, they have not yet been employed to improve imaging.

J. *Reconstruction of Electron Micrographs by Light Optical Methods*

The reconstruction method proposed by Gabor (1948–1951) can be explained by transfer theory as follows (Hanszen, 1970a). The electron micrograph of the object (126), with the transparency (127), is illuminated in the same way as is done with monochromatic light in the diffractometer. Except for a fixed scaling factor, the light optical system used for reconstruction should have the same generalized transfer function as the electron microscope. In this case, the reconstructed image has the intensity distribution

$$F_I''(x'') = 1 + \text{const} \int\limits_0^\infty B^2(R \neq 0) \cos\left[2\pi R x'' + \xi_I(R)\right] dR \qquad (132)$$

and therefore, a positive definite transfer function $B^2(R)$ or $\mathscr{B}^2(\mathscr{R})$: see Fig. 21d. The effect of the unavoidable frequency gaps of the resulting transfer functions should, however, not be underestimated.

The micrograph on the photographic plate has an amplitude and a phase structure. If the amplitude structure is bleached, a pure light-optical phase object remains, and the light-optical step can be directly described by one of the phase-contrast transfer functions of the family $\mathscr{B}(\mathscr{R})$. The only experimental requirement is to adjust the generalized light-optical defocusing to the generalized defocusing in the electron microscopical step. If the phase structure in the unbleached plate is suppressed by means of appropriate immersion, a pure amplitude object remains, the imaging of which is described by an amplitude-contrast transfer function of the family $\mathscr{D}_I(\mathscr{R})$ if no other measures are taken. This function, however, can be converted into the corresponding phase-contrast transfer function $\mathscr{B}(\mathscr{R})$ by means of a $\lambda/4$-plate, inserted into the zero order of the diffraction spectrum. In this case the reconstruction, which leads to contrast of uniform sign, is also described by (132); for details see Hanszen 1970a.

Better results should be obtained by *single-sideband holography*. When one side of the diffraction pattern in the microscope is cut off by an opaque mask, the only imperfections of the micrograph are lateral displacements (136) of the spatial frequencies (see transfer function (II.7b) in column D of Table II). The image will be perfectly reconstructed, when these displacements are compensated in the light optical process. This can be done by a masking which leads to the conjugate complex transfer function (Table II; D, 7b) (see the model experiments by Hanszen, 1969b; 1970b).

K. *Information Gain by Evaluating Focal Series: Using Complementary Objective Diaphragms*

If the image is not improved by masking methods, there will be frequency gaps in the transfer functions of Fig. 12 which in many cases may be serious. In order to interpret unknown structures, the evaluation of only a single micrograph may therefore lead to misinterpretations.

The frequency gaps move with defocusing; this means that the information, missing in one micrograph, can be substituted by information from other micrographs of a focal series. Conventional micrographs are less appropriate for information replenishment since, according to Fig. 12, contrast alternations create intricate conditions. Reconstruction should rather be based on micrographs like Fig. 21a, b or on reconstructions as in Fig. 21d. Starting from micrographs like those in Fig. 21a was proposed by Hanszen (1966b) and verified (1970a) by light-optical model experiments; see Fig. 23. The image improvement of a phase object is interpreted in Fig. 21e: the first micrograph is produced using zone plate I and the defocusing $\Delta = 1$. The frequency gap is filled, in this micrograph, by a second micrograph in which zone plate II and the defocusing $\Delta = \sqrt{5}$ are used.

By varying Δ only, it is impossible to obtain two phase-contrast transfer functions such that all minima of the first function coincide with the maxima of the second one. Therefore, it is impossible to obtain an image containing the whole frequency spectrum by reconstructing it from only two original micrographs. At worst, the number of required plates photographed at carefully calculated Δ-values is equal to the number of the frequency gaps to be removed. Unfortunately, the contrast in the reconstruction decreases with the number of micrographs used, because the undiffracted beam which produces the background is needed for constructing *each* image (see Section VD).

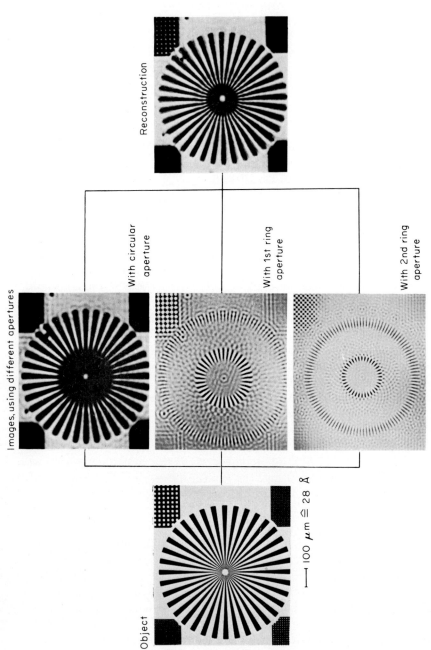

FIG. 23. Transfer improvement by complementary objective diaphragms, demonstrated by means of a light-optical amplitude object. In order to eliminate the spatial frequencies suffering contrast alterations, the objective aperture has to be severely limited (image, shown above); the frequencies lost in the first image can be restored by the method described in Section VIK (images at the centre and the bottom); (Hanszen, 1968, where the captions of Figs. 1 and 3 have to be exchanged).

VII. Contrast by Means of Phase Shift: Contrast Transfer Functions for Oblique Illumination

A. *Without Masking*

Calculations for axial illumination are easy, because the spatial frequency R can be identified with the reduced pupil coordinate S related to the axis point of the pupil plane, and therefore $\mathscr{R}$ with $\mathscr{X}_P$. This is not true for oblique illumination. In this case the wave aberration must be separated, according to (107) and (108), into an odd component $W_u(R)$ and an even one $W_v(R)$. Writing $\xi = 2\pi \mathscr{W}_u(R)$ and $\zeta = 2\pi \mathscr{W}_v(R)$, it can be seen from Table II.7a,b.C that the contrast transfer functions are complex. The even component of the wave aberration determines the amount and the odd component the phase of the contrast transfer function. Therefore, in addition to the effects known from axial illumination, lateral displacements of the spatial frequencies occur in the image. For this reason, it is not advantageous to operate with oblique illumination without aperture limitation.

B. *Elimination of One Side of the Diffraction Pattern*

If a symmetrical objective aperture is used and the illumination is oblique enough to place the zero diffraction order near the edge of the objective aperture, the simple relations of (Table II.7.D) exist. We give the results in generalized coordinates. The coordinate $\mathscr{Q}$ of the undiffracted beam coincides with the coordinate $\mathscr{X}_{Pe}$ of the edge, and the generalized spatial frequency, as defined in (98), is

$$\mathscr{R} = \mathscr{X}_P - \mathscr{X}_{Pe} = \mathscr{X}_P - \mathscr{Q}. \tag{133}$$

For the phase shifts χ, we can write

$$\chi = 2\pi[\mathscr{W}(\mathscr{X}_P) - \mathscr{W}(\mathscr{X}_{Pe})] \tag{134}$$

and, according to (II.7.D) for the contrast transfer functions,

$$\mathscr{D}_I = (1/2) \exp\{2\pi i[\mathscr{W}(\mathscr{X}_p) - \mathscr{W}(\mathscr{X}_{Pe})]\};$$
$$\mathscr{B} = \exp\{2\pi i[\mathscr{W}(\mathscr{X}_P) - \mathscr{W}(\mathscr{X}_{Pe}) + (1/4)]\}. \tag{135}$$

The absolute value of the transfer factor, therefore, is the same for all spatial frequencies and independent of defocusing. Using (27), the lateral displacement is $\delta \mathscr{X}'$ of the image structures with respect to the object structures is

$$\delta \mathscr{X}'(\mathscr{R}) = -[\mathscr{W}(\mathscr{X}_P) - \mathscr{W}(\mathscr{X}_{Pe})]/\mathscr{R};$$

for amplitude transfer or

$$\delta \mathscr{X}'(\mathscr{R}) = -[\mathscr{W}(\mathscr{X}_P) - \mathscr{W}(\mathscr{X}_{Pe}) - (1/4)]/\mathscr{R} \tag{136}$$

for phase transfer respectively. This is dependent on spatial frequency and defocusing. From (136) we learn that the essential image properties can already be derived from the wave aberration curves.

The phase-contrast transfer properties will now be discussed with the aid of Fig. 24. We assume defocusing to be $\varDelta = \sqrt{2}$, by which the shape of the wave aberration curve is given, and the coordinates of the diaphragm edge to be $\mathscr{X}_{Pe} = \pm 1{\cdot}56$. Furthermore we assume the undiffracted beam to intersect the pupil plane at $\mathscr{X}_{Pe} = -1{\cdot}56$. As can be seen from the diagram, the highest transferred spatial frequency is twice that in the case of central illumination. Spatial frequencies, for which $\mathscr{W}(\mathscr{X}_P) - \mathscr{W}(\mathscr{X}_{Pe}) = n - 1/4$ with $n = 0, \pm 1, \pm 2 \ldots$ holds, do not suffer lateral displacements.

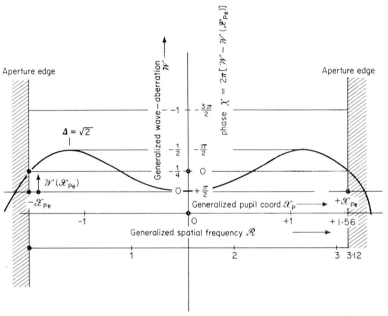

Fig. 24. An example showing the phases χ occurring with oblique illumination. The zero diffraction order touches the aperture edge.

In order to reduce the lateral displacements of the other frequencies as far as possible, tilting of the illumination and defocusing should be adjusted in such a way:

(1) that the most important frequency interval is covered by an extremum of the wave aberration,
(2) that this extremum has the height $(n - 1/4)$, compared with the wave aberration at the zero order.

For further details see Hanszen and Morgenstern (1965), Hanszen and Trepte (1971b, c); for details about single-sideband holography see Hanszen (1969b; 1970b).

C. *Practical Applications*†

The discussion of contrast transfer under oblique illumination was brief because of its limited range of application. As we have seen in the last section, the transfer limit can be raised by the factor of 2 in one coordinate direction. Unfortunately, however, transfer is drastically reduced in the perpendicular direction.

One-sided oblique illumination is therefore only suitable for investigating very special objects, e.g. one or a small number of lattice planes having spacings so small that it is impossible to separate them by axial illumination (see Dowell, 1963; Komoda *et al.*, 1964–1967). In order to obtain more information about crystal lattices from electron micrographs than from electron diffraction patterns, it should be expected that lattice images indicate the position of each lattice plane in a reliable way. It has been established, however, that lateral displacements occur depending on spatial frequency (i.e. reciprocal lattice spacing) and defocusing. Only if (136) vanishes for the considered lattice spacing $1/\mathscr{R}$ is the information obtained on the lattice position correct (especially with respect to the crystal boundaries). Unfortunately, it is difficult to find out if this requirement is fulfilled. Lateral displacements always occur when the rotational symmetry of the optical system is disturbed, as is the case for one-sided oblique illumination. The symmetry can be restored by using conical oblique illumination; see the general considerations of Hanszen and Morgenstern (1965). This illumination, however, is a particular case of partial coherent illumination (Hanszen and Trepte, 1971b, c); see also Section XA.

VIII. The Problem of Point Resolution

A. *The Concepts of Contrast Transfer and Resolution Limit*

So far the concept of "resolving power" has been avoided. If the imaging properties of an electron microscope are described by a "resolution limit", one may be led to think that all structures larger than the "smallest resolved distance" are transferred perfectly and all smaller structures not at all. By means of transfer functions a very detailed description of the imaging properties can now be given.

Contrast transfer theory is based on "spatial frequencies" as the object elements to be analyzed, while older theory is based on "object points" as elements; see Section IVC. The two concepts differ from each other in a similar way as do the corpuscular and wave interpretations of matter. An infinite spectrum of spatial frequencies is related to

† For single-sideband holography, see Section VIJ.

an object point. For this reason, in accurate discussions, the reciprocal spatial frequency should not be identified with the distance between two object points as is often done. In fact, a similar complementarity between point-image and transfer function can be given as is known from quantum theory (see e.g. Röhler (1967), p. 34f).

B. *Definition of Point Resolution*

Two object points are said to be resolved when their Airy-disks overlap in such a way, that the intensity in the saddle between the two maxima is 75% or less of the intensity in the maxima. In most calculations of the intensity distribution in the image of the two disks, incoherent illumination is assumed (see Glaser 1956, p. 371). Figure 25 shows, however, that in electron microscopy, where illumination is practically coherent, the overlapping is more complicated. Thus we should not identify the half-width of the diffraction disk of a *single* atom with the "point resolution" of the electron microscope. Point resolution can only be determined by calculating the overlapping of the wave function in *both* disks and taking interference effects into account. Shape and size of the diffraction disks, and therefore also the resolution, depend on aperture, spherical aberration, and defocusing. In a paper by Scherzer (1949) the influence of these parameters was treated, at least in all fundamentals.

C. *Transfer-Theoretical Treatment of Point Resolution*

We now study the image formation of an object having a homogeneous background interrupted by two small, equal phase details (shortly called "object points") at a distance x apart. In the diffraction pattern, the shape of these details is described by the wave function at the pupil points $R_s = k/x$ with $k = \pm 1;2;\ldots$. That the object detail repeats only twice is due to the fact that each spectral point is a "sampling point" with a *finite* width; see Section VG. In order to resolve the point distance x in the micrograph, it is sufficient to use only the zero diffraction order and the spectral points with $k = \pm 1$. In order to produce positive phase contrast, it is necessary to have a wave aberration $\mathscr{W} = n - 1/4$ with $n = 0; \pm 1 \ldots$ at these points. Furthermore, it is necessary to maintain this value of the wave aberration in an extended interval around the pupil points considered, because the information of having only two object points, which is stored in the width of the sampling points, must also be transferred with phase contrast into the image. This condition is best fulfilled for the generalized pupil coordi-

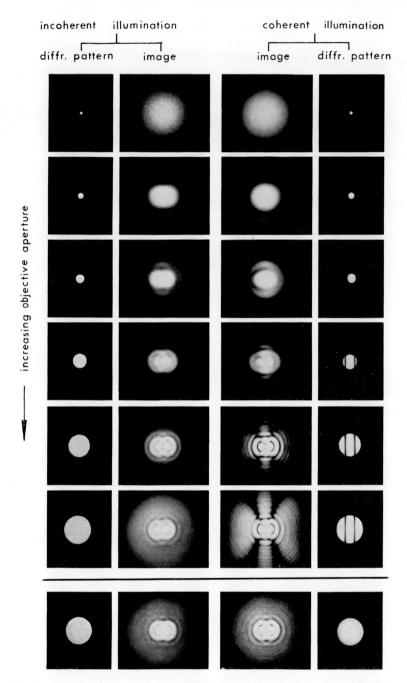

incoherent illumination | coherent illumination

diffr. pattern | image | image | diffr. pattern

increasing objective aperture

Fig. 25. Light-optical model experiments demonstrating point resolution. Object: diaphragm with two holes of 9 μm diameter at a distance of 18 μm; $\lambda_c = 6328$ Å; $C_S = 1660$ mm; $\Delta = 0$. Left column: using incoherent illumination, the intensities are to be added. Right column: using coherent illumination, the wave functions are to be added. The Fresnel fringes appearing are *spurious structures*. Row at the bottom: double exposure; in each case, one of the holes was covered.

nate $\mathscr{X}_P = 1$ and a defocusing $\varDelta = 1$; then n equals zero. Relating $\mathscr{X}_P$ to the distance x of the object points according to (115), we obtain the condition for resolving this distance,

$$x = \sqrt[4]{C_s \lambda^3} \tag{137}$$

which complies with the resolution limit of the older theory up to a factor of about $1/2$. The new description is more reliable, because (137) was derived under the assumption of coherent illumination and under the requirement of producing phase contrast. Its statements, however, are much more restricted than the statement of older resolution theory because there is no doubt that phase contrast transfer of two object details having distances smaller *or* larger than (137) is made questionable by the new theory. This means that even in the simple case of weak objects it is impossible to describe the image quality by a single numerical value, e.g. the resolution limit. Instead, we need the contrast transfer functions as evaluated in previous chapters.

D. *Practical Consequences*

At best, a resolution limit x_{min} can be given if a transfer function exists for which all period lengths $\epsilon \geq x_{min}$ are transferred with sufficient contrast. Since an electron microscopist is not interested in imaging the "coarse" structures known from light microscopy, the unavoidable frequency gap of the phase-contrast transfer functions in the vicinity of $\mathscr{R} = 0$ (see Fig. 12) raises no problems. Thus, the resolution limit of the electron microscope is given in practice by the high-frequency end of the main transfer-interval in Fig. 12a. Consequently, in generalized coordinates, $\mathscr{E}_{min} = 1/\mathscr{R}_{min} = 1/1\cdot45 = 0\cdot69$. The resolution limit of the normal objective defined in this sense would be $x_{min} = 4\cdot6$ Å.

As has already been pointed out, an alternating transfer function such as given in Fig. 12c is generally used to obtain high resolution micrographs. In this case, the resolution limit is usually quoted to be $\mathscr{E}_{min} = 1/1\cdot79 = 0\cdot56$ (i.e. $x_{min} = 3\cdot8$ Å for the normal objective). In light optics, however, resolution of this kind would be called "spurious resolution". Röhler (1967) writes in his monograph (p. 61): We should speak of spurious resolution in this case, as such micrographs "normally cannot be used for imaging problems because the shape of non-periodic objects is incorrectly reproduced". This criterion of light optics illustrates the problematic situation of present electron microscopy as well as the need to develop reconstruction methods and microscopes with better transfer functions. In order to elucidate the present situation, we refer to the experimental results of Thon (1967), especially p. 450, showing that all micrographs of a focal series show the same resolution. The indication of a resolution limit by means of point resolution tests is, therefore, a very unreliable criterion for the quality of an image.

IX. Aperture Contrast

A. *Comparison Between Contrast by Phase Shift and Aperture Contrast*

We have seen, that in the case of *weak objects* a symmetric objective aperture serves only the purpose of eliminating those frequencies which are transferred to the image with the wrong sign of contrast; it has, however, no influence on the contrast transfer factor of the spatial frequencies that appear in the image (see Fig. 26). With coherent illumination, therefore, no enhancement of resolution can be achieved by giving the objective aperture an optimal size, as is possible with incoherent illumination.

The aperture stop, however, does raise the contrast of strong objects. This effect is well known as "scattering-absorption contrast". Unfortunately, the aperture diaphragm eliminates the highest spatial frequencies from the image. Instead of them, phase structures of larger dimension become visible in the image. The contrast transfer conditions for imaging a phase edge are discussed in detail by Hanszen *et al.* (1963; p. 483).

In the case of *weak objects*, no aperture contrast is possible and the

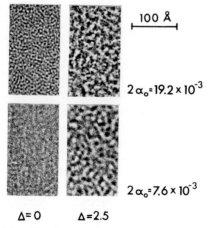

$2\alpha_0 = 19.2 \times 10^{-3}$

$2\alpha_0 = 7.6 \times 10^{-3}$

$\Delta = 0$ $\Delta = 2.5$

Fig. 26. Influence of the objective aperture $2\alpha_0$ on the image of a weak object (carbon foil). Micrographs taken with the "normal" objective described in Section IIB. The objective aperture only limits the spatial frequency spectrum in the image without influencing the transferred frequencies.

contrast caused by phase shifts can be described by linear theory. In the case of *strong objects*, aperture contrast is possible, but it does not follow the linear theory. Since in the latter case the imaging depends on the object, only characteristic examples can be discussed (see Hanszen *et al.* (1963), p. 481ff; and Hanszen and Morgenstern (1965), pp. 219f and 223). In the following section some supplementary considerations about the strong phase object (49) are given.

B. *An Example for Aperture Contrast*

In (49) the diffraction spectrum of a strong single-spatial-frequency phase object was given. We saw that the Fourier coefficients for $|k| > 1$ are significant only for the amount of modulation, but not for the value of the frequency in the object. For this reason, we want to find out whether we can eliminate the diffraction orders $|k| > 1$. If we do so by using an aperture stop, the image-related transparency in the diffraction plane is

$$\tilde{F}'(R = 0) = J_o(\tilde{a});$$
$$\tilde{F}'(R = \pm R_1) = iJ_{+1}(\tilde{a}) \exp\left[2\pi i \mathscr{W}(R_1)\right], \tag{138}$$

where $\mathscr{W}(R_1)$ describes the influence of spherical aberration and defocusing. From this, we obtain the wave function and the intensity transparency in the image:

$$F'(x') = J_o(\tilde{a}) + 2iJ_1(\tilde{a}) \exp\left[2\pi i \mathscr{W}(R_1)\right] \cos 2\pi R_1 x' \tag{139}$$

$$\left.\begin{aligned}
F'_I(x') =\ & J_o^2(\tilde{a}) + 2J_1^2(\tilde{a}) - 4J_o(\tilde{a})J_1(\tilde{a}) \sin 2\pi \mathscr{W} \cos 2\pi R_1 x' + \\
& \qquad\qquad\qquad\qquad + 2J_1^2(\tilde{a}) \cos 2\pi 2 R_1 x' \\
=_{df} \quad A(\tilde{a}) \quad + \quad B(\tilde{a}, \mathscr{W}, R_1) \quad + \\
& \qquad\qquad\qquad\qquad + C(\tilde{a}, 2R_1)
\end{aligned}\right\} \tag{140}$$

The background A of the image is modulated by two terms B and C. Even in the case $\mathscr{W}(R_1) \equiv 0$—where no contrast formation by phase shifts is possible—the modulation term C remains. It therefore expresses the aperture contrast. Unfortunately, the image intensity does not contain the object frequency R_1, but rather the double frequency $2R_1$, (see Fig. 27, curve a). The decisive requirement of contrast theory, that each frequency of the image has to correspond to the same frequency in the object, is now violated. Moreover, the intensity modulation in the image is not proportional to $\tilde{a}$; thus the contrast transfer depends on the object. The modulation in the image completely disappears whenever $J_1(\tilde{a}) = 0$; that is when $\tilde{a} = 3\cdot83; 7\cdot02$, etc.

As a result of the wave aberration the basic frequency can be transferred into the image as well (see term B in (140) and Fig. 27, curve b). This transfer also depends on the object. It disappears for $J_o(\tilde{a}) = 0$

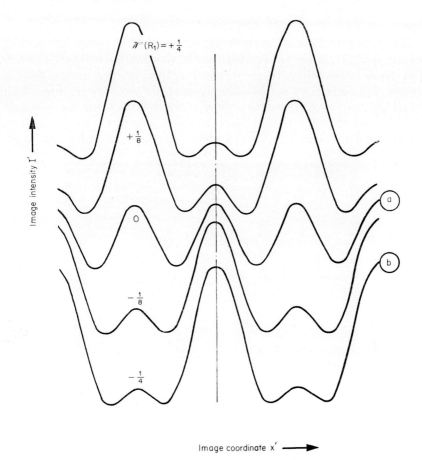

$\mathscr{W}'(R_1) = +\frac{1}{4}$

$+\frac{1}{8}$

0

$-\frac{1}{8}$

$-\frac{1}{4}$

Image intensity I'

a

b

Image coordinate x'

FIG. 27. Intensity distribution in the image of the strong phase object $F(x) = 1 + i\tilde{a}$ $\cos 2\pi R_1 x$, limiting the diffraction pattern to $-R_1 < R < +R_1$: defocusing, expressed by $\mathscr{W}(R_1)$, varies. (Transfer-theoretical interpretation of a diagram given by Komoda, 1964.) The micrographs corresponding to curves a and b are shown in Fig. 28.

and for $J_1(\tilde{a}) = 0$. Consequently, there are object thicknesses for which the modulation in the image vanishes completely. Nevertheless, at some specific values of defocusing only the frequency $2R_1$ is present in the image. It is impossible, however, to have only the basic frequency in the image.

If a phase object consists of numerous spatial frequencies $R_1; R_2 \ldots$ R_k, it may happen that a low diffraction order of one frequency coincides with a higher order of another frequency. If the corresponding area of the pupil plane is masked, the transfer of *several* frequencies is affected and image interpretation is difficult.

C. *Spurious Structures*

The theoretical results of (140) and Figure 27 have been confirmed by experiment; Komoda (1964). Figure 28a shows a micrograph which is "in focus", according to geometrical optics. It does not show the lattice spacings 12·6 Å actually present in the object, but "half spacings" of 6·3 Å. Structures of this kind (sometimes inaccurately called "defocusing artifacts") are *spurious structures*. The defocused micrograph in Fig. 28b, on the other hand, shows natural lattice spacings.

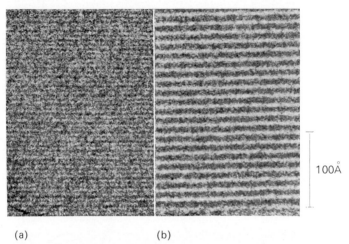

(a)   (b)

Fɪɢ. 28. Electron micrographs showing the same portion of a Cu-phthalocyanine crystal, taken with axial illumination and symmetric objective aperture but slightly different focusing. The corresponding curves in Fig. 27 are marked by a and b (Komoda, 1964.)

It is difficult to define the concept "spurious structures" in electron microscopy; for particular considerations see Hanszen (1969a). It is convenient to reserve this term for smaller details, i.e. for spatial frequencies appearing in the image that are higher than those existing in the object. Knowing the radius R_e of the object aperture, it is easy to say that all structures $x' < 1/R_e$ appearing in the micrograph of a phase object are spurious structures†.

Dark field images may also lead to objectionable spurious structures in the micrograph; for a detailed discussion see Hanszen (1969a). A drawing, given by Hanszen (1967; Fig. 4c and b), illustrates how spurious structures are produced when imaging strong phase objects with aperture contrast or in dark field.

† This statement does not hold for amplitude objects, where the object *intensity* is usually the reference quantity, see Hanszen and Morgenstern (1965), p. 223, footnote 1.

X. Contrast Transfer for Partially Coherent and for Incoherent Illumination

A. *Partially Coherent Illumination*

As previously mentioned, coherent illumination, i.e. illumination by a point source, is an idealization. In reality, sources of finite diameter are used, and different points of the source radiate incoherently. If the source is not too large, illumination is *partially coherent* and the effective transfer function can be calculated by taking the mean value of the transfer functions for each point of the source. Numerical calculations of these transfer functions and experiments with a light optical model take a long time; a publication on them is under way (Hanszen and Trepte, 1971b, c). Here we give only some general results. For small illumination apertures, the zero points and the maximum contrast values of the main transfer intervals remain unchanged. For larger apertures, the zero points move and the contrast always decreases. As shown in Fig. 29, there are values of defocusing for which contrast transfer functions can be produced with a uniform sign, but with a rather low contrast transfer factor, by widening the illumination

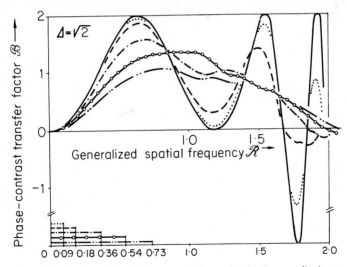

Fig. 29. A family of phase-contrast transfer-functions $\mathscr{B}(\mathscr{R})$. Parameter is the generalized illumination aperture $\mathscr{Q}$ (see the scale at the abscissa); defocusing is $\varDelta = \sqrt{2}$. The frequency gap disappears when the illumination aperture is increased. $\mathscr{Q} = 0.18$ corresponds to an illumination semi-aperture of $1 \cdot 10^{-3}$ for the normal objective.

aperture. In the case of a normal objective, the semi-illumination angles needed are not much larger than 1.10^{-3}.

A ring condenser gives a special kind of partially coherent illumination. Such a ring-shaped source can be regarded as composed of pairs of point sources located under every azimuth at equal distances from the axis and radiating incoherently. It was shown by Hanszen and Morgenstern (1965) that, contrary to one-sided oblique illumination, this illumination eliminates lateral displacements. On the basis of the formulae given there, contrast transfer functions for conical illumination are obtained by integrating over the azimuth θ. Under optimal conditions with this type of illumination, transfer functions having uniform sign up to high spatial frequencies are obtained. However, contrast is strongly reduced for all frequencies; see Hanszen and Trepte (1971c).

B. *Incoherent Illumination*

An illumination is said to be incoherent, when the illumination aperture is so large that the undiffracted beam fills the whole objective aperture. With incoherent illumination, phase structures cannot be shown. Hence, such an illumination is advantageous when the phase component of a mixed object is to be suppressed. In order to increase the amplitude component, a low voltage microscope combined with an energy filter has to be used. For such an instrument, see Wilska (1965).

With incoherent illumination, the intensity transparency of the object is linearly transferred into the intensity distribution of the image regardless of the modulation depth of the object. Here it may be sufficient to give only a few details pertinent to electron microscopy; for particulars see Hanszen, Rosenbruch and Sunder-Plassmann (1965).

When the undiffracted beam fills the entire pupil the beams diffracted by the spatial frequency R_k in the object have also the cross section of the pupil. However, their centres are displaced by $\pm R_k$. Only those portions of these beams which are hatched in Fig. 30a pass through the aperture and participate in image formation. For coherent illumination the frequency R_k is no longer transferred, as in the case drawn in Fig. 30a at the right, because the centre of one circle lies outside the radius of the other. For incoherent illumination this frequency appears, although weakly, in the image, because the hatched area in the figure is not yet reduced to zero. The amplitude contrast transfer function for an aberration free objective (Fig. 30c; curve $h_i/h_a = 0$) is nearly triangular; it ends at $\mathcal{R}_e = 2$, which is double the generalized radius of the aperture. When we use the same aperture and incoherent illumination, therefore, the highest transferred spatial

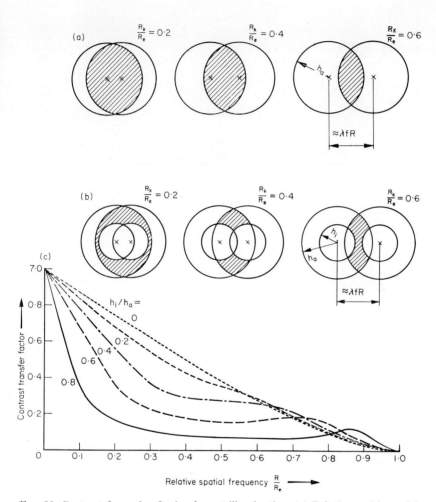

FIG. 30. Contrast formation for incoherent illumination. (a) Relative positions of the undiffracted beam and of the beam which is diffracted at the spatial frequency R_k in the pupil plane; R_k is increasing from left to right. (c) Contrast transfer-functions of the aberration-free objective, using ring apertures in the objective having the same outer radius h_a but different inner radii h_i, as shown in (b). The area of the pupil plane hatched in Fig. 30b is responsible for contrast transfer (Linfoot, 1957).

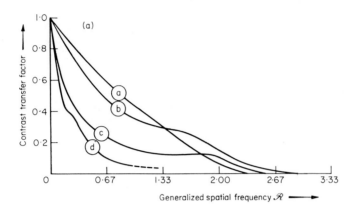

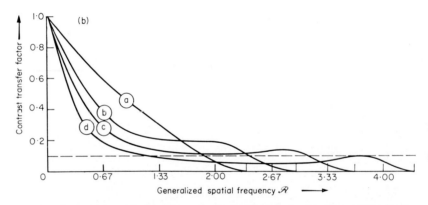

Fig. 31. Contrast transfer-functions of a real objective for incoherent illumination. The generalized radii of the objective aperture are denoted by $\mathscr{H}$ (Hanszen, Rosenbruch and Sunder-Plassmann, 1965).

(A) Circular objective apertures:

    (a) $\mathscr{H}_a = 1\cdot17$; $\varDelta = 0\cdot70$
    (b) $\mathscr{H}_a = 1\cdot50$; $\varDelta = 1\cdot08$
    (c) $\mathscr{H}_a = 1\cdot84$; $\varDelta = 1\cdot30$
    (d) $\mathscr{H}_a = 2\cdot17$; $\varDelta = 1\cdot90$

(B) Ring apertures:

    (a) $\mathscr{H}_a = 1\cdot17$; $\mathscr{H}_i = 0\cdot00$; $\varDelta = 0\cdot70$
    (b) $\mathscr{H}_a = 1\cdot50$; $\mathscr{H}_i = 0\cdot68$; $\varDelta = 1\cdot45$
    (c) $\mathscr{H}_a = 1\cdot84$; $\mathscr{H}_i = 1\cdot12$; $\varDelta = 2\cdot50$
    (d) $\mathscr{H}_a = 2\cdot17$; $\mathscr{H}_i = 1\cdot74$; $\varDelta = 4\cdot00$

frequency in the image of an amplitude object, with arbitrary modulation depth, is twice as high as in the image of a weak object using coherent illumination†.

If a circular aperture is used, and the diameter of it is so small that the spherical aberration gives no wave aberration larger than $\lambda/4$ at any point inside the aperture, the true transfer function coincides with that for perfect imaging; see Fig. 31a, curve a. There are no practical focusing problems in this case. If a larger aperture is used the influence of the spherical aberration should be compensated by a slight underfocusing in order to keep the wave aberration within the hole as small as possible. In this manner, transfer functions can be obtained which are extended to higher spatial frequencies without appreciable loss of contrast at low frequencies; see curve b, Fig. 31a. The improvement declines rapidly if the aperture is increased further, see curves c and d. It is possible, however, to raise the contrast transfer factor close to the resolution limit by using ring apertures in the objective. Unfortunately the contrast transfer factor decreases for low frequencies in this case. Figures 30b, c demonstrate the improvement of transfer by ring apertures, using an aberration-free objective and a fixed outer radius of the aperture ring. Figure 31b demonstrates the same for a lens with aberration, using apertures with adapted ring radii and adapted defocusing, in order to obtain optimal transfer functions. The primary condition is to keep the changes of the wave aberration below $\lambda/4$ within the free opening of the ring. No appreciable improvement in transfer can be expected when using zone plates with several rings.

XI. Test Objects

In order to determine the contrast transfer properties of an electron microscope, test objects are needed having continuous spatial frequency spectrum in the interval of interest. In light optics, radial gratings (as in Fig. 23) are mainly used for this purpose. There are no such test objects for electron microscopy.

We have already seen in Section VIE that a thin carbon foil is often a suitable test object, because it has a "white" spatial frequency spectrum with sufficiently intense diffracted waves (Thon, 1965–68). The diffractogram of an electron image of such a foil gives far-reaching information about the contrast transfer functions (see for example (129) and Fig. 18). Unfortunately, a carbon foil has no regular structures and contrast transfer properties can hardly be determined visually from the random image structure.

† For incoherent illumination, there are no objections to covering the central portion of the objective aperture.

An example of an object having a regular structure and containing all spatial frequencies is given by a phase edge. The spectrum of such an object is weak, however, and therefore it is difficult to record and to evaluate. Information on contrast transfer in this case can be obtained from the Fresnel fringes appearing in the image of the object. It has been shown (Hanszen and Morgenstern, 1966), that Fresnel fringes are less characteristic for electron microscopical contrast transfer functions when the objective aperture is limited. In this case, the shape of the fringe system depends on the aperture, but the contrast transfer functions do not. Better information can be expected in experiments without an aperture stop.

If a biprism interference field (see Möllenstedt and Düker, 1956) replaces the object in an electron microscope, wave aberration and contrast transfer cannot be obtained from the fringe *contrast* in the image (Lenz, 1965b). However, one can derive them from the *lateral displacements* (Hanszen, 1967). Experiments of this kind appear to be very difficult to carry out.

Artificial objects such as proposed by Boersch (1943) can serve as test objects also (see Hanszen (1967)). Such experiments have not yet been done. For information concerning transfer tests by means of an enlarged light optical model for the electron microscope, see Hanszen (1968–1971); an example is given in Section VIK.

XII. Historical Remarks

The references in the preceding chapters concern only papers contributing further details; historical aspects were not considered. Therefore, a short historical evaluation is given in the following.

The results of light optical imaging and transfer theory can be assumed to be known; see the summary by Born and Wolf (1964) and the monograph by Röhler (1967) with numerous bibliographical references. Special mention should also be made of the publication of Zernike (1942) about image formation with special relation to phase contrast, and its contrast theoretical interpretation by Menzel (1958 and 1960) and Hauser (1962), which extended the theory to problems of particular interest in electron microscopy.

The wave optical imaging theory of the electron microscope began with the papers of Boersch (1936–1943) on the importance of Abbe's diffraction pattern and his recognition of the fact that electron microscopical objects behave as phase objects (Boersch, 1947). Proposals were already made in these papers for phase shifting foils in the diffraction plane, and the phase shifting influence of spherical aberration was discussed.

The theory of electron microscopical resolution (Glaser, 1943) turned only slowly away from incoherent illumination. Focusing conditions were first included by Scherzer (1949); Figs 3 and 4 of his paper gave examples, although unrecognized ones, of phase-contrast and amplitude-contrast transfer-functions. Further details about imaging theory were presented by v. Borries (1949), Uyeda (1955) and Haine (1957), considering partially the concept of phase objects. These articles as well as some more recent ones (Vyazigin and Vorobev, 1963; Eisenhandler and Siegel, 1966b, Reimer, 1966, 1969; Heidenreich *et al.*, 1965–1967) are more concerned with the problem of resolving atomic structures than with instrumental transfer. Moreover, the distinction between "contrast formation by phase shift" as an instrumental method and "phase-contrast imaging" (i.e. imaging of phase structures in the object as intensity structures in the image) as a requirement for image formation is not always strictly observed. The proposal for and the realization of zone plates as objective apertures (Hoppe; Langer and Hoppe, 1961–1967; Lenz, 1963; Eisenhandler and Siegel, 1966a; Siegel *et al.*, 1966; Möllenstedt *et al.*, 1968) spurred new studies about image formation. The question, how such a masking influences contrast formation, could be answered by transfer theory (Hanszen *et al.*, 1963–69; Lenz, 1965a, b; Morgenstern, 1965). This theory closely connects the contrast and resolution problems, which earlier were treated rather separately. For several years it has been known (v. Borries and Lenz, 1956; Lenz and Scheffels, 1958; Faget *et al.*, 1960; Fert, 1961; Fagot *et al.*, 1961; Farrant and McLean, 1965; Dowell and Farrant, 1965; Hahn, 1965) that object structures the size of which depends on defocusing can be emphasized in the image. In investigations of this kind Albert *et al.* (1964) and Thon (1965) also considered the influence of spherical aberration, which was neglected in the previous papers. Independently of transfer theoretical considerations they came to nearly the same results as that theory. Thon (1965–1968) introduced the carbon foil as electron microscopical test object and used successfully the diffractometer method, known from X-ray investigations.

To quote all publications on the problem of resolving lattice spacings in crystals would lead us too far. The papers of Komoda (1964–1967) should be mentioned, however. A theoretical treatment of these experiments brought results which are discussed in transfer theory from a general point of view.

The possibility of reconstructing electron micrographs by holography was first recognized by Gabor (1948–51); see also Haine and Mulvey (1952). Transfer theory is capable of outlining the possibilities of this method by describing it in terms of the spatial frequency concept and comparing the method with other procedures for image improvement.

Contrast theory enjoys increasing popularity among the designers of electron microscopes, who need an imaging theory which is independent of the object. Electron microscopists have been more reserved, however. It is important to them, of course, to know which misinterpretations of electron micrographs are possible, under which conditions they may occur, and how we can avoid them. It was the author's desire to work out these points for the readers of this article.

ACKNOWLEDGEMENTS

I wish to express my gratitude to Professor Alvar P. Wilska, Professor Jan B. Le Poole and Professor Kurt W. Just in Tucson, Arizona, as well as to Dr. Rolf Lauer and Dr. Lutz Trepte in Braunschweig, Germany, for reading the manuscript and for many useful remarks.

REFERENCES

Albert, L., Schneider, R. and Fischer, H. (1964). *Z. Naturforschg.* **19a,** 1120.
Albert, L. (1966). *Optik* **24,** 18.
Archard, G. D. (1958). *Rev. Sci. Instr.* **29,** 1049.
Boersch, H. (1936). *Ann. Phys.* (5) **26,** 631; (5) **27,** 75.
Boersch, H. (1938). *Z. Techn. Phys.* **19,** 337.
Boersch, H. (1943). *Phys. Z.* **44,** 202.
Boersch, H. (1947). *Z. Naturforschg.* **2a,** 624.
Boersch, H. (1948). *Optik* **3,** 24.
v. Borries, B. (1949). *Z. Naturforschg.* **4a,** 51.
v. Borries and Lenz, F. (1956). Electron Microscopy: Proc. Stockholm Conf. (Almqvist and Wiksell, Stockholm), p. 60.
Born, M. and Wolf, E. (1964) "Principles of Optics", 2nd Edition, Pergamon Press, Oxford.
Dowell, W. C. T. (1963). *Optik* **20,** 535.
Dowell, W. C. T. and Farrant, J. L. (1965). Int. Conf. on Electron Diffraction and Crystal Defects, Melbourne. p. I P-2.
Eisenhandler, C. B. and Siegel, B. M. (1966a). *J. appl. Phys.* **37,** 1613.
Eisenhandler, C. B. and Siegel, B. M. (1966b). *Appl. Phys. Lett.* **8,** 258.
Faget, J. Fagot, M. and Fert, Ch. (1960). Electron Microscopy: Proc. Delft. Conf., (Nederlandse Vereniging voor Electronenmicroscopie, Delft), Vol. I, p. 18.
Fagot, M., Ferré, J. and Fert, Ch. (1961). *C.r. hebd. Seanc. Acad. Sci. (Paris)* **252,** 3766.
Farrant, J. L. and McLean, J. D. (1965). Conf. on Electron Diffraction and Crystal Defects, Melbourne. p. I P-5.
Fert, Ch. (1961). *J. Phys. Rad.* **22,** 26S.
Gabor, D. (1948). *Nature* **161,** 777.
Gabor, D. (1949). *Proc. R. Soc.* **A 197,** 454.

Gabor, D. (1951). *Proc. Phys. Soc.* **B 64**, 449.

Glaser, W. (1943). *Z. Phys.* **121**, 647.

Glaser, W. (1956). *In* Handb.d.Phys., (ed. S. Flügge), Springer, Berlin, Vol. 33, 123 ff.

Hahn, M. (1965). *Z. Naturforschg.* **20a**, 487.

Haine, M. E. (1957). *J. Sci. Instr.* **34**, 9.

Haine, M. E. and Mulvey, T. (1952). *J. opt. Soc. Am.* **42**, 763.

Hanszen, K.-J., Morgenstern, B. and Rosenbruch, K. J. (1963) *Z. ang. Phys.* **16**, 477.

Hanszen, K.-J., Rosenbruch, K. J. and Sunder-Plassmann, F. A. (1965). *Z. ang. Phys.* **18**, 345.

Hanszen, K.-J., and Morgenstern, B. (1965). *Z. ang. Phys.* **19**, 215.

Hanszen, K.-J., and Lauer, R. (1965). *Optik* **23**, 478.

Hanszen, K.-J. (1966a). *Z. ang. Phys.* **20**, 427.

Hanszen, K.-J. (1966b). Electron Microscopy: Proc. Kyoto Conf. (Maruzen, Tokyo), Vol. 1, p. 39.

Hanszen, K.-J. (1966c). Physikertagung München, Fachberichte, p. 94.

Hanszen, K.-J. and Morgenstern, B. (1966). *Optik* **24**, 442.

Hanszen, K.-J. (1967). *Naturwiss.* **54**, 125.

Hanszen, K.-J. (1968). Electron Microscopy: Proc. Rome Conf. (Tipografia Poliglotta Vaticana, Rome), Vol. 1, p. 153.

Hanszen, K.-J. (1969a). *Z. ang. Phys.* **27**, 125.

Hanszen, K.-J. (1969b). *Z. Naturforschg.* **24a**, 1849.

Hanszen, K.-J. (1970a). *Optik* **32**, 74.

Hanszen, K.-J. (1970b). *Microscopie électronique; 7me congrès international, Grenoble.* Vol. **I**, p. 21.

Hanszen, K.-J. and Trepte, L. (1970). *Microscopie électronique; 7me congrès international, Grenoble.* Vol. **I**, p. 45.

Hanszen, K.-J. and Trepte, L. (1971a). *Optik* **32** (in press).

Hanszen, K.-J. and Trepte, L. (1971b). *Optik* **33** (in press).

Hanszen, K.-J. and Trepte, L. (1970c). *Optik* **33** (in press).

Hauser, H. (1962). *Opt. Acta.* **9**, 121.

Hawkes, P. W. (1968). *Brit. J. appl. Phys.* (2), **1**, 131.

Heidenreich, R. D. and Hamming, R. W. (1965). *Bell System Tech. J.* **44**, 207.

Heidenreich, R. D. (1966). *Bell System Tech. J.* **45**, 651.

Heidenreich, R. D. (1967). *J. Electron Microscopy* **16**, 23.

Hopkins, H. H. (1961). Proc. Conf. on Opt. Instr. and Technol. London, ed. K. J. Habell (Chapman and Hall, London), p. 480.

Hoppe, W. (1961). *Naturwiss.* **48**, 736.

Hoppe, W. (1963). *Optik* **20**, 599.

Hoppe, W., *et al.*, (1969). *Siemens Review* **36**, 24.

Jeschke, G., and Niedrig, H. (1970). *Acta Cryst.* **A, 26**, 114.

Komoda, T. (1964). *Optik* **21**, 93.

Komoda, T. and Otsuki, M. (1964). *Jap. J. appl. Phys.* **3**, 666.

Komoda, T. (1966). *Jap. J. appl. Phys.* **5**, 603.

Komoda, T. (1967). Proc. 25th EMSA Meeting (Claitor, Baton Rouge), p. 234.

Kunath, W. and Riecke, W. D. (1965). *Optik* **23**, 322.

Langer, R. and Hoppe, W. (1966). *Optik* **24**, 470.

Langer, R. and Hoppe, W. (1967). *Optik* **25**, 413 and 507.

Lenz, F. and Scheffels, W. (1958). *Z. Naturforschg.* **13a**, 226.

Lenz, F. (1963). *Z. Phys.* **172,** 498.

Lenz, F. (1964). *Optik* **21,** 489.

Lenz, F. (1965a). *Lab. Invest.* **14,** 808.

Lenz, F. (1965b). *Optik* **22,** 270.

Linfoot, E. H. (1957). *Optica Acta* **4,** 12.

Linfoot, E. H. (1964). "Fourier Methods of Optical Image Evaluation", Focal Press, London, p. 15.

Lohmann, A. and Wegener, H. (1955). *Z. Phys.* **143,** 413.

Menzel, E. (1958) *Optik* **15,** 460.

Menzel, E. (1960). "Optics in Metrology", Pergamon Press, Oxford, p. 283.

Möllenstedt, G. and Düker, H. (1956). *Z. Phys.* **145,** 377.

Möllenstedt, G., Speidel, R., Hoppe, W., Langer, R., Katerbau, K.-H., and Thon, F. (1968). Electron Microscopy: Proc. Rome Conf. (Tipografia Poliglotta Vaticana, Rome), Vol. 1, p. 125.

Morgenstern, B. (1965). *Z. Naturforschg.* **20a,** 972.

Nagendra Nath, N. S. (1939). *Akust. Z.* **4,** 263.

O'Neill, E. L. (1963). "Introduction to Statistical Optics", Addison Wesley Publishing Corp., Reading, Mass.

Reimer, L. (1966). *Z. Naturforschg.* **21a,** 1489.

Reimer, L. (1969). *Z. Naturforschg.* **24a,** 377.

Röhler, R. (1967). "Informationstheorie in der Optik", Wiss. Verl. Ges., Stuttgart.

Scherzer, O. (1949). *J. appl. Phys.* **20,** 20.

Shannon, C. E. (1949). *Bell System Techn. J.* **27,** 379, 623.

Siegel, B. M., Eisenhandler, C. B. and Coan, M. G. (1966). Electron Microscopy: Proc. Kyoto Congress (Maruzen, Tokyo), Vol. 1, p. 41.

Thon, F. (1965). *Z. Naturforschg.* **20a,** 154.

Thon, F. (1966a). *Z. Naturforschg.* **21a,** 476.

Thon, F. (1966b). Electron Microscopy: Proc. Kyoto Congress (Maruzen, Tokyo), Vol. 1, p. 23.

Thon, F. (1966c). Physikertagung, München, Fachberichte, p. 101.

Thon, F. (1967). *Phys. Blätter, Mosbach* **23,** 450.

Thon, F. (1968a). Electron Microscopy: Proc. Rome Conf. (Tipografia Poliglotta Vaticana, Rome), Vol. 1, p. 127.

Thon, F. (1968b). Thesis, Tübingen.

Tsujiuchi, J. (1963). *Prog. Optics* **2,** 131.

Uyeda, R. (1955). *J. phys. Soc. Japan* **10,** 256.

Vyazigin, A. A. and Vorobev, Y. V. (1963). *Bull Akad. Sci. USSR Phys. Ser.* **27,** 1103.

Wilska, A. P. (1965) *Lab. Invest.* **14,** 825.

Zernike, F. (1942). *Physica* **9,** 686 and 974.

Image Processing for Electron Microscopy: I. Enhancement Procedures

ROBERT NATHAN

*Jet Propulsion Laboratory, California Institute of Technology,
Pasadena, California, U.S.A.*

I. Introduction	. . . .	85
II. Problems of High Resolution	. . . . . .	86
A. Spherical aberration and instrument instability	. . .	86
B. Contrast	. . . .	86
C. Specimen damage	. . . .	87
D. Image sensor noise	. . . .	87
III. Computer Image Processing	. . . . . .	87
A. Geometric correction	. . . . . .	89
B. Photometric correction	. . . . . .	89
C. Random noise removal	. . . . . .	95
D. System noise removal	. . . .	97
E. Scan line noise removal	. . . . . .	101
F. High-frequency attenuation correction	. . . .	101
IV. Enhancement of Periodic Images	. . . . .	102
V. High Resolution by Computer Synthesis	. . . .	113
Acknowledgements	. . . .	119
References	. . . .	119
Appendix	. . . .	120

I. Introduction

TODAY's electron microscope is still not capable of atomic resolution for bio-organic compounds. It is shown here that the digital computer with associated video camera scanner equipment can be used in conjunction with the microscope's best performance to give improved and ultimately perhaps atomic resolution. Examples of computer application are limited to space photography, where the computer enhancement methods originated, and to a micrograph of negatively stained catalase. Work is proceeding in our laboratories to couple successfully a sensitive video camera system to an electron microscope as a prerequisite to the computer-microscope marriage.

A theoretical discussion of the combination of bright and dark field images is included, which should lead to a synthetic aperture for the

Work prepared under contract No. NAS 7–100
National Aeronautics and Space Administration

objective lens in order to obtain 1 Å resolution. Some of the practical problems in performing these operations are also discussed.

II. Problems of High Resolution

One Ångstrom unit resolution should reveal atomic geometry. There are several reasons why existing electron microscopes have not yet attained this capability when examining biological molecules: spherical aberration and instrument instability, lack of contrast for biological material, specimen damage by the electron beam and vacuum, and image sensor noise.

A. *Spherical Aberration and Instrument Instability*

Spherical aberration limits the useful aperture of the electron objective lens. The error due to this defect causes an object point to be spread over an image area and makes it appear out of focus. In a more technical sense spherical aberration affects the phases of higher spatial frequency components in a calculable manner, causing that scattered portion of the beam which passes further from the lens center not to be deflected accurately back to the position it would have if it remained close to the axis of the lens. The information of the higher spatial frequencies is recorded but not in an obviously useful manner.

Recently, several microscope manufacturers have announced instruments which have highly stabilized lens currents and high voltage, so that even though spherical aberration may still be present to the extent of limiting resolution to 3 Å or 4 Å, the image is stationary to 1 Å or less for a period of many seconds. From this stationary image it may be possible to recover 1 Å resolution by computer transformation to the Fourier or spatial frequency domain. A correction for phase error can be performed by the computer, and then a transformation would be made back to the real plane to reconstruct a higher resolution image. Discussion of this type of image processing must be postponed, however, until some experimental experience has been obtained with one of these better instruments.

For the present discussion it will suffice to accept the effects of spherical aberration and impose a requirement of a limited objective aperture, limiting apparent resolution to 3 Å per single recorded image.

B. *Contrast*

Any attempt to use a computer to overcome aberration limitations is directly dependent upon contrast even though the instrument is stable to better than 1 Å. The simple definition of resolution must also be questioned in terms of contrast.

If the ratio of signal-to-noise is high enough, it should be possible to record information which discloses two peaks separated by a very small dip in intensity. The visibility of this dip is directly dependent upon the signal-to-noise ratio and it will not in general agree with the arbitrary definition of resolution established by Rayleigh which, for incoherently illuminated opaque object points, involves a dip of 25%. A computer may be used to improve the signal-to-noise ratio and thereby the visibility of the image. An example is described in Section IV. A basic difficulty is that unstained material must be used if atoms are to be resolved, with the consequence that contrast is minimal in bright field imaging, although it can be improved by dark field operation.

C. *Specimen Damage*

Another problem related to contrast occurs in relation to total exposure time. When the exposure time is too short, an insufficient number of electrons have interacted with the specimen and sensor to form a non-uniformly illuminated image, giving rise to photon or "shot" noise. The difficulty with increasing the exposure time is that the electron beam begins to damage organic specimens through inelastic scattering. This damage gives rise to a need for more sensitive sensors. It may partly be offset by operating at very high voltages.

D. *Image Sensor Noise*

The ability to record small contrast differences without adding more noise is very difficult. In the past the prime recording media have been photographic glass plates or film. Recently, electronic recording has begun to compete with photography, especially when compared to the real time viewing of a phosphor screen.

It is expected that electronic means of recording weak images (especially those taken in dark field) will become the prime mode of viewing unstained organic molecules. In pursuing this approach an SEC vidicon is being installed in our microscope. It is hoped that this instrument will be capable of responding meaningfully to a single photon arising from electron impact on a phosphor surface.

III. Computer Image Processing

Techniques for handling television camera images have been under development in our laboratories for the past ten years. These techniques have been applied primarily to television signals received from spacecraft cameras, recording images of the surfaces of the Moon and Mars. Because we have not yet achieved success in using our television camera system to record superior electron microscope images, the

pictures chosen to illustrate the computer application to image enhancement and clean-up will be chosen mostly from the space mission images.

The problems arising from the use of television cameras and their
solution form the main body of this chapter. It is the function of
the video-data-handling system to reproduce the original scene of
transmitted television pictures as faithfully as possible in terms of
resolution, geometry, photometry, and perhaps color. The difficulty
lies in overcoming limitations imposed by the noise, distortions, and
information bandwidth of the system. These corrections are performed
by computer after the pictures have been digitized. The pictures in
cleaned-up form can be enhanced in contrast and used for detailed
visual photo-interpretation.

Once the pictures have been corrected, information can be extracted
from them. Since the pictures are now in digital form, some of the
analyses can be performed by the computer.

There are several significant differences between taking a picture
with a film camera and a television vidicon camera. Assuming that the
lenses are not the limiting factor, the differences appear in the manner
in which the image projected onto the receiving surface is sensed.
Spectral and dynamic sensitivity and linearity differ. Grain size limits
film resolution, and scanning-beam spot size limits vidicon resolution.
Geometric fidelity is worse in the vidicon scanning camera than in
film. Noise in transmission is unique to electrically encoded pictures.

There are several other problems unique to film, but emphasis here
is upon those weaknesses of television systems which add to the photo-
interpretive difficulties.

Several years ago, when the Ranger† effort was first proposed, no
known methods existed of performing by analog means alone all the
desired operations of clean-up, calibration correction, and information
extraction on video data. The most practicable approach to the solution
of these problems available at that time was to digitize the data and
perform these operations on a computer. The next problem was the
conversion of analog video data to and from digital form. A determined
effort was undertaken by the video-processing group at the Jet Propulsion Laboratory to digitize the data directly from photographs produced
from an analog signal. Although it was possible to recover most of what
was on the film, there was already too great a system loss from the film
recording itself. However, if the signals were recorded on magnetic tape
at the time of transmission, the analog video could be digitized directly
from the tape, and ground recovery losses became minimal.

---

† Ranger was the first successful hard landing spacecraft to take close-up photographs
of the lunar surface.

After the analog tapes were converted to digital tapes, the remaining major problem was reduced to creating the computer programs which would perform the corrections, enhancements, and analyses.

The last step in the sequence was the conversion of the digital tapes to an accurate visual presentation. A discussion of the hardware follows in the next chapter, by F. C. Billingsley.

The following series of corrections evolved as a result of working with the pictures themselves. (Other photo or video systems may or may not require these operations.)

Geometric correction: physical straightening of photo image.

Photometric correction: correction of non-uniform brightness response of vidicon sensor.

Random-noise removal: superposition and comparison.

System-noise (periodic) removal: elimination of spurious visible frequencies superimposed in image.

Scan-line-noise removal: correction of non-uniform response of camera with respect to successive scan lines.

Sine-wave correction: compensation for attenuation of spatial high-frequency components.

A. *Geometric Correction*

The first calibration to be applied must be geometric in order to ensure the proper registration of other calibrations. This correction is determined from preflight grid measurements as well as postflight reseau measurements.

The geometric correction is measured from the distorted image of the calibration grid, which has about ten to fifteen rows per picture height and width. The corresponding video elements between these intersections are shifted by a linear interpolation to the corresponding original position. If it appears by visual inspection that the change between grid points warrants more than a single interpolation because of severe nonlinearity, then more correction points may be chosen between rows. While these shifts could be determined prior to flight, in practice the measurements are made after success is assured. In fact, calibration and reseau-shift information are combined into one geometric correction. (Fig. 1.)

B. *Photometric Correction*

Examination of the photometric response to a uniformly lit field along a single scan line for each of several illuminations (Fig. 2) shows that the response is not uniform in either sensitivity or magnitude.

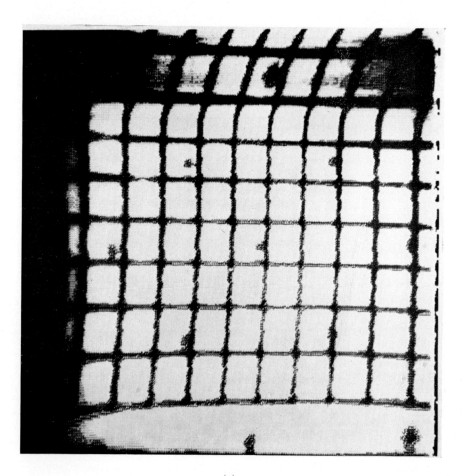

(a)

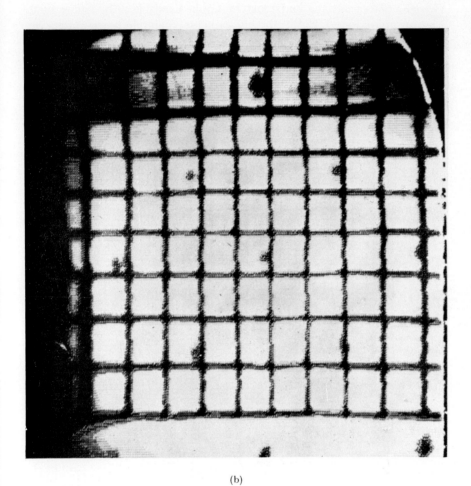

(b)

FIG. 1. (a) Image of a uniform grid as seen by an early Ranger camera; (b) Corrected grid after moving intersections back to a square array. (Note that some distortion remains in the third row as a result of extreme nonlinear distortion. Reference points could have been selected in a finer mesh to create better results.)

Photometric measurements are made for each line over the entire picture frame. The calibration data are unique for each point of the vidicon-camera surface and must be applied individually. Since there were so many points in the Ranger cameras, a simple linear interpolation was used to adjust the actual data lying between calibration brightnesses. An example of the intensity correction is seen in Fig. 3 for a Mariner† photograph.

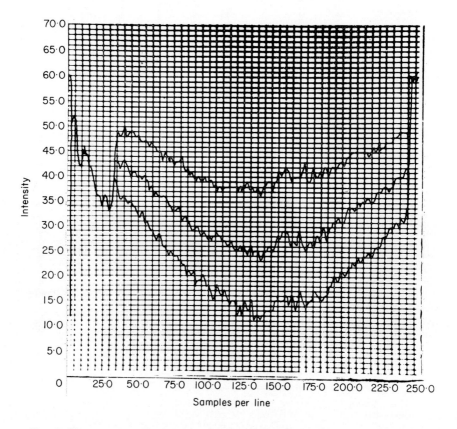

FIG. 2. Photometric calibration. (Abscissa represents distance along one particular scan line—in about the middle of a video frame. Ordinate shows voltage response to three levels of light from a uniformly lit screen; white is down.)

† Mariner is the name for the first successful unmanned Mars spacecraft.

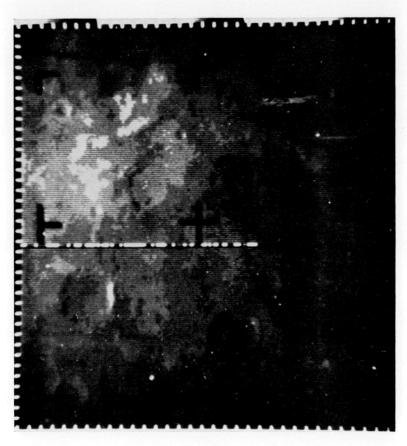

FIG. 3(a)

(b)

FIG. 3. (a) Mariner frame 11; (b) Mariner frame 11 after preliminary, experimental field-flattening correction and contrast enhancement. Scale 200 Km on edge of frame.

C. *Random Noise Removal*

Most of the noise discovered in the Ranger pictures had not been anticipated. The programs for its removal were written after the data were received.

There were, however, two classes of noise which had been anticipated and for which programs were written in advance of the actual spacecraft flight. This noise was caused by a poor signal-to-noise ratio, which created random points of bad data. In one case, the random noise gave rise to the appearance of "snow". However, this extreme change in the data can be detected, and the affected points can be replaced by the average of the neighbouring points. If the amount of snow is extreme, the theoretical picture resolution is degraded by this method of clean-up; but without it, the picture would be too hard to interpret. (Fig. 4.)

The second class of random noise is less apparent. However, it can be detected by superimposing pictures with overlapping areas of view. This process requires a very accurate registration of data, which, in turn, involves adjustments in translation, rotation, and magnification. The magnitude of these matching parameters can be determined visually, but a computer program has been developed which registers at least two small corresponding sectors in two pictures and determines their translation differential. For local regions, a translation correlation calculation is reasonably accurate and independent of small amounts of rotation and magnification. The vector differences between the two regions are sufficient to enable the computer to calculate the three parameters of translation, rotation, and magnification for matching the whole frame.

Once the pictures are matched, one way to improve the image is by simple averaging of the repeated areas. A more powerful approach utilizes the trustworthiness of each contribution. This reliability factor is derived from the history of that point—either from its magnification or calibration adjustment, or from the validity of the measurement in terms of the noise recognized in the individual frame. This judgment associates a weight with each point, which is then incorporated into the averaging.

In addition, after the average has been computed, a comparison of the original points can be made against the neighbouring points and the average. If the deviation of an original point from the average is too high, then that point can be omitted and the remaining points reaveraged. This method of majority logic is far superior to that of $\sqrt{N}$ improvement in the signal-to-noise ratio derived by straight

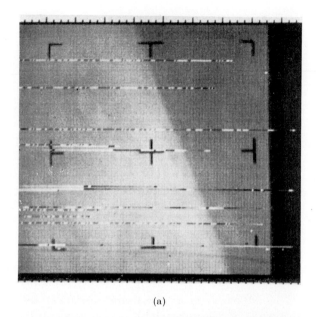

(a)

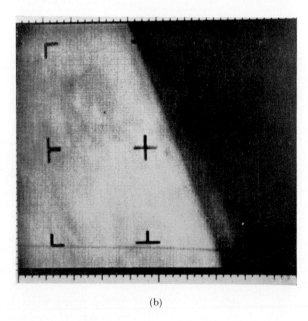

(b)

FIG. 4. Mariner frame 1 before (a) and after (b) clean-up. Edge of planet Mars. Scale 200 Km on lower edge of frame.

averaging (where N is the number of averaged frames). An example of improvement by simple averaging is illustrated later in the discussion of a catalase electron micrograph.

D. System Noise Removal

The film records of the first Ranger mission were such an overwhelming success that no further improvement appeared to be possible. The indication that improvement of the results was possible was the suspicion that some loss of resolution must have taken place in the ground film recorder because of its finite recording-beam size. A concentrated effort was made to make the data directly from the magnetic tape, with the result that the picture obtained did indeed retrieve the resolution lost by the prime film record.

Examination of this new picture disclosed a systematic frequency superimposed upon the original image (Fig. 5). Closer inspection indicated that this noise, even though superficially of a single frequency, did in fact drift in phase throughout the picture to such an extent that no single application of the formula

$$N(x, y) = N_0 \cos 2\pi \ (hx + ky + \Delta)$$

(where N is the magnitude of noise at coordinates x and y in the picture, Δ represents the phase shift, and h and k are the horizontal and vertical frequency components) would match the noise at all times.

The parameters N_0, h, k, and Δ were therefore not unique. The vertical and horizontal frequency components could be selected reasonably well in a local region; amplitude N_0 and phase Δ remained to be chosen. At any particular point, the noise could be considered as a sum of cosine and sine components of the original noise, each with zero phase shift relative to that point; i.e. it was necessary to determine only the cosine component of the noise. (Note that a sine component of zero phase at the origin is zero.) This determination can be made by performing a cross-correlation of the picture against the function $N \cos 2\pi \ (hx + ky)$, where N is a normalizing factor and h and k are chosen approximately by visual examination of the picture. The calculation becomes

$$\rho(x_0, y_0) = N \sum_{x=-r}^{r} \sum_{y=-s}^{s} B_0 \ (x + x_0, y + y_0) \cos 2\pi \ (hx + ky)$$

$$\frac{1}{N} = \sum_{x=-r}^{r} \sum_{y=-s}^{s} \cos^2 2\pi \ (hx + ky) \tag{1}$$

where $B_0 \ (x_0, y_0)$ is the original image brightness and $\rho(x_0, y_0)$ then gives the magnitude of noise contributing to that point. It should be

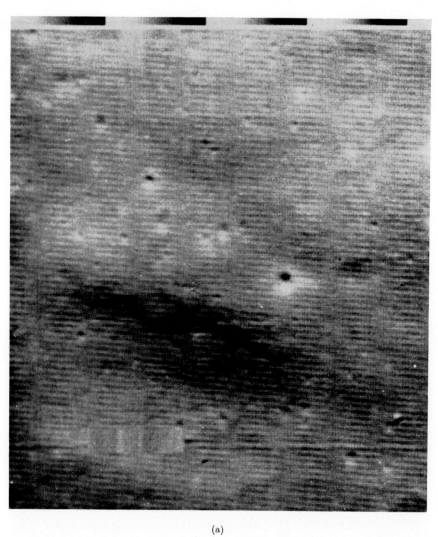

(a)

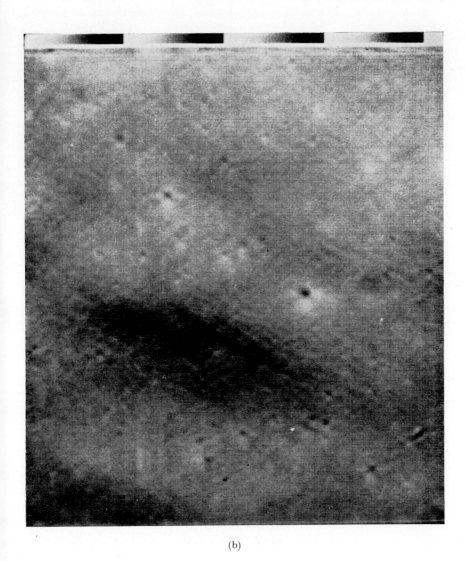

(b)

FIG. 5. (a) Digitized data but unprocessed with respect to noise removal. (Reseau mark has been removed; noise from television-camera erase cycle is visible.); (b) Result of subtracting noise from (a). (Photograph from Ranger VII.)

noted that r and s are chosen somewhat arbitrarily to accommodate the computer time taken in these calculations. It should also be mentioned that the function is stored in memory as a table and not recalculated for each point.

The correction to the picture is simply

$$B(x, y) = B_0(x, y) - \rho(x, y) \tag{2}$$

It becomes very useful to generalize these calculations in terms of the Fourier or frequency transform. For simplicity, the discussion can temporarily be kept to one dimension without immediate loss of generality. The Fourier transform of a real function in x (either time or distance), $P(x) \rightarrow A(h)$ to a frequency domain h, is

$$A(h) = \int_{-\frac{1}{2}}^{\frac{1}{2}} P(x) \cos 2\pi(hx + \Delta)dx \tag{3}$$

where $A(h)$ is the amplitude of each component of the original picture with frequency h. Where x is discrete, the integral becomes a summation. The original picture can then be represented in the frequency domain as a set of vectors whose direction normal to the base line indicates the phase angle Δ and where A is the length of the vector for each h. This vector can point in any direction between the real and imaginary planes.

Let us consider the probable envelope of the A-vectors in the real plane only as being random but distributed (roughly uniformly) over all possible frequencies. Systematic noise, however, as found in these pictures, is clustered very heavily around a single frequency. A filter peaked near this frequency is all that is needed to clean out the noise, but if the noise is not exactly at a single frequency, then too sharp or accurate a filter will not remove all of it. Yet, too broad a filter removes too much of the picture. Subjective judgment and consideration of computer time now become factors as various trials are made to determine the optimum filter.

The easiest digital filter to design would be a very sharp one, consisting of essentially a delta function in frequency and an infinite cosine wave of a single frequency in the real domain.

The filter next in complexity as well as effectiveness would be

$$\frac{\sin 2\pi[(h - h_0)n]}{2\pi(h - h_0)n} \tag{4}$$

which in the real domain consists of a square truncation of a cosine wave of frequency h_0.

The chosen filter is of the form

$$\frac{\sin^2 2\pi[(h - h_0)n]}{4\pi^2(h - h_0)^2n^2} \tag{5}$$

which is a triangular truncation of a single frequency h_0, and the truncation factor n is related to the sharpness of cutoff.

When this filter is subtracted from the original data, a notch results at the dominant noise frequency. Mathematically, the filtered frequency is positive and negative, which gives rise to a cosine transform of zero phase (no sine component). Rather than multiplying the Fourier transform of the picture by this notch filter in the frequency domain and then inverse-transforming back to the real dimension, it is more practical to perform a convolution operation of the inverse transform of the filter and the picture. The convolution operation is identical to eq. (1) for the sharp truncation and becomes eq. (6) for the triangular truncation.

$$\rho_{TR}(x_0, y_0) = N \sum_{x=-r}^{r} \sum_{y=-s}^{s} B_0(x + x_0, y + y_0) \cos 2\pi(hx + ky)$$

$$\times \left[\frac{r - |x|}{r} \right] \left[\frac{s - |y|}{s} \right]$$

$$\frac{1}{N} = \sum_{x=-r}^{r} \sum_{y=-s}^{s} \cos^2 2\pi(hx + ky) \left(\frac{r - |x|}{r} \right) \left(\frac{s - |y|}{s} \right) \qquad (6)$$

E. Scan-Line Noise Removal

The treatment of other kinds of noise requires bringing the discussion back to two dimensions in both the real and frequency domains. Among other things, television pictures are different from film in that they are scanned in some particular direction. Because not every scan line is carefully reproduced, noise is generated as a series of frequencies at right angles to the scan. In the two-dimensional frequency domain, these noises appear as high frequencies on the vertical axis.

After some mathematical manipulation, the filter which will remove these frequencies can be described as follows. Take the average value of the scene brightness in the region of the point to be corrected. Compare it with the average of the scan line containing this point and apply the difference between the scene average and the line average as a correction to the point. Application of the truncation logic of Section III.D to this filter results in a gentler response to areas more remote from the point. (Fig. 6.)

F. High Frequency Attenuation Correction

The camera scan beam is finite in size and somewhat Gaussian in shape. If it scans a scene which has a resolution finer than the beam spot, there will be significant loss in the transmitted resolution; the higher frequencies will be severely attenuated, if not lost completely.

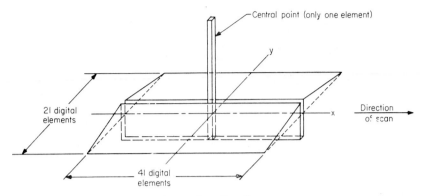

FIG. 6. Scan-line filter.

In the frequency domain (using one-dimensional logic to begin with), the desired system response (modulation transfer function) would be unity for all frequencies out to the upper limit cutoff (Fig. 7a). Calibration measurements of the actual frequencies show the response illustrated in Fig. 7b. If, for each frequency h the reciprocal of the response is plotted, then the curve plotted in Fig. 7c results.

To avoid overemphasis of high-frequency noises, the upper bound of the curve is arbitrarily chosen not to exceed 5. The product of the actual- and inverse-response curves (Fig. 7d) gives a flat response out to the point where the original response has fallen to 1/5 its original value. The filtering program can again be applied using the Fourier inverse of the inverse response. The convolution of this inverse function and the brightness of the original photograph enhances the higher frequencies to a point equivalent to the original scene (Fig. 8).

IV. ENHANCEMENT OF PERIODIC IMAGES

Thus far in this chapter there has been a discussion of some problems which affect resolution and some computer methods for manipulating image data. Emphasis has been given to the idea that more information can be made visible as improvement occurs in the signal-to-noise ratio.

In light microscopy it has long been accepted that a human observer can see more than can apparently be recorded photographically. It may be that visual sensitivity to contrast is greater, but there are some other factors of operation which are more subtle: integration of object characteristics by the human memory as the microscope is adjusted through a focal series, aspect changes as lighting is varied, stereovision, and historical comparison. The first three observation methods are different ways of improving signal-to-noise ratio by building a mental

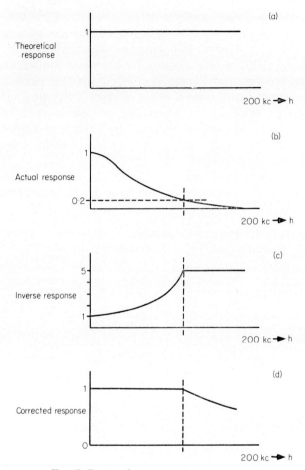

Fɪɢ. 7. System frequency-response curves.

image of the desired object and rejecting the remaining or background information. The last, historical comparison, is the building up of an image from the examination of a series of related but not necessarily identical objects and thus represents the expertise of a particular observer. This latter ability falls into that vague area called pattern recognition, and by definition, if the methods used to deduce the morphology of an object are not quantitative, then the image remembered must be called intuitive. It is the purpose of this chapter to focus attention upon those methods which when fully developed will lead to quantitative image reconstruction from a series of less than perfect recorded images.

For the non-periodic image it is expected that eventually three-dimensional reconstruction of images from a series of photographs of a slowly rotating specimen will improve tremendously the present signal-to-noise problem now existing for the usual biological specimen. (Algorithms are presently being developed for the reconstruction of three-dimensional detail from a series of skull X-ray projections.)

The fact that images exist where the information is available in a form where it can be separated from the background noise must be examined first. One such class of objects are periodic structures and the best of those are crystals.

In the case where the image is known to be periodic as in a crystal, an extreme improvement can be obtained in the signal-to-noise ratio

(a)

(b)

FIG. 8. Lunar rock as seen by Surveyor camera (a) before; (b) after enhancement.

by translating the image by the amount of the periodicity and superimposing the image upon itself. The assumption that is made is that the
noise is random relative to the translational distance while the signal
is additive. The result is an improvement in the signal-to-noise ratio by
an amount approximately equal to the square root of the number of
superpositions.

As a test of this assumption an electron micrograph of catalase
as recorded by Professor Fernandez-Moran (1966) was taken from the
literature and scanned by our equipment (see Fig. 9). [Catalase has
received a significant amount of attention in the literature (Kiselev
et al., 1968; Klug and Berger, 1964; Labaw, 1967; Longley, 1967).]

Fig. 9. Raw Catalase Micrograph—Negatively stained.

For the sake of sharpness for publication the brightness of the published image was essentially binary (i.e. the image was completely black and white) and therefore suffered somewhat in transmitting all the information that could be made available from the original micrograph. The image was rotated by the computer using our geometric correction program so that the periodicity appeared parallel to an edge. (See Fig. 10.) Because of an unexpected peculiarity which appeared upon performing the superposition, it was decided to indicate the effect of the superposition by reproducing here a series of superpositions which show an unambiguous buildup of the basic periodic structure. (See Fig. 11.) Examination of the final averaged image gives an approximation to the crystallographic two-dimensional space group pgg (i.e. two glide planes at right angles). But it is only an approximation! Realization finally occurred that this image must be essentially

FIG. 10. Raw Catalase—but rotated so unit cell is parallel to picture edge. (Note insert in center shows average image.) ⁷

a monomolecular layer and that the picture was taken slightly off the normal to the pseudo crystal face. The delightful result is that, by duplicating one of the averaged images and reproducing it after a rotation of 180°, we can see a stereo image! (See Fig. 12.)

A Fourier transform of the averaged image discloses diffraction maxima out to 20 Å resolution (see Fig. 13). Even though the higher frequencies were enhanced (Fig. 14), the information about the even higher frequencies has been permanently lost by the objective aperture.

Included in this series of experimental displays is an image where the intensities are contoured. More detail may sometimes appear by this means. (Fig. 15).

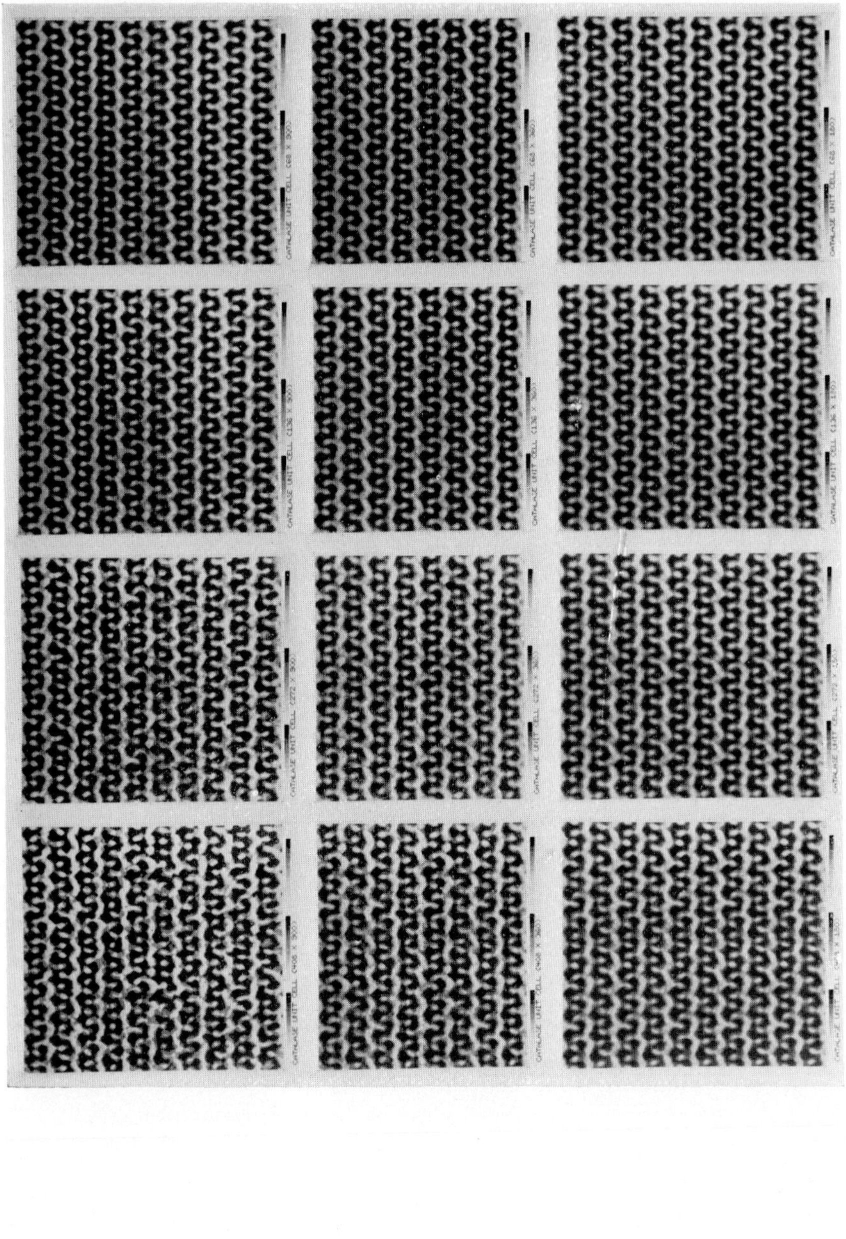

Fig. 11. Sequence of catalase overlays. Number of samples and lines for each repeat unit as follows. Top row from left to right: 408×900, 272×900, 136×900; middle row: 408×360, 272×360, 136×360, 68×360; bottom row: 408×180, 272×180, 136×180, 68×180. Final product appears in lower right. By sheer accident the size of one picture element is equal to one

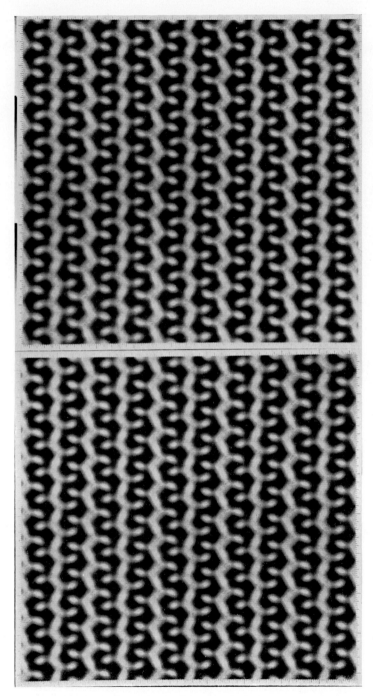

FIG. 12. Stereo pair is result of rotating image by 180° and placing it next to original.

ROBERT NATHAN

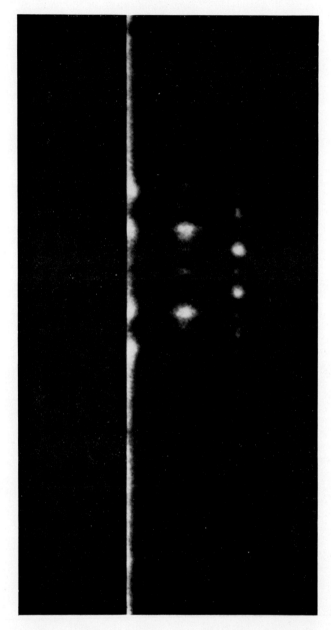

FIG. 13. Fourier transform of average catalase image.

FIG. 14. High frequency enhancement of averaged image.

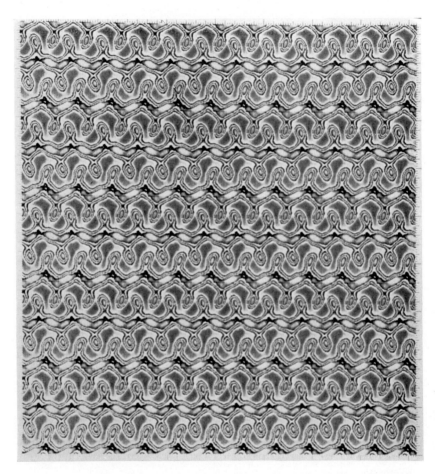

FIG. 15. High order digital bits removed and contrast stretched on remaining bits gives artificial contouring and brings out detail.

V. High Resolution by Computer Synthesis

X-ray diffraction methods do resolve 1 Å distances on simple molecules, but only in rare instances for moderately complex substances do these techniques work.

X-ray and electron diffraction methods of atomic structure determination use a diverging diffraction pattern which arises from the crystal under examination when illuminated by an incoming beam. (See Fig. 16.) Figure 17b shows the diffraction pattern which is recorded

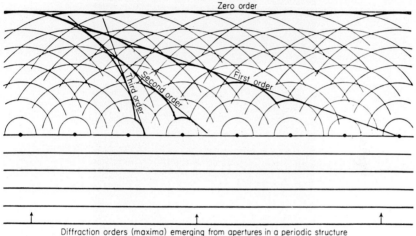

Diffraction orders (maxima) emerging from apertures in a periodic structure

Fig. 16. A plane wave, "illuminating" a periodic structure from below, gives rise to interference combinations called 0th, 1st, 2nd, etc. orders, each of which then travels as a wave front moving perpendicular to a new wave crest.

on film when light is passed through the pattern shown in Fig. 17a. Because of the divergence there can be no interaction between the various parts of the diffraction pattern. When this pattern is intercepted by a photographic film, all that is recorded is the intensity of the diffraction pattern, while the phases of the pattern's maxima are completely lost. But a knowledge of these phases along with the intensity of the maxima can be used by a computer to prepare a Fourier synthesis of the electron density in the crystal for the purpose of finding the atomic arrangement in the crystal.

In the case of electron diffraction, if an electron lens is placed between the crystal and the photographic film, the lens can be made to cause those diffraction maxima passing through the lens aperture to

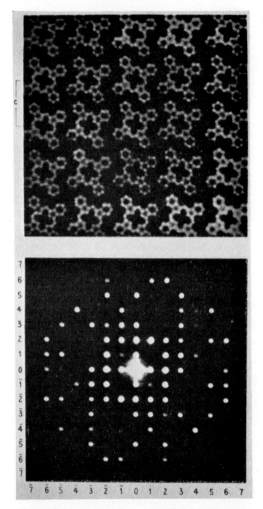

FIG. 17. Phthalocyanine—Two dimensional. Above: Phthalocyanine molecules. Below: Diffraction pattern.

converge and interfere in a calculable manner depending upon their relative phases. If the electron lens had an effective aperture of about five to ten times its present capability, then resolutions of 1 Å would be possible. But there appears to be a theoretical limit arising from spherical aberration which stops the lens from being used to collect more diffraction maxima to produce a higher resolution image. If the crystal and incident beam are tilted relative to the lens or if the lens aperture in the rear focal plane of the objective lens is shifted laterally, then different diffraction maxima can be made to interfere in a way that reveals their relative phases. (See Fig. 18.)

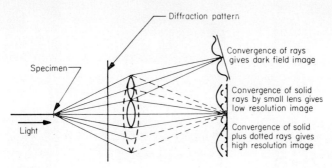

FIG. 18. Comparison of small aperture lens, both on and off optical axis and large aperture lens on axis in system illuminated by monochromatic parallel light.

Simulation of this method has been performed at visible wave lengths on larger targets on an optical bench. Figure 19b shows a high resolution image whose repeat unit is indicated by the square. The diffraction pattern of this image is shown in Fig. 19a. Suppose now a poor lens (one with a limited aperture) were used and allowed only those diffraction maxima indicated in Fig. 19c to pass through its opening. Because the higher frequency components were not included, the resultant image in Fig. 19d has lost its sharpness and some of the image details are missing. Figure 19e shows some of the off-axis diffraction maxima which can be made to go through the restricted lens upon shifting the lens to one side of the optical axis. An image is formed in Fig. 19f which no longer resembles the original object but which can be analyzed to produce phase information on the higher order diffraction maxima which produced it. A Fourier analysis can also be performed on the central-axis image Fig. 19d. A reconstitution of the low resolution image from its Fourier components is shown in Fig. 20 a and b. If to the Fourier components for the low resolution image are added the Fourier components from the off-axis diffraction maxima (from Fig. 19e), the computer synthesizes a high resolution image in Fig. 20 c and d resembling the original object Fig. 19b. (Note the sharpening of edges as higher frequency components are added.) This kind of computer reconstitution is used in X-ray diffraction and the kind of final product which is to be expected is indicated in Fig. 21, which shows the electron density map of a simple compound, hexamethylbenzene.

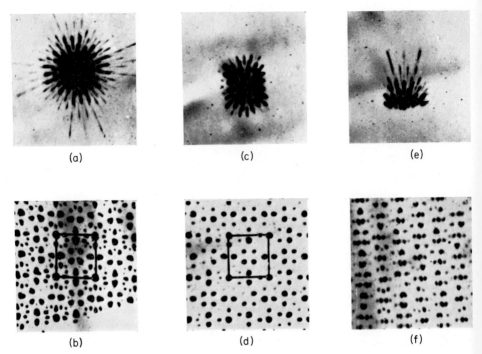

Fig. 19. Optical Bench Simulation of Problem (a) Large aperture gives full diffraction pattern; (b) High resolution possible if able to use large aperture, does not exist for electron microscope; (c) Restricted aperture allowing only central diffraction maxima to pass; (d) Low resolution image resulting from Fig. 19c; (e) Restricted aperture allowing only off-axis diffraction maxima to pass; (f) Low resolution "dark field" image resulting from Fig. 19e. Note the lack of resemblance to image formed in Fig. 19d.

As the instrument stands now, a crystalline organic specimen has been stable enough to give rise to the electron diffraction pattern shown in Fig. 22.

Mathematically, the above method is well understood. (Cowley, 1953; Moody, 1968; Nathan, 1953, 1954). What problems are encountered when attempts are made to apply it? The electron beam which is used in the microscope is very energetic, and when it is applied in high concentration onto an organic specimen for examination, the specimen melts from the heat. Also, organic substances which are of interest are low in contrast, which has led to the practice of shadowing or immersing the specimen in some heavy metal such as platinum or tungsten. Since these methods only give superficial information something must be done to get images from the original organic materials. The solution may be to install a sensitive television camera directly in the microscope

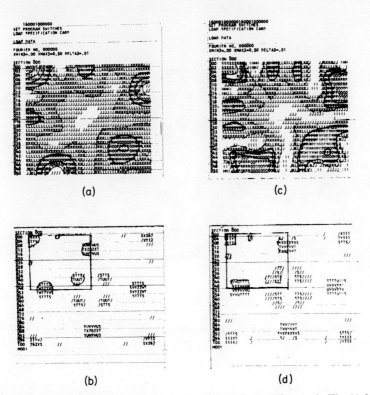

Fig. 20. Computer Reconstruction From Fourier Analysis of Images in Fig. 19 d and f. (a) Reconstruction of 1/4 cell repeat unit as seen in Fig. 19d (low resolution image); (b) Reconstruction of full unit cell from Fourier components obtained from analysis of Fig. 19d. Note loss of image detail; (c) Reconstitution of 1/4 cell taking phase and magnitude of Fourier components from both on-axis (Fig. 19d) and off-axis (Fig. 19f) images to form single high resolution image resembling Fig. 19b; (d) Full unit cell simulation of high resolution (see Fig. 20c). Note increased sharpness and appearance of non-round shape.

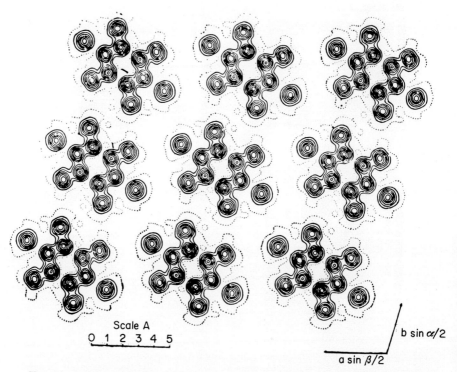

FIG. 21. Atomic resolution obtained by X-ray methods for simpler substances. This resolution is the goal of our present effort on the electron microscope, but it is expected to be applied to more complex organic substances.

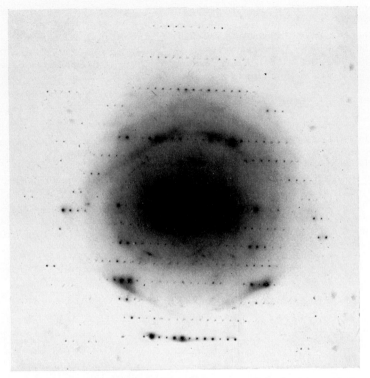

Fig. 22. Diffraction pattern of indanthrene olive crystal obtained in our instrument.

column and place the operation of the microscope under computer control in order to accelerate specimen manipulation during the picture-taking sequence. Such modifications have been initiated. It is planned to feed the television image directly (after digitizing) into the computer memory for analysis and subsequent synthesis.

ACKNOWLEDGEMENTS

I would like to acknowledge the work performed by Thomas C. Rindfleisch for the computer programming and computations performed on the catalase image enhancement.

REFERENCES

Cowley, J. M. (1953). *Proc. Phys. Soc. Sec. B.*, **66**, 1096–1100.
Fernandez-Moran, H. (1966). "Membrane Ultrastructure in Nerve Cells: Intensive Study Program in the Neurosciences", Boulder, Colorado.
Kiselev, D. J., De Rosier, D. and Klug, A. (1968). *J. Molec. Biol.* **35**, 561–566.
Klug, A. and Berger, J. E. (1964). *J. Molec. Biol.* **10**, 565–569.
Labaw, Louis W. (1967). *J. Ultrastruct. Res.* **17**, 327–341.

Longley, W. (1967). *J. Mol. Biol.* **30**, 323–327.

Moody, M. F. "Method for Eliminating Spherical Aberration and Preserving Phase Contrast Information in Electron Microscopy", Internal Publication, The Rockefeller University. (Received, 1968.)

Nathan, R. (1953–1954). Yearly Summary, Dept. of Chemistry and Chemical Engineering at the Calif. Inst. of Technology p. 18.

APPENDIX

The following are simplified examples of (1) a two-dimensional scan-noise line filter and (2) a one-dimensional high frequency recovery filter.

The computer presently uses a 21×41-element array; obviously, the 3×3 array (used in the line filter) does not work as well as the larger matrix but is still effective.

The high-frequency enhancement filter operates identically to the periodic-noise filter, but the example presented here has the added complexity of illustrating the determination of the frequencies to be enhanced and the extent of enhancement provided.

Example 1. Let us construct a simple scan-line filter F of a 3×3 array:

$$
F(x,y) = n_1 \begin{vmatrix} 1 & 1 & 1 \\ 1 & 1 & 1 \\ 1 & 1 & 1 \end{vmatrix} - n_2 \begin{vmatrix} 0 & 0 & 0 \\ 1 & 1 & 1 \\ 0 & 0 & 0 \end{vmatrix}
$$

$$
+ \begin{vmatrix} 0 & 0 & 0 \\ 0 & 1 & 0 \\ 0 & 0 & 0 \end{vmatrix}
$$

$$
= \begin{vmatrix} \tfrac{1}{9} & \tfrac{1}{9} & \tfrac{1}{9} \\ \tfrac{1}{9} & \tfrac{1}{9} & \tfrac{1}{9} \\ \tfrac{1}{9} & \tfrac{1}{9} & \tfrac{1}{9} \end{vmatrix} - \begin{vmatrix} 0 & 0 & 0 \\ \tfrac{1}{3} & \tfrac{1}{3} & \tfrac{1}{3} \\ 0 & 0 & 0 \end{vmatrix}
$$

$$
+ \begin{vmatrix} 0 & 0 & 0 \\ 0 & 1 & 0 \\ 0 & 0 & 0 \end{vmatrix}
$$

$$
= \begin{vmatrix} \tfrac{1}{9} & \tfrac{1}{9} & \tfrac{1}{9} \\ -\tfrac{2}{9} & \tfrac{7}{9} & -\tfrac{2}{9} \\ \tfrac{1}{9} & \tfrac{1}{9} & \tfrac{1}{9} \end{vmatrix}
$$

or

F $\diagdown$ z	-1	0	1
-1	$\tfrac{1}{9}$	$\tfrac{1}{9}$	$\tfrac{1}{9}$
0	$-\tfrac{2}{9}$	$\tfrac{7}{9}$	$-\tfrac{2}{9}$
1	$\tfrac{1}{9}$	$\tfrac{1}{9}$	$\tfrac{1}{9}$

where $1/n_1$ = the number of elements in the filter array ($n_1 = 1/9$), and $n_2 = n_1$ × the number of rows in the filter array ($n_2 = n_1 \times 3 = 1/3$).

As a test application of filter F on a received noisy signal B, consider a 6 × 6 array of numbers representing brightness A. To A add some scan-line noise N to produce received image B.

A

y \ x	1	2	3	4	5	6
1	3	4	6	4	3	1
2	5	7	3	2	1	2
3	6	1	6	3	2	1
4	5	3	3	4	4	3
5	3	1	8	3	1	4
6	4	2	3	6	2	3

+

N

y \ x	1	2	3	4	5	6
1						
2	−1	−1	−1	−1	−1	−1
3	3	3	3	3	3	3
4						
5	−1	−1	−1	−1	−1	−1
6						

=

B

y \ x	1	2	3	4	5	6
1	3	4	6	4	3	1
2	4	6	2	1	0	1
3	9	4	9	6	5	4
4	5	3	3	4	4	3
5	2	0	7	2	0	3
6	4	2	3	6	2	3

Perform a convolution of filter F and array B to give approximate recovery R back to A:

$$R(x_0, y_0) = \sum_{x=-1}^{1} \sum_{y=-1}^{1} B(x_0 + x, y_0 + y) \, F(x, y)$$

$$\text{for each} \begin{cases} x_0 = 2 \text{ to } 5 \\ y_0 = 2 \text{ to } 5 \end{cases}$$

$R\diagdown^x_y$	2	3	4	5
2	7	4	4	2
3	1	7	3	3
4	4	4	5	4
5	0	7	2	1

The error of recovery is $E = R - A$.

$E\diagdown^x_y$	2	3	4	5
2	0	1	2	1
3	0	1	0	1
4	1	1	1	0
5	-1	-1	-1	0

Effectiveness may be measured by comparing average noise $|\overline{N}|$ against error matrix $|\overline{E}|$ over the range $x = 2$ to 5 and $y = 2$ to 5.

$$|\overline{N}| = \frac{1}{16} \sum_{x=2}^{5} \sum_{y=2}^{5} |N(x,y)| = 1{\cdot}25$$

$$|\overline{E}| = \frac{1}{16} \sum_{x=2}^{5} \sum_{y=2}^{5} |E(x,y)| = 0{\cdot}75$$

Hence, a decided improvement is shown for a very small filter.

Example 2. Let us consider a one-dimensional scene in x with brightness $A(x)$.

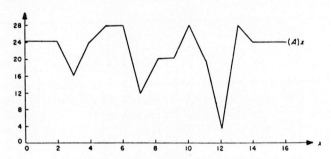

Let A be scanned by a beam with the response shape $S(x)$:

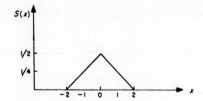

The transmitted brightness is a convolution of A and S to form $B(x)$:

$$B(x) = \sum_{x=-1}^{1} A(x_0 + x)\, S(x) \quad \text{for each } x_0 = 1 \text{ to } 15$$

There is a visible drop in resolution from A to B.

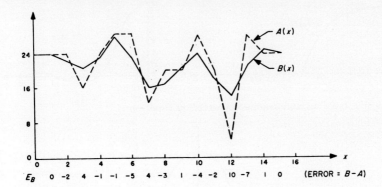

The Fourier transform of $S(x)$ is $(\sin 2\pi h)^2/(2\pi h)^2 = T_S\,(h)$.

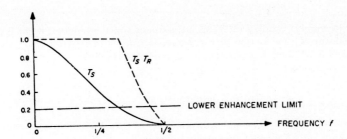

Now, let us take the reciprocal of T_S,

$$T_R(h) = \frac{1}{T_S}$$

but $T_R \leqslant 5$ is an arbitrary upper bound to avoid noise enhancement.

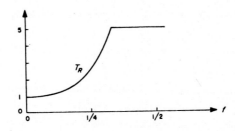

Let us subtract T_R from 5 and take an inverse Fourier transform to real space.

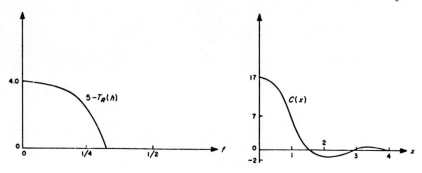

We now have an unnormalized correction function C.

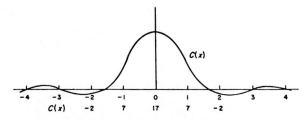

In order to convert C into a filter, an adjustment must be made for the fact that the transform of this fuction was subtracted from 5 in the frequency domain. Two constant, K_1 and K_2 must be determined for the filter.

$$F(x) = K_1\, \delta(0) - K_2\, C(x)$$

where $\delta(0)$ is a delta function $\begin{cases} = 1 \text{ for } x = 0 \\ = 0 \text{ for } x \neq 0. \end{cases}$

The convolution of $F(x)$ and a brightness of very high frequency will cause $C(x)$ to drop out, and an enhancement factor of 5 will result (see T_R at high frequency).

Therefore,

$$5 = F = K_1\, \delta(0)$$

$$K_1 = 5$$

The convolution of $F(x)$ and a constant brightness of magnitude 1 should give an enhancement factor of 1 [see $T_R(0)$].

$$1 = \sum_{x=-2}^{2} 1\, F(x) = \sum_{x=-2}^{2} [5\, \delta(0) - K_2\, C(x)]$$

$$= 5 - K_2[\, -2 + 7 + 17 + 7 - 2\,]$$

Therefore,

$$K_2 = \frac{4}{27}$$

$$F(x) = 5\, \delta(0) - \frac{4}{27}\, C(x)$$

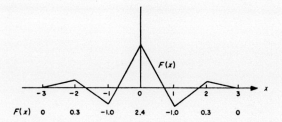

$F(x)$	0	0.3	-1.0	2.4	-1.0	0.3	0

When a convolution is performed between $F(x)$ and $B(x)$, the result $R(x)$ represents a reconstruction of $A(x)$ to the degree permitted by the enhancement of the higher frequencies, as indicated by $T_S\, T_R$.

$$R(x_0) = \sum_{x=-2}^{2} B\,(x_0 + x)\, F(x)$$

for $x_0 = 1$ to 15

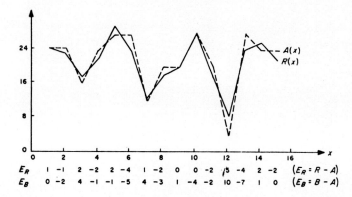

| E_R | 1 | -1 | 2 | -2 | 2 | -4 | 1 | -2 | 0 | 0 | -2 | 5 | -4 | 2 | -2 | $(E_R = R - A)$ |
| E_B | 0 | -2 | 4 | -1 | -1 | -5 | 4 | -3 | 1 | -4 | -2 | 10 | -7 | 1 | 0 | $(E_B = B - A)$ |

Compare the error E_B caused by the scanning beam against the error E_R remaining in the reconstructed image.

$$|\overline{E_B}| = \frac{1}{15} \sum_{x=1}^{15} |E_B| = \frac{45}{15} = 3{\cdot}0$$

$$|\overline{E_R}| = \frac{1}{15} \sum_{x=1}^{15} |E_R| = \frac{30}{15} = 2{\cdot}0$$

This is a distinct improvement. Visual comparison of the plots of A vs B and A vs R also shows that R matches A much better than does B.

Image Processing for Electron Microscopy: II. A Digital System

F. C. BILLINGSLEY

Jet Propulsion Laboratory,
California Institute of Technology,
Pasadena, California, U.S.A.

I.	Introduction	127
II.	General Considerations	129
III.	Resolution	130
	A. Scanner resolution	131
	B. Data quantity	134
	C. System design	135
IV.	Quantization	135
	A. Digital step distribution	136
	B. Film grain	138
	C. Quantization noise	140
	D. System noise	141
	E. Detector noise	141
V.	Electronic Cameras	145
VI.	Data Recording	149
	A. Electronic recording	149
	B. Conversion to output film	150
	C. Display console	153
VII.	Computer System	153
	A. Hardware	153
	B. Software	156
	C. A complete system	158
References		158

I. INTRODUCTION

THE preceding discussion by Nathan of methods for image processing has assumed that this processing will be done by digital computer. The design of a complete system must allow for the possibility of data recording either upon film for subsequent scanning, or by some form of television system. In addition, it is not at all clear *a priori* that computer processing of the images is necessarily the best. Indeed, for images which have initially been recorded upon film a distinct contender is processing by coherent optical means. At the moment, however, less than half of the operations listed in Fig. 1 can be performed in this

Area	Technique	Used for
GENERATION	Computer-Originated	Test Targets Graphical Displays
	Computer Substitutions	Insert Windows in Pictures Insert Good Data for Bad
INTENSITY MANIPULATION	Intensity Calibration of Systems	Photometry
	Non-linear Lookups	Film Curve Corrections Grey Scale Alterations
	Chromaticity Calculations	Color Shift, Balance, Alteration
GEOMETRIC MANIPULATION	Geometric Calibration of Systems	Good Geometry Needed for Stereo
	Reprojection	Convert Slant Pix to Ground Projection
	Rubber Sheet Stretching	Overlay Match of Two Pix
	Independent X and Y Adjustments	Aspect Ratio Corrections
SPATIAL FREQUENCY OPERATIONS	Spatial High Frequency Boost	Correct for Detail Losses in System
	Spatial Low Frequency Reduction	Minimize Broad-Brush Shading Remove Effects of Glare
	Single Frequency Filtering	Remove Coherent Noise
ANALYSIS	Fourier Transform	Analysis in Spatial Frequency Plane
	Image Light Distribution	Star Cluster Analysis
	Pattern Extraction	Counting Blood Cells, Autos, Stars, etc. Analyzing Shapes of Objects
	Convolution	Filtering, Correlation
MULTI-PICTURE	Subtraction	Change Detection Stereo Information Extraction
	Addition	Averaging, Noise Removal
	Multiplication	Spatial Domain Filtering
	Division	Normalizing

FIG. 1. Some uses for image processing.

manner, and other means such as computer processing must be used for the remainder.

Dr. T. Huang (1967), discusses these two prime contenders, digital processing or coherent optics. In general it can be stated that the main advantages of a coherent optical system are its efficient storage of the large amount of data in a picture and high processing speed due to the implicit parallel processing of all elements simultaneously. Coherent processing has the disadvantages of being rather touchy to operate properly and hence subject to anomalies, its inability to handle non-linear problems and the relatively high cost of large aperture systems.

Digital processing has the advantages of being extremely flexible, of being rather trouble-free and easy to use once the programming system has been worked out, has the ability to handle problems which are non-linear in both intensity and geometry, and can do these with an accuracy limited only by the user's knowledge of the incoming data. Although the possible large cost of a digital computer system is a potential disadvantage, this cost can be traded for processing speed by using a smaller computer which will require longer to do the processing.

To gain the advantages of digital processing requires high-quality signals with low noise and associated accurate analog to digital conversion equipment. This equipment for either scanning film or digitizing analog magnetic tape signals is fairly expensive and requires appreciable time to operate. Thus immediate turn-around is not practical except at a complete Image Processing Laboratory. The remainder of this chapter will discuss some of the considerations involved and assumes that the data will be later reduced to digital form for computer processing.

II. General Considerations

In designing an imaging system, a number of things must be considered more or less simultaneously. Some of these will be found to have rather practical hardware limitations due to the current state of the art, while others may be open-ended in this respect, but limited by cost of implementation.

Briefly the considerations to be covered in detail below are:

Resolution. The sharpness of the image as recorded on film or as presented spatially to the imaging device will determine the required spatial density of sampling points.

Quantizing Accuracy. The finer the digital steps used to characterize the points the better. This will ultimately be limited by noise.

Recording Considerations. Any film used for recording will contribute noise and therefore is to be avoided if possible. However, some forms of image tube sensors cause even more problems and, therefore, a trade-off between methods is needed.

Data Processing. Minimization of computer time suggests that a small picture is desirable but this limits either resolution or area coverage. Another trade-off. Design of a data handling system geared to picture processing will greatly facilitate this procedure.

Conversion to output film. Film recording equipment specifications must be stringent as quite small deviations will cause noticeable artifacts.

These points are discussed in more detail below and a rationale is developed which may either be copied or used as a guide by the experimentor in the design of his own system.

These considerations apply to image processing in general, regardless of the specific field of application. For example, the concept and understanding of the requirements of image processing have been developed for the processing of pictures as returned from NASA space vehicles to the Jet Propulsion Laboratory, but in general these same concepts, and to a very large extent even the same specific programs, have beeen utilized to process pictures from the electron microscope, light microscope and radiographs from the medical field, industrial X-rays, traffic study photographs, photographs of wind tunnel models and analysis of star clusters.

III. RESOLUTION

Nathan has discussed the various causes of limitations in resolution obtainable in practicable equipment. Knowing the equipment magnification, the quoted—or desired—resolution in the object domain may be converted to the image domain, as this is the quantity of importance to the image recording system. Though a number of different criteria and rules of thumb have been used to describe resolution and image sharpness, the best quantitative measure of the image sharpness is the modulation transfer function, "MTF", which measures the loss in high spatial frequency components of the image, i.e., fine detail. A good general discussion of this is given by Perrin (1960). Since MTF does not result in a convenient single number, various rules of thumb are used to reduce this to a number. "Resolution" is the normally quoted parameter, often taken as the distance between the central peak and the first minimum in the image of a point object. Alternatively, it has been defined as that spatial frequency in cycles per millimeter at which the high frequency information has been reduced to some low fractional value (e.g. 5%) of its original value.

"Limiting resolution" in the case of a diffraction limited optical system is the lowest spatial frequency at which no contrast is visible, and may be related to the Airy disc radius† (O'Neill, 1963). In converting a picture to sample form the important criterion is that two samples must be taken at the highest frequency at which there is picture information. This is the Nyquist criterion. Sampling at this spacing or more often will insure that no picture information is lost, and that the picture can theoretically be reconstructed perfectly (O'Neill, 1963). Failure to sample this often will result in loss of picture information at best, and may result in a severe beat pattern or moiré which will interfere with picture analysis.

A. Scanner Resolution

The finite size of the scanning spot will further limit the high frequency response of the overall system. In particular, practical flying-spot scanner systems and camera tubes may have apertures (i.e. spot size) of 10 to 25 μm. While in principle these apertures may be imaged onto the film at reduced size, the projection lens itself will limit this process by producing a spot which is larger than expected.

An approximation of this effect sufficiently accurate for our purpose may be devised as follows:

Approximate the half-amplitude diameter of the Airy disk by:

$A = \lambda F$ $\lambda =$ wavelength, micrometres (μm)

$F = F$-number

$A =$ approximate half amplitude of Airy disk, μm.

This is the impulse response of the lens. The output image is the convolution of this with the CRT scanning spot, which is essentially Gaussian. This rather messy convolution may be approximated by an r-m-s combination since the impulse response of practical lenses is approximately Gaussian.

$SA = \sqrt{(D \times M)^2 + A^2}$ $SA =$ half-amplitude diameter of spot on film (scanning aperture), μm

$D =$ half amplitude diameter of CRT spot, μm

$M =$ magnification, film/CRT.

† See Eastman Kodak Co., *Modulation Transfer Data for Kodak Films* Pamphlet P–49. EKCo Sales Service Divn., Rochester, N.Y. 14650.

Figure 2 shows this approximation for a perfect lens and for $\lambda \approx 0.5\,\mu$m. Real lenses, being less than perfect, will result in further enlargement of the projected spot size and consequently less resolution.

To find the number of independent scan spots across one scan line, assume a 7″ CRT with a useful diameter = 6″, with a 10.8×10.8 cm square raster.

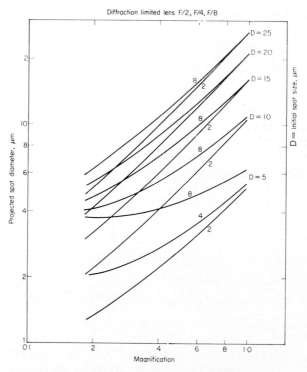

Fig. 2. Projected spot diameter of CRT spot.

Then assuming equal SA size over the entire raster,

$$N = 108,000\ \frac{M}{SA}$$

N = number of pixels/line
= number of lines

This equation is plotted in Fig. 3 for various F-numbers. It can be seen that small CRT spots are seriously degraded by lenses smaller than F/2, especially as small rasters on film are attempted.

Actual scanner resolution in the image or film plane can now be

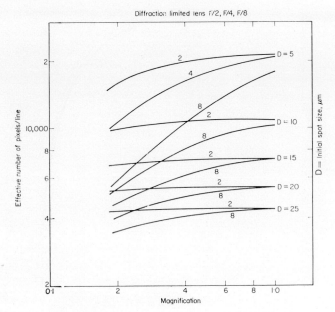

FIG. 3. Effective number of pixels on 7″ CRT.

determined from the given spot size as projected on film. This is calculated by treating the CRT spot as a low pass filter, i.e. an aperture whose transmittance varies as a Gaussian or $\cos^2$ function of radius.

The response for a Gaussian spot is:

$$\frac{A(\omega)}{A(o)} = e^{-(0.30\beta)^2} \qquad\qquad \beta = \frac{D}{P_\omega}2\pi$$

$$D = \text{Half-Amplitude Diameter}$$
$$P_\omega = \text{period of sinusoid}$$
$$\text{at frequency } \omega$$

The response for a $\cos^2$ spot is:

$$\frac{A(\omega)}{A(o)} = \frac{J_3(\beta)}{\beta^3} + 3\frac{J_4(\beta)}{\beta^4} - (15 - \pi^2)\frac{J_5(\beta)}{\beta^5} \ldots \text{ where } J_n(\beta) \text{ is a Bessel}$$

function of order n. Figure 4 gives the curve for the types of apertures as produced by real flying spot systems or camera tubes as well as for Gaussian and $\cos^2$ spots. Because of the similarity of the curves, no great error is caused by using whichever one makes the mathematics of a particular analysis easier.

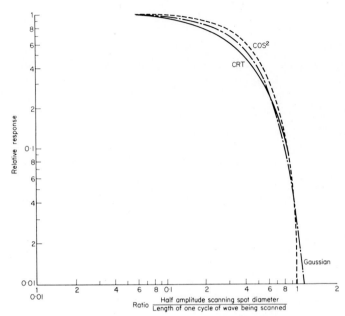

FIG. 4. Spatial frequency response of CRT.

B. *Data Quantity*

Although, from the viewpoint of minimizing picture information loss and minimizing the possibility of moiré, it is desirable to use as many picture elements (pixels) as possible, minimization of computer operating time dictates that the number of pixels be reduced to a minimum. The practical number of pixels per line is bounded by several parameters:

1. One standard 2400 foot digital magnetic tape recorded at 800 samples per inch will hold a picture about 4,000 elements square, and this therefore becomes a convenient break point. If the picture is formatted to contain one picture line per digital record, this results in about 4,000 records containing about 4,000 pixels each.

2. The film may be scanned with a flying spot scanner. Ultimate limits of a flying spot scanner system seem to be about 7,000 to 8,000 pixels per line. A more practical limit for a work-a-day piece of hardware is about 4,000 pixels per line. If the picture is scanned with a television camera of some sort, the practical limit is approximately 1,000 pixels per line. The new return-beam vidicons promise to raise this limit to greater than 4,000 pixels/line.

3. Since the processing time goes up as the square of the number of pixels per line, it behooves the user to minimize this quantity to minimize computer time.

4. Because of the characteristics of some of the processing programs, it is convenient to think of the number of pixels per line as a power of 2; that is, 512, 1024, 2048, or 4096. We at JPL have done almost all of our processing at 1024 or below.

C. *System Design*

The first invariant is the highest spatial frequency at which information occurs in the image plane, as this determines the maximum permissible element spacing. This frequency is used with Fig. 4 to determine the projected spot size required to avoid further loss of high frequency information. This spot size is used in Fig. 2 with a chosen F-number and the manufacturer's quoted CRT spot size to determine the magnification to be used. This magnification is then used with Fig. 3 to determine the effective number of picture elements per line which can be obtained in the system. This procedure is iterated until a satisfactory combination is obtained.

For example, assume an X-ray film having the highest spatial frequency content at 40 cycles/mm. To avoid high frequency loss, let us require the relative response to be ≥ 0.7 at this frequency. This gives

$$\beta_{40} = \frac{SA}{\text{period of } 40 \ \text{~/mm}} \approx 0.3 \text{ from Fig. 4 for CRT scanning}$$

$$SA = 0.3 \times \frac{1}{40} \text{ mm} = 0.0083 \text{ mm} = 8.3 \ \mu\text{m}$$

Assume a CRT having $D = 20 \ \mu\text{m}$

try an F/2 lens:

From Fig. 2 $\qquad M = 0.37$

From Fig. 3 $\qquad N \approx 5500$ pixels/line or number of lines

This is the largest N which can be expected, assuming a perfect lens and optimum adjustments. This number of pixels will cover an area on the film or in the image of

10.8×0.37 cm $\approx 4.0 \times 4.0$ cm.

IV. QUANTIZATION

It is found that the limiting factor in image processing is noise. This is particularly true for the high frequency enhancement which is used to sharpen fine details. For this reason, in the design of any imaging system, one must consider closely the noise performance of the system, especially in the manner in which the noise interacts with the visual appearance of the picture and perturbs the subsequent digital processing. Since these two considerations lead to somewhat different noise requirements, they must be discussed separately.

The effect of two-dimensional noise upon the human observer has attracted the attention of a large number of workers (see, for example Schade, 1951–1955; Schade, 1956; Stultz and Zweig, 1959; Stultz and Zweig, 1962; Hacking, 1962; Huang, 1965). The visual effects of noise are tied up intimately with the desired resolution, appearance, contrast, surroundings, brightness, and other aspects of the picture as well as the magnitude, spectral content and other aspects of the noise itself. As a very general rule of thumb (and recognizing the likelihood of oversimplification) a 40 db (100/1) signal-to-noise ratio will give a reasonably clean picture. More noise will degrade the ability to differentiate by eye two areas of different brightness (or in the case of film, of different transmittances) (Heyning, 1966; Altman, 1967). Although the visual effects are of importance, they are secondary to data considerations when image processing is to be attempted, and will therefore not be discussed in detail here.

Since film systems and most television cameras are not linear, the effect of noise will be different depending upon the brightness of the part of the picture on which the noise is superimposed. Since the effective amount of granularity or noise is a function of the picture element size, the question of granularity is tightly coupled to that of resolution (Levi, 1958; Zweig et al., 1958; Jones, 1961; Eyer, 1962; Levi, 1963; Hacking, 1964).

The effects of noise on digital reconstruction and enhancement are determined from the statistics of the amount of perturbation caused by the noise. Accordingly, the effects of noise on quantization accuracy are derived, and the resulting curves used to estimate the performance of the system.

A. *Digital Step Distribution*

A given pixel will be converted to digital form by quantizing it into some number of steps for ultimate digital processing. The quantization is normally in terms of equal increments of light, or in the case of film scanning, of equal increments of film transmittance. On a high-quality print or a film transparency the eye can readily discern a difference in quality as produced by four bit (16 level) and five bit (32 level) quantizing, but cannot readily see any extra quality at 6 bits. In order to maintain five-bit quantization of the output of the digital processing, the input quantizing must be at least 6 bits or 64 levels.

There is a fundamental limitation in the maximum density (D) which can be digitized to equal transmittance (T) steps. Since transmittance and density are related by

$$D = \log \frac{1}{T}$$

equal transmittance steps are unequal density steps, with the density steps being large at the high density end. Figure 5 gives a curve of this for 6-bit quantizing. In particular, the last three (6-bit) digital levels occur at densities of 1·5, 1·8, and infinity. Since normal films quite often have densities above two, special precautions must be taken to reduce the peak density to 1·8 or less for that part of the picture containing information.

Alternatively, if measurements must be made in density regions above that corresponding to the normal first step of linear transmittance

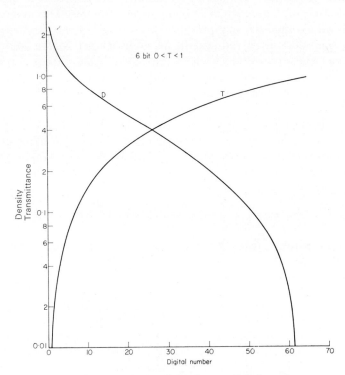

Fig. 5. Transmittance and density vs. digital number.

intervals, the corresponding transmittance intervals must be made smaller (i.e. the first step must be sub-divided). The smaller steps make additional demands on the noise performance of the system, especially if accuracy proportional to the continually diminishing transmittance is required. This will be discussed in more detail below.

In some examples of image processing, such as that shown by Oppenheim *et al.* (1968) and that discussed by Shelton *et al.* (1967), the step spacing may not be linear in transmittance, but rather may be linear in density or have some other distribution. However, the quantity that is seen by the measuring device is light, which is linearly related to film transmittance. Transmittance, therefore, is the film parameter of importance, although the user may choose to think in other terms. Similarly, the noise considerations are in terms of transmittance noise rather than the density noise which is of importance to considerations of visual effect (Altman, 1967).

B. *Film Grain*

The image on any normal silver halide film is produced from the superposition of a number of opaque silver granules. The macroscopic effect is an average density produced as a result of the local exposure, but as the aperture is reduced in an effort to get higher and higher resolution, the silver grains, which appear as black patches on a clear background, now begin to become visible. This is the film granularity, which is usually stated in terms of the rms value of the statistical fluctuation in density (σ_D), when measured with an aperture of 48 μm, and at a density $D \approx 1$.

From a consideration of probability of overlap of the silver grains it can be shown that the noise (σ_T) at any value of transmittance (T) may be derived from a known value of noise (σ_{T1}) at a transmittance T_1 (O'Neill, 1963):

$$\sigma_T = \frac{\sigma_{T1}}{\sqrt{T_1 (1 - T_1)}} \cdot \sqrt{T (1 - T)}$$

The maximum σ_T from this expression occurs at $T = 0.5$, and therefore represents the worst-case noise. The maximum value of this worst-case noise is chosen to be:

$$\sigma_{T_{max}} = 1/3 \times \text{size of one digital step}$$

Then, for 6-bit digitizing over the range $0 < \mathrm{T} < 1$:

$$\sigma_T = \frac{1/3 \times 1/64}{\sqrt{0.5 (1 - 0.5)}} \cdot \sqrt{T (1 - T)}$$

$$= 0.0104 \sqrt{T (1 - T)}$$

"Standard Conditions" at which granularities are quoted are approximately at $T = 0.1$ $(D = 1)$.
This substitution gives:

$$\sigma_{T = 0.1} \approx 0.003$$

This is the worst-case σ_T, permissible at $T = 0.1$. This must be converted to σ_D, (i.e. σ measured in density units) since normal specifications quote σ_D.
For a density well above fog level (O'Neil, 1963)

$$\frac{\sigma_T}{T} \approx \frac{\sigma_D}{0.4343} \qquad \sigma_D = \text{rms granularity in density units}$$

$$\therefore \sigma_D = 0.4343 \frac{\sigma_T}{T}$$

$$= 0.0133 \quad \text{This is the allowable maximum at } T = 0.1, \text{ measured with the aperture as determined by resolution requirements.}$$

But it is found that (O'Neill, 1963)

$$\sigma_D \sqrt{A} = \text{const for apertures} >> 10 \times \text{grain diameter}$$

$$A = \text{area of aperture}$$

Conversion to various diameters is then done by:

$$\frac{\sigma_{D1}}{\sigma_{D2}} = \frac{SA_2}{SA_1}$$

Therefore

$$\sigma_D \big| 48\,\mu\text{m} = \sigma_D \big|_{SA} \cdot \frac{SA}{48} = \frac{0.0133}{48} \cdot SA \approx 0.00025 \cdot SA$$

Thus, with this criterion the required granularity is:

$$\sigma_D = 0.00025 \times SA$$

Where

σ_D = required rms density granularity at $D = 1$ and aperture = $48\,\mu\text{m}$

SA = scanning aperture diameter in μm

Perusal of film specifications will show that normal film has a granularity considerably higher than this value and hence the system will be film-limited. For this reason it is desirable where possible to go to direct camera pickup of the light image rather than use film as an intermediary step.

C. *Quantization Noise*

A derivation and discussion of the effects of noise on quantization is given by Friedman (1965), and the resulting curve is shown in Fig. 6. Briefly, the quantizer transforms the magnitude of the signal into a discrete number of steps. In a noiseless system there is no ambiguity in the designation of a particular signal level as a certain digital number. However, in the presence of noise (assumed to be random), it is the signal-plus-noise which is quantized, and the level of the signal alone in somewhat uncertain from inspection of the digital number.

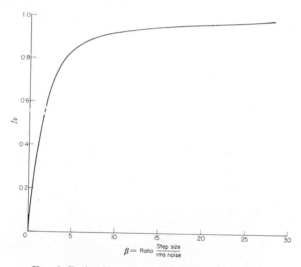

FIG. 6. Probability of correct digitization vs. S/σ.

We can develop a rule of thumb for the determination of β as follows:

Let us start by defining the statement "digitizing to M useful bits" to mean that as a result of noise, the $(M + 1)^{\text{th}}$ bit is correct with a probability $= 0.5$. From Fig. 6, it is seen that a value of $\beta = 1.4$ is required for the 7th bit where $\beta =$ step size/rms noise. This gives a $\beta = 2.8$ for the 6th bit.

Alternatively, system performance may be approached in terms of noise introduced by the digitizer. The value of the noise power introduced by the act of digitizing is given by (Walli, 1964)

$$\sigma^2{}_{\text{Dig}} = \frac{1}{12} \text{ step size}$$

We can define a balanced system as one in which the quantizing noise signal is equal to the rms sum of all prior noise sources in the system.

Thus σ prior to digitizing $= \dfrac{1}{\sqrt{12}} \times$ step size

or

$$\beta = \sqrt{12} = 3 \cdot 56$$

We will use as a general rule of thumb that the rms noise into the digitizer should be $= 1/3$ step size $(\beta = 3)$ as representing a reasonable consensus from these criteria.

D. *System Noise*

If we consider the requirements in terms of digitizing a film, the required black/white signal-to-noise ratio (S/σ) may be derived as follows:

Define $\sigma = 1/3 \times$ size of one digital step $= 1/3 \times \Delta T$ $(\beta = 3)$

$$\Delta T = \text{size of one digital step}$$
$$= 1/64 \text{ for 6-bit, } 0 < T < 1$$
$$S = \text{black-white signal} \triangleq 1; \text{ where } \triangleq \text{ means equal by definition, or some equivalent notation.}$$

The density and transmittance at the first step away from $T = 0$ are:

$$T_1 = \Delta T = 1/64 = 1/N \qquad \text{Where } N = \text{number of steps}$$

$$D_1 = \log \frac{1}{T_1} \qquad\qquad D_1 = \text{density of first step away from } T = 0.$$

Therefore,

$$S/\sigma = \frac{1}{1/3 \times 1/N} = 3N$$

S/N in db is expressed by:

$$S/\sigma|_{\text{db}} = 20 \log S/\sigma$$
$$= 20 (\log N + \log 3)$$

But $T_1 = 1/N$ and $D_1 = \log 1/T_1 = \log N$

$$\therefore \quad S/\sigma|_{\text{db}} = 20 (D_1 + \log 3) = 20 (D_1 + 0 \cdot 477)$$

For a 6-bit system $D_1 = 1 \cdot 8$ and therefore $S/\sigma = 45 \cdot 7$ db.

E. *Detector Noise*

Another prime source of noise is the detector itself, typically a photomultiplier tube in the case of a flying spot scanner system. For determination of the system bandwidth in terms of the photodetector noise, we will consider a system (Fig. 8) in which the measuring light, after attenuation by the film being measured, falls on a photodetector having an efficiency $\epsilon < 1$. The electrons released will be counted, and the count will exhibit a statistical fluctuation. Associated with the

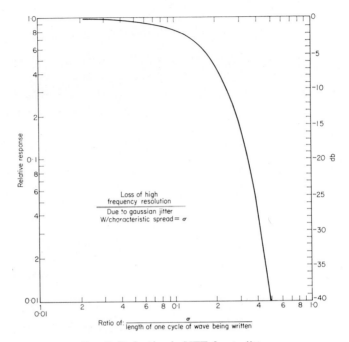

FIG. 7. Reduction in MTF due to jitter.

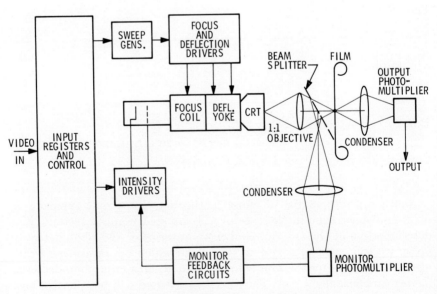

FIG. 8. Block diagram of film recorder/scanner.

photocathode will be an electron multiplier showing multiplication noise. This multiplication noise will not be included in the derivations but may be taken into account later with a multiplicative factor. Since load resistor thermal noise is generally small compared to the other noise sources, it will be ignored. For purposes of derivation, the detector will be allowed the full noise allowance (i.e. $\beta = 3$).

From considerations of the shot noise produced in a photo-multiplier tube, it can be shown that (Van der Ziel, 1954, 1957)

$$I_S = 2 \frac{G}{G-1} G^n e \left(\frac{S}{\sigma}\right)^2 B$$

Where I_S = signal current, amps
G = stage gain
n = number of stages
e = charge on electron
  = $1 \cdot 6 \times 10^{-19}$ coulomb
S/σ = signal/noise as before
B = bandwidth, Hz

Each element (film, CRT, amplifier, etc.) in the system will cause a loss in high frequency information, or roll-off. Generally, the combination of the various roll-offs will limit the total system bandwidth such that the high frequency attenuation curve is more or less Gaussian. Assuming that the band-limiting roll-off is Gaussian, maximum S/σ is transmitted through the channel when (Schwartz, 1959)

$$B'\tau \approx 1/3$$
B' = cutoff bandwidth, MHz
τ = optimum pulse width, μsec

However, at the termination of the pulse, the output pulse top is still rising, so that pulse width errors cause amplitude errors. A compromise which does not sacrifice much S/σ and yet is less critical to τ variations is

$$B'\tau \approx 1/2$$

The sampling time is therefore:

$$T = \frac{e\,(S/\sigma)^2\,G^{n+1}}{I_S\,(G-1)}$$

Since the digitizing step size $= \dfrac{\text{total signal}}{\text{number of steps}} = \dfrac{S}{N}$,

$$B = \frac{S/N}{\sigma}, \text{ and}$$

$$S/\sigma = N\beta.$$

thus $T = \dfrac{eN^2\,\beta^2\,G^{n+1}}{I_S\,(G-1)}$ N = number of equal-size digital steps

The lowest β will be found at $T = 1$ (open aperture), and will improve as T decreases.

To get a feel for real numbers, the following parameters will be assumed. These are typical of practical photomultipliers when used in fine spot CRT scanners.

$$G = 3 \qquad\qquad I_S = 1\text{mA at open aperture}$$
$$n = 10$$

This gives:

$$T = 14 \cdot 2 \times 10^{-12} \beta_0{}^2 N^2 \text{ seconds} \qquad \beta_0 = \text{required } \beta$$
$$\text{at } T = 1$$

and an optimum system bandwidth

$$B = 0 \cdot 0352 \times 10^{12} \frac{1}{\beta_0^2 N^2} \text{ Hz}$$

For $N = 64$ and $\beta = 3$,

$$T = 14 \cdot 2 \times 10^{-12} \times 9 \times 64^2 = 1/2 \ \mu\text{sec}$$

Allowing a dead time between pulses of $20 \ \mu\text{s}$ for spot motion, phosphor decay, and digital readout (a practical value), the resultant maximum digitizing rate is about 50,000 samples per second.

This sampling rate, together with the size of the picture, now determines the total sampling time required for a complete picture. For example, approximately 20 seconds is required to completely digitize a 1024×1024 picture sampled at 50,000 samples per second.

For digitizing to $D = 3$, $N = 1024$, giving $T = 125 \ \mu\text{s}$. The time penalty required for high-density digitizing is thus easily seen.

The β at some partial value of transmittance, T, is given by:

$$\beta_T = \beta_0 \frac{1}{\sqrt{T}}$$

and the optimum system bandwidth is

$$B_{opt} = 0 \cdot 0352 \times 10^{-12} \frac{1}{\beta_T^2 N^2 T}$$

The allowable bandwidth is a function of the film transmittance and hence if the bandwidth and sampling time were to be varied in accordance with the transmittance, the maximum total digitizing rate could be achieved. The accompanying reduction in pulse width at partial T allows a shorter total sampling than would be the case if the system bandwidth were fixed at the low value required to give adequate S/σ at the required minimum β step.

"Constant confidence" scanning is usually accomplished by allowing the scanning aperture to dwell on a given picture element while integrating the output signal until it reaches a predetermined threshold value. The length of time required for this integration is measured. This time is linearly proportional to the reciprocal of the film transmittance or the image light value. This system implicitly produces uneven time intervals between sampling pulses and side effects such as nonsynchronous data recording must be designed for. It produces a constant signal-to-noise ratio, but gives a variable β.

$$\beta_{\text{const confid}} = \frac{SNR}{N} \cdot \frac{1}{T}$$

All of the discussion thus far implies division of the transmittance range into equal incremental steps, since most picture processing is most easily visualized in this domain. This is not necessary, however, and sometimes other constraints may dictate that these steps are unequal. Shelton *et al.* (1967) discuss a different approach to constant confidence scanning using digital steps having unequal size but equal reliability. This scheme makes it possible to achieve more levels within a given range than either the equal transmittance steps or equal density step schemes, but has the disadvantage of introducing non-linearity into the digital computation. Whether or not this is tolerable depends on the accuracy to which the computation is desired and on the specific picture content. No general rule can be given.

V. Electronic Cameras†

It was implied above that the introduction of a film into the system will result in appreciable extra noise and is therefore to be avoided if possible. In those instances for which the direct image digitizing time as derived above is not satisfactory, initial photographing of the image onto film with subsequent film scanning may be the only alternative. Where sufficient light for film exposure is available, scanning the photographic film for digitizing purposes to achieve high quality results should be done at slow-scan rates. Scanning at video rate with a TV camera tube can derive only low quality data. Commercial TV cameras such as CCTV camera systems generally utilize a vidicon camera tube. The operating illumination range known as dynamic range is low for such camera tubes and is limited by the signal-to-noise ratio (S/σ) obtainable (Weimer, 1960). The reproduced image from such tubes will possess much noise especially in those areas of the image which are dimly lit. Image quality is further degraded by spatial shading and

† Contributed by John Volkoff, Jet Propulsion Laboratory.

blemishes which are added to the reproduced image by vidicon camera tubes.

For low level light conditions, such as is available from an electron microscope, electronic cameras must be utilized for image recording. Typical performance parameters for some candidate electronic cameras are shown in Table I. These camera tubes have been selected with the intention of being applicable for an electron-microscope image recording system.

Illumination is usually expressed as a flux having units of lm/ft^2. However, it is sometimes advantageous to estimate the performance of a camera tube from a total quanta aspect especially for slow scan operation. For this purpose, Table I includes the illumination density in units of $lm\ sec/ft^2$. The high level illumination listed in Table I corresponds to the saturation point of the tube. The threshold level of illumination is set by a S/σ of unity. These two levels define the dynamic range capability of the tube. The term resolution is used to define the tube's ability to distinguish an image element on the tube face plate. Values listed in the table are those corresponding to a modulation transfer function response of 10%.

The maximum luminance available at the viewing screen (37 × 50 mm²) of an electron microscope is typically 10^{-3} ft Lamberts (Sadashige, 1967). This is equivalent to an illumination of 10^{-3} lumens/ft² at the camera tube face plate. As can be seen from Table I, the camera tubes that can perform at this illumination level are the Secondary Electron Conduction (SEC)-Vidicon[†], the Image Orthicon[‡], and an image intensifier camera tube system.

The SEC-Vidicon is capable of storing a charge pattern for long periods of time (24 hr) without noticeable degradation of the stored image. However, this tube inherently adds shading and blemishes to the reproduced image. At the light level of 10^{-3} lm/ft², the performance of the SEC-Vidicon at video scan rate will be of lower quality than that described in Table I. The S/σ and resolution will be degraded, the dynamic range will decrease to 10, and shading effects in the image pattern will be increased. The estimated S/σ will be 25 and the resolution at the center of the image will be 18 cycles/mm. At the edge of the image the resolution will be less than 16 cycles/mm.

However, the SEC-Vidicon has the capability of integrating and storing the charge produced for an extended period of time. By time integration an image at a relatively low illumination level may be stored

[†] Westinghouse Electric Corp., WL–30691, *SEC Camera Tube*, No. TD86–817, March 1968.
[‡] RCA, *Image Orthicons*, Catalogue No. CAM–800, March 1968.

TABLE I

Storage Type TV Cameras

Camera Tube	Sensitivity† (high light illumination on face plate)		Dynamic Range	Gamma	S/σ at high light (pk-pk/rms)	Center Resolution at high light ($\sim$/mm)
	Flux at 30 fr/sec ($1m/ft^2$)	Density ($1m$ sec/ft^2)				
Vidicon (video mode)	10^{-1}	$3 \cdot 10^{-3}$	10	·7	80	37
Vidicon (slow-scan mode)	10^{-2}	$3 \cdot 10^{-4}$	9	·4	430 (ultimate)	34
Vidicon (Surveyor V S/C)	60	2	32	·7	100	42
SEC-Vidicon	10^{-2}	$3 \cdot 10^{-4}$	30	·7	50	26
Image Orthicon (3″)	$2 \cdot 5 \cdot 10^{-2}$	$8 \cdot 10^{-4}$	20	·75	30–60	25
Image Orthicon ($4\frac{1}{2}$″)	$6 \cdot 10^{-2}$	$2 \cdot 10^{-3}$	25	·75	90	25
Image Orthicon (high sens.)	10^{-5}	$3 \cdot 10^{-7}$	10	·75	3	25
Image Intensifier-Vidicon	$10^{-6} - 10^{-2}$	$3 \cdot 10^{-8} - 3 \cdot 10^{-4}$	30	·7	80	20
Image Intensifier-Image Orthicon	$10^{-7} - 10^{-2}$	$3 \cdot 10^{-9} - 3 \cdot 10^{-4}$	25	·75	75	14

† Based on white light. The photocathode's spectral sensitivity must be applied to estimate the tube's response to other light.

to an ultimate charge level. Subsequent scanning of this charge pattern will result in a reproduced image having the qualities listed in Table I.

A range of sensitivity levels of five orders of magnitude is available in Image-Orthicon camera tubes having magnesium oxide targets. Somewhat more complex to operate than the SEC-Vidicon, the Image-Orthicon can reproduce an image of good quality at illuminances greater than 10^{-2} lm/ft^2, but at low illumination levels, low quality results because of the very low S/σ produced. When directly applied to an electron microscope, degradation of the image will occur at the optical interface. Assuming that the optical interface light transmittance is 10%, the estimated resolution of the reproduced image will be about 24 cycles/mm, S/σ will be 20 at high light, and the dynamic range will most probably not exceed 10.

Luminous intensification for an image camera tube is often used to increase sensitivity of the tube. This can be done but with deterioration of image quality. Resolution is most affected. The center resolution capability of an image intensifier tube is usually better than 40 cycles/mm and if fiber optics are used, resolutions up to 200 cycles/mm are possible. Pronounced resolution losses occur toward the edges of the image. To overcome this requires a large diameter image intensifier tube. The tube can be considered to be noiseless if operated in a thermally cold environment. The luminous gain can be varied over an order of magnitude. However, image graininess is characteristic of intensifiers. The estimated performance of two intensification camera tube systems is listed in Table I.

Scanning of photographic film should be performed at a scan rate compatible with the quality of picture required. As slow a frame rate as possible should be used to minimize the noise. Because of the inherently poor S/σ performance of electronic systems at the high bandwidth required for high rate scanning (for example, commercial rate video), the temptation to use a video camera system must be avoided where ultimate digitization and processing is the goal. A non-storage type camera tube whose performance is based on quanta rate can derive accurate photometric data from the image for digital processing. The image dissector camera tube, basically a photo-multiplier tube, can be constructed to perform as a very high quality scanning camera tube.

The dynamic range of an image dissector depends upon the scan rate. A decrease in scan rate results in an increase in the dynamic range capability of the tube. High S/σ and high resolution can be obtained from an image dissector† (Eberhardt, 1966). Resolution capability is a

† Westinghouse Electric Corp. WI-1300. *Image Dissector*, No. TD-86-876. May 1968.

constant parameter based on an aperture dimension. The maximum illumination is limited by photo-cathode characteristics. A resolution of 36 cycles/mm and dynamic range comparable to digitizing to 256 gray levels can reliably be achieved with an image dissector camera tube.

VI. Data Recording

A. *Electronic Recording*

Scanning of either the original image or a film with a scanning system is ultimately required and the electronic image must be disposed of. Once the image has been converted to electrical form by a camera of some sort, the signals should be FM-recorded on magnetic tape (if it cannot be quantized in real time) to minimize degradations in the S/σ ratio. Irretrievable data loss in both noise and in resolution occurs in the recording of a signal onto film and hence film recording of an electronic camera signal with the intent of future film scanning should be avoided.

In the best of all possible worlds, the data will be digitized directly during the image scanning process, thus avoiding all intermediate forms of recording. At moderate digitizing rates (for example between 10 and 100 kilosamples per second, with the exact number depending upon the computer system used) the input data can be handled directly by computer by recording it on either magnetic tape or disk simultaneously with its receipt.

If on-line digitizing cannot be done, FM-recording of the analog signal is recommended rather than direct recording because of the amplitude stability which FM-recording affords. Modern instrumentation tape recorders can record in a wide band FM mode a data bandwidth of 400 KHz at a tape speed of 120 inches per second, and at proportionally lower bandwidth at reduced speed.

At the present state of the art, there is no device for recording large quantities of samples at high sample rates. Therefore, means to reduce the rate to within the recording capabilities of practical equipment are often required. Therefore, it is desirable to replay the analog magnetic tapes at a fraction of the original speed to get the digitizing rate down to the rate which can be handled by the system.

An immediate problem is encountered if for some reason the input data has been analog recorded on a helical scan or rotary head magnetic tape recorder, as these machines cannot be slowed down for replay. If this type of machine has been used, and if the required data rate is above that which the computer system can handle with simultaneous recording, the input signal may be dubbed onto a longitudinal tape recorder for subsequent replay at slow speeds. This dubbing process will add more noise to the signal and results in further degradation.

All magnetic tape recorders suffer from some amount of time-base instability caused by variations of the tape speed. Attempts to run analog tape at extremely slow speeds result in an increase in the amount of wow and flutter and other problems involved in tape playback. To combat the time base instability and to provide a real time reference for subsequent digitizing, it is recommended that a pilot tone (a constant frequency steady-state signal) be recorded on an adjacent tape track. Zero crossings of this pilot tone will be used to key the digitizer and will relegate further time base instabilities to second order effects.

The severity of the time base instability can be analyzed from the impulse response point of view. At some time after the sync pulse for each line a single impulse spot is recorded. The vertical line of delta functions so generated will be straight and sharp in the absence of all jitter, or will be wiggly or diffuse in the presence of jitter. The spot position density function vs horizontal position is the impulse response of the group of lines taken as a whole. The effective horizontal frequency response for this group of lines is then the one-dimensional Fourier transform of this impulse response (Harger, 1965).

It has been shown that the probability density function for time base errors is approximately Gaussian for the error in a magnetic tape recorder (Chao, 1966). For Gaussian jitter,

$f(x) = \exp\left[-1/2\,(x/\sigma)^2\right]$, and the transform is

$F(\omega) = \exp\left[-1/2\,(\sigma \cdot \omega)^2\right]$, when normalized.

Letting $\omega = \dfrac{2\pi}{P_\omega}$, where P_ω is the period of the sinusoid,

$$F(P_\omega) = \exp\left[-2\pi^2\,(\sigma/P_\omega)^2\right]$$

This is the equation plotted in Fig. 7.

For other than Gaussian jitter, of course, the transform of the appropriate function must be taken.

The horizontal MTF will in general be different for different parts of the picture. Note that, in general, for the case of a picture reproduced from longitudinal tape, resolution degrades progressively along the scan line as the time base displacement errors accumulate.

B. Conversion to Output Film

The output film recording device generally will take the form of a cathode-ray tube flying spot device. Precisely the same set of limitations apply as were previously outlined for a cathode-ray tube flying spot film scanner, and in fact the same cathode-ray tube and optics system will probably be used for both purposes.

In addition to the resolution requirements as previously outlined for the scanning of film, the geometrical accuracy in the placement of the recording spot is of added importance. Slight displacement of the spot during film scanning will result in reading of a portion of the film slightly misplaced from the intended position and will cause some loss of resolution. This same displacement of the spot during recording will cause quite visible lines, streaks, beat patterns and other visible degradations. We have found that the visible effects are particularly bad if the displacements are systematic, as would be caused for instance by a small amount of 60 Hz hum. Therefore additional effort must be made to ensure that both the actual displacements and the rate of change of displacement with position are minimized. Spot displacements of $0 \cdot 1$ of the pixel to pixel spacing and a rate change of displacement of $0 \cdot 01$ will allow satisfactory pictures to be produced.

Cathode-ray tubes are normally quite non-linear in output light with respect to input drive, and therefore will produce a distorted gray scale unless corrective efforts are made. One method which has been used at the Jet Propulsion Laboratory is to monitor part of the light being produced by the CRT by using a beam splitter which diverts part of the light into a photomultiplier tube. This is a feedback input to an operational amplifier which closes a servo loop around the CRT and includes the light path, insuring that the light is a reasonably linear function of the input voltage. The linearity depends on how much loop gain can be used, with the normal servo problem of attaining a sufficiently high loop gain while keeping the loop stable.

The loop must remain closed to well above the highest frequency components of the picture if it is to be useful. This in turn necessitates the use of a phosphor of the P-16 type which decays to less than 10% response after 1/10 of a microsecond. With this scheme the loop can be kept closed past 1 MHz.

A block diagram of a typical CRT recorder is given in Fig. 8 and a photograph of the JPL recorder is shown in Fig. 9.

As indicated in the previous analysis, if the spot diameter can be kept small compared to the size of the information to be recorded, loss of high frequency data can be minimized. However, if the spot is reduced in size in an effort to get better high-frequency resolution, line-to-line ripple begins to show which may be more detrimental to the viewer than the loss of resolution caused by the larger spot (Schade, 1956). Thus, the number of lines in the picture influences the useful horizontal

Fig. 9. Picture of JPL film recorder/scanner.

resolution. Since the raster structure disappears when the half-amplitude diameter is approximately equal to the line spacing, this condition is generally selected as the operating point.

C. *Display Console*

The development of processing techniques and the determination of the parameters to be applied during the processing are quite subjective, and require the continued attention of the analyst to the results of his processing. This is an iterative procedure in which the analyst examines his latest results and submits new processing in accordance. Upon eventual successful processing, the process is "locked up" and the job run. During the development time, however, the numerous iterative cycles are quite time-consuming.

The time required to obtain a picture from the computer system will be minimized if a data display is available. This display must be of high enough quality to present the details required by the analyst and may take the form of either hard copy printout or cathode-ray tube display. For the former case the computer will output the picture once to a display equipped with a Polaroid camera.

A live display must present the picture to the analyst in non-flickering form, which requires that it be renewed at least 20 times per second. Renewal may be from computer memory if the computer memory is large enough to hold an entire picture. Otherwise some form of storage such as a rotating disk is required which may be repeatedly scanned at display rates. These are the only current forms of display which are entirely satisfactory. As an alternate to them, but producing pictures with degraded resolution and gray scale, storage tubes or image converter tubes may be used to display the gross aspects of pictures.

VII. Computer System

A. *Hardware*

The development of the processing algorithms and the accomplishment of some moderate amount of actual processing can be done on almost any general purpose computer. It will be found, however, that the smaller computers will be quite inefficient for picture processing. This is so because the smaller machines do not have adequate capability for the large amount of pixel manipulation which must be done to handle the large arrays present in pictures. Therefore, if the microscopist is concerned with either fast turnaround from the computer, or with processing either large pictures or large numbers of small pictures he must eventually consider going to a medium or large sized computer.

Since the designs of the hardware system and the data handling software are closely interrelated, consideration of the hardware design must take into account the pecularities of processing pictures.

A large amount of the picture processing can be done with 6-bit pixels. Since manipulation is on a pixel-by-pixel basis it is convenient to store each pixel as a 6-bit character in a character oriented machine. By character oriented, we mean a computer which can address data and perform calculations to the character level rather than being able to handle only longer words. For more sophisticated processing, 8-bit pixels are used, which suggests that the pixels be stored as 8-bit bytes and that the machine be able to address bytes. If a character or byte oriented computer as appropriate is not available, the pixels will normally be stored one pixel per computer word and processed in this condition. This use of a computer, while possible, is not economical as considerable time and memory capability is wasted in manipulating the vacant bit positions.

Since even the normal large size computer installations do not have adequate memory for storing an entire picture, some form of picture roll-through is required. In this mode a few lines of the picture at a time are read into core, processed and then read out to allow room for more lines for further processing. In addition to the core space required for the program storage, capacity must be provided for pixel storage. Some of the processing algorithms operate on one pixel at a time, and could conceivably work with a picture memory as small as a few pixels. Other programs such as the two dimensional convolution program for filtering require simultaneously a number of picture lines at least as high as the convolution matrix size, and for efficient operation should have room for several times this amount. But for a picture of 1024 pixels per line the total memory size for picture storage rapidly gets out of hand. Fortunately the state of the art in computer memories is rapidly improving, and memory of 65,000 to 130,000 bytes is now quite practical.

A large number of the processing programs will be found to be CPU bound. That is, the larger part of the processing time will be caused by central processing unit manipulations rather than input/output operations. For this reason and especially in view of the large number of operations to be performed on one picture, as fast a CPU as can be afforded should be used. Since a large number of computers in all sizes are available with cycle times of approximately one microsecond, this speed or faster is recommended.

At least two tape units will be required with the system since the pictures will normally be handled and stored on the magnetic tape.

For those operations requiring more than one input picture simultaneously (for example, multipicture averaging) a large amount of line interleaving and picture shuffling will be required, unless one magnetic tape unit and/or data set on disk is simultaneously available for each input picture. The extra cost of the extra tape drives must of course be traded off against the savings in processing time.

Whether or not magnetic disks are useful will depend upon the type and mix of the processing to be done. We have found it convenient to group the pixels in one record per line, with a complete picture occupying one file or data set. The lines are normally recorded sequentially along the recording medium. For those processes in which sequential line access is adequate, obtaining a picture from tape is entirely satisfactory. However, for those processes in which a number of lines simultaneously must be accessed or for which non-sequential pixels from several lines must be used, magnetic tape is very inconvenient. In this case, it will be better to serially read a complete picture from tape onto disk and then to randomly access the required pixels or lines from the disk. If the disks are organized in a cylindrical mode† access time will be further reduced. At least one disk drive is recommended.

Again depending upon the quantity of pictures to be processed, some special purpose peripheral devices may be useful. A number of computer manufacturers now have available a hardware box which will rapidly and efficiently perform the sequence of multiplication and additions required in the convolution filter process. Use of such a box can reduce the processing time by as much as 10/1. Consider a convolution using a 15 by 15 pixel matrix: 225 multiplications must be performed and their results summed for each point in the picture. For a 1024 × 1024 picture, allowing 15 microseconds for each multiplication/summation, without the special purpose hardware, one hour per filter is required. The saving in reducing this to between 6 and 10 minutes is apparent.

Development of the Cooley-Tukey algorithm for fast Fourier transforms has been a material factor in making this type of manipulation practical. An FFT on a 1024 × 1024 picture using the Cooley-Tukey algorithm requires 8 to 10 minutes per transform. Hardware boxes are now becoming available which claim to reduce this transform time by a factor of 10/1 over software.

It can thus be seen that image processing can be done on almost any size computer from the table top model up to the largest installa-

---

† *Introduction to IBM System/360 Direct Access Storage Drives and Organization Methods,* IBM student text, C20–1649–1.

tions. Processing time on the large machines will be much less than that on a small one, and a larger memory and more rapid processing speeds will allow much more ambitious algorithms to be performed than would be attempted on the small machine.

B. *Software*

As with the hardware, a large amount of picture processing can be done with normal general-purpose computers and the standard Fortran batch processing software. This is not particularly efficient, however, as the Fortran compilers with their associated I/O routines are often much slower than machine language programming. For this reason, with large numbers of pictures to be processed, special attention must be given to reducing or by-passing the inefficient parts of the Fortran language and/or doing programming in machine language. In addition, an image processing laboratory should make available to the analyst an efficient data handling system for the development of algorithms and for the recording, processing and display of pictures.

To facilitate the use of the system by the analyst and to enable him to rapidly call for new processes, an image processing software system must be designed which is based upon English language commands. This system should require a minimum of programming knowledge and data inputs from the analyst, and perform automatically as much of the input/output processing and routine bookkeeping as possible. Such a supervisory language has been developed for the Image Processing Laboratory of the Jet Propulsion Laboratory (Efron, 1968). This language has been designed to allow the analyst to easily and quickly process one or a string of pictures through one or more processes automatically. This is done with a fairly simple set of commands which, together with the required numerical parameters, sre submitted as a card deck.

Only a supervisory program is permanently resident in core. The various processing programs as required are read from the disk libraries by these commands to the core, thus minimizing the amount of core which must be reserved for programs. The system contains special-purpose input/output routines optimized for picture handling which are used instead of the normal machine or Fortran input/output routines. This saves considerable space in the program libraries and obviates the necessity of writing these for each processing program. As a result, the writing of processing programs (which may be either assembly language or Fortran) is considerably simplified.

The entire programming system is available from COSMIC, the NASA software repository at the University of Georgia at Athens, Georgia.

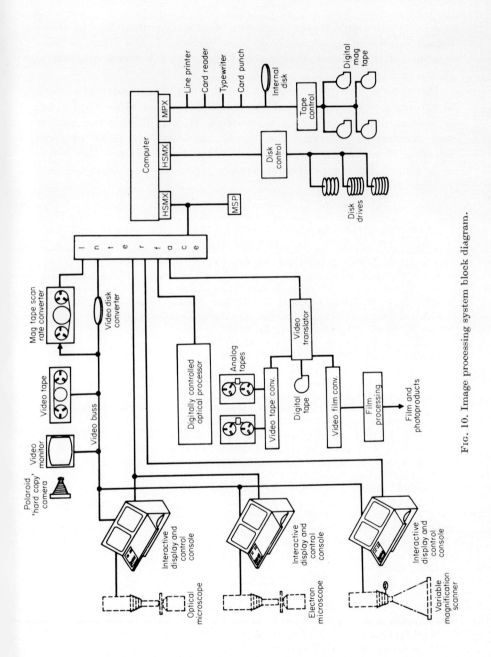

Fig. 10. Image processing system block diagram.

C. *A Complete System*

A complete digital image processing laboratory embodying the above considerations either presently or to be included in the future is being established at the Jet Propulsion Laboratory. A block diagram is shown in Fig. 10.

The system has been designed to allow rapid computer processing of the images, direct entry of the image data from films, hard copy, or optical devices such as a microscope. It will provide the analyst with a computer driven video display and the capability of rapid selection of processing by keyboard entry. The computer also serves as a control for the peripheral devices when they are operated in a system mode.

REFERENCES

Altman, J. H. (1967). *J. SMPTE*, **76**, 629–634.
Chao, S. C. (1966) *Flutter and Time Errors in Instrumentation Magnetic Recorders*, IEEE Trans. on Aerospace and Electronic Systems, *AES-2*, 214–223.
Eberhardt, E. H. (1966). *Signal-to-Noise Ratio of Image Dissector*, ITT Industrial Laboratories, Fort Wayne, Indiana, Technical Note No. 101.
Efron, E. (1968). *Phogram. Engn.*, **34**, 1058
Eyer, J. A. (1962). *Photo. Sci. Engn.*, **6**, 71–74.
Friedman, H. D. (1965). *Proc. IEEE.*, **53**, 658.
Hacking, K. (1962). *J. Br. IRE.*, **23**, 307–310.
Hacking, K. (1964). *J. SMPTE*, **73**, 1015–1029.
Harger, R. O. (1965). *Appl. Optics*, **4**, 383–386.
Heyning, J. M. (1966). "The Human Observer", Proc. Sem. on the Human in the Photo-Optical System, SPIE, New York City.
Huang, T. S. (1965). *IEEE Trans. Information Theory*, IT-11, 43–53.
Huang, T. S. (1967). Combined Use of Digital Computers and Coherent Optics in Image Processing, SPIE Computerized Imaging Techniques Symp. Washington, D.C.
Jones, R. C. (1961). *J. Opt. Soc. Am.*, **51**, 1159–1171.
Levi, L. (1958). *J. Opt. Soc. Am.*, **48**, 9–12.
Levi, L. (1963). *Photo Sci. Engn.*, **7**, 26–28.
O'Neill, E. (1963). "Introduction to Statistical Optics", p. 25,83, Addison-Wesley, London and New York. (See also p. 163 and Chapter 7.)
Oppenheim, A. V., Schafer, R. W. and Stockham, T. G. (1968). *Proc. IEEE*, **56**, 1264–1291.
Perrin, F. H. (1960). *J. SMPTE*, **69**, 151; **69**, 239.
Sadashige, K. (1967). *Appl. Optics*, **6**, 2179–2190.
Schade, Otto H. Sr. (1951–1955). Image Gradation, Graininess and Sharpness in Television and Motion Picture Systems. Part I. *J. SMPTE*, **56**, (Feb. 1951), 137–171: Part II. *J. SMPTE*, **58**, (March, 1952), 181–222; Part III. *J. SMPTE*, **61**, (August, 1953), 97–164; Part IV. **64**, (November, 1955), 593–617.
Schade, Otto H. Sr. (1956). *J. Opt. Soc. Am.*, **46**, 721–739.

Shelton, C. F., Herd, H. H. and Leybourne, J. J. (1967). Gray-level Resolution of Flying Spot Scanner Systems, SPIE Photo Optical Systems Seminar, Rochester, N.Y.

Stultz, K. F. and Zweig, H. J. (1959). *J. Opt. Soc. Am.*, **49**, 693–702.

Stultz, K. F. and Zweig, H. J. (1962). *J. Opt. Soc. Am.*, **52**, 45–50.

Schwartz, M. (1959). "Information Transmission, Modulation and Noise", McGraw-Hill, New York.

Van der Ziel, A. (1954). "Noise", Prentice-Hall, N.J., U.S.A.

Van der Ziel, A. (1957). "Solid State Electronics", Prentice-Hall, N.J., U.S.A.

Walli, C. R. (1964). Quantizing and Sampling Errors in Hybrid Computation, *Proc. Fall Joint Computer Cong.*, **25**, 545–558.

Weimer, P. K. (1960). *Adv. Elect. Elect. Phys.*, **13**, 387–437.

Zweig, H. J., Higgins, G. C. and MacAdam, D. L. (1958). *J. Opt. Soc. Am.*, **48**, 926–933.

SPIE: Society of Photo-optical Instrumentation Engineers, 1716 So. Catalina Ave., Redondo Beach, Calif. 90277.

SMPTE: Society of Motion Picture and Television Engineers, 9 E. 41 St., N.Y.C., 10017.

IEEE: Institute of Electrical and Electronic Engineers, Box A, Lennox Hill Sta., N.Y.C., 10021.

Mirror Electron Microscopy

A. B. BOK, J. B. ʟᴇ POOLE, J. ROOS ᴀɴᴅ H. ᴅᴇ LANG

Philips P.I.T.E.O. Eindhoven, Netherlands and
Technische Physische, Dienst, TNO–TH Delft, Netherlands
(with Appendices by H. de Lang, H. Bethge
and J. Heydenreich, and M. E. Barnett)

I.	Introduction	161
II.	Contrast Formation in a Mirror Electron Microscope with Focused Images	167
	A. Principle	167
	B. Calculations	168
	C. Model A	171
	D. Model B	182
	E. Model C	189
	F. Comparison with the image contrast for a mirror projection microscope	198
	G. Magnetic contrast	205
	H. Concise list of symbols used in Section II	206
III.	Description and Design of a Mirror Electron Microscope with Focused Images	207
	A. Description	207
IV.	Results and Applications	218
	A. Results	218
	B. Applications	226
V.	Appendix	227
	A. A scanning mirror electron microscope with magnetic quadrupoles	227
	B. Calculation of the influence of the specimen perturbation on the phase of the reflected electron beam in a mirror electron microscope (H. de Lang)	233
	C. Practice of mirror electron microscopy (H. Bethge and J. Heydenreich)	237
	D. Shadow projection mirror electron microscopy in "straight" systems (M. E. Barnett)	249
	References	259

I. Introduction

Fʀᴏᴍ the birth of electron optics (about 1920) till the late 1950's most efforts in the field of electron optics were concentrated on the theory, the design and development of the nowadays widely available transmission electron microscope. A guaranteed point resolution of less than 0·5 nm is considered normal for high quality instruments.

Since the transmission electron microscope provides information about the internal structure of an electron transparent specimen, this technique does not allow for the direct investigation of surfaces of solids. Two useful alternatives are either putting both the illuminating and imaging system at a glancing angle with the surface to be examined (reflection electron microscopy) or the application of the replica technique. The indirect observation of a surface by means of a replica permits a resolving power up to 5 nm.

The increasing interest in direct observation of surfaces of solids or investigation of surface phenomena has resulted during the past 15 years in the development of the following types of electron microscopes:

1. scanning electron microscope;
2. emission electron microscope;
3. reflection electron microscope;
4. mirror electron microscope.

Before going into more detail concerning the mirror electron microscope a brief description of the other types of microscopes is presented.

A. *Scanning electron microscope*

In a scanning electron microscope (Knoll, 1935) a primary electron beam, emitted from a heated tungsten filament, is focused into a fine electron probe on the specimen and made to scan on a raster—similar to television techniques—on the surface by a deflection system. Electrons liberated from the specimen by the focused primary beam are detected by a photomultiplier tube with a scintillator mounted on top. The photomultiplier output signal is used to modulate the brightness of the electron beam in a cathode-ray tube, which is scanned in synchronism with the electron probe. The resolution—being of the order of 20 nm in favourable operating conditions—depends upon the diameter of the electron probe, the accelerating voltage, the detector system and the type of specimen.

B. *Emission electron microscope*

In an emission electron microscope the specimen acts as a self-illuminating object. Electrons are liberated from the specimen by either heating of the specimen (thermionic emission), electron and ion bombardment of the specimen (secondary emission) or quantum irradiation of the specimen (photo-emission). The image is usually formed by a combination of two or three electron lenses. The obtainable resolution—mainly determined by the energy spread of the emitted electrons and the strength of the electrostatic field at the specimen surface—amounts to about 20 nm. In the case of thermionic and photo emission the image contrast is mainly dominated by the local work function of the specimen surface. Since the successful application of photo-emission, by means of ultra-violet radiation, this type of microscope has become of great importance.

C. *Reflection electron microscope*

Reflection electron microscopy is rarely used nowadays. The first experiments (Ruska, 1933) did not show very promising results until von Borries and Janzen, 1941, suggested that the large energy spread of the scattered electrons could be reduced by having the illuminating and imaging system at a glancing angle with the specimen. The remaining energy spread still requires a small aperture in order to minimize the dominating chromatic aberrations. Since the reflected electrons are scattered over a wide angle, a small angular aperture of the accepted beam has to be selected. This gives an image barely bright enough to be focused at the necessary magnification.

D. *Mirror electron microscope*

Contrary to the techniques mentioned above the specimen is neither struck by electrons nor emits electrons. An accelerated electron beam enters the retarding field of an electrostatic mirror. Application to the mirror electrode of a potential, which is slightly more negative than the accelerating voltage, causes the electrons to be reflected from an equipotential plane closely in front of the mirror electrode, which is in this technique the specimen surface. The electron trajectories near the point of reversal, in front of the specimen, are highly sensitive to deviations from flatness of the reflecting equipotential plane. These deviations are either caused by electrostatic or topographic perturbations at the physical specimen surface.

The possibility of converting on a microscopic scale electrostatic and, to a certain amount, magnetic potential distributions into a directly observable image has given access to new information in phenomena such as diffusion of metals, contact potentials, surface conductivity and magnetic properties.

The first experimental results of Hottenroth (1937) and the calculations of Henneberg and Recknagel (1935) clearly revealed the feasibility of mirror electron microscopy. Hottenroth showed that the manner of formation of non-focused images of the mirror electrode closely resembles that of the light optical "Schlieren" method. Following his experiments the research in this field was mainly directed towards the

application of this technique for visual observation of surface phenomena. Numerous articles, especially by Mayer (1957a, b, 1959a, b) and
Spivak *et al.* (1955, 1959, 1962) are published about different kinds of
applications with this type of microscope.

Little attention has been paid to optimizing the imaging technique
of the mirror electrode. It was Le Poole (1964c) who pointed out that
the attainable resolving power for this type of microscope could be
improved considerably by forming a focused image of the mirror
electrode onto the fluorescent screen. Also it became evident that
mirror electron microscopes with rotationally symmetric lenses require
a separation of the illuminating and reflected beam in order to obtain
a focused image of the specimen with sufficient field of view. In instruments with rotationally symmetric lenses and without beam separation
(Bethge *et al.*, 1960; Forst and Wende, 1964; Barnett and Nixon, 1967a, b)
the specimen is illuminated through a central hole in the final screen.
The field of view, which disappears entirely when the mirror electrode
is exactly conjugated to the final screen, can only be increased by
defocusing. The formation of contrast in a microscope with focused
images is achievable in a way similar to the transmission electron
microscope by having an aperture in the objective lens. To avoid a
new limitation of the field of view and to have at the same time normal
incidence of the illuminating beam this aperture must be in the back
focal plane of the objective lens. Simultaneous focusing of the specimen
and a sufficiently large field of view can be obtained by the separation
of the illuminating and reflected beam with a magnetic prism or the
use of magnetic quadrupoles.

Concerning this latter method a scanning mirror electron microscope
with magnetic quadrupoles has been designed and constructed (Bok
et al., 1964). A brief description of this instrument is given in Section
V.A.

Although beam separation with a magnetic prism has been applied
earlier by several experimenters (Orthuber, 1948; Bartz *et al*, 1956; Hopp,
1960; Schwartze, 1967), it was never used (except for the instrument of
Schwartze (1967) in combination with a contrast aperture in the back
focal plane of the objective lens.

The absence of this aperture means that for in-focus images the
contrast disappears. Here defocusing provides the necessary image
contrast.

From geometrical considerations it follows that for an imaging
mirror microscope special attention has to be paid to the condenser
system. Seen from the objective aperture the electron source must
appear as large as the required field of view.

This leads to the conclusion that all mirror electron images, obtained so far, except for some pictures made by Schwartze (1967) are point projection, out of focus, images.

When a point projection image of the mirror electrode is to be made, an electron probe is formed in front of the specimen. The electrons then reflect from a paraboloid of revolution which is the envelope of all parabolic electron trajectories in the retarding field. Contrary to this, in a microscope with focused images and an objective aperture in the back focal plane, all electrons reflect from a flat equipotential plane normal to the z-axis.

The effect of the reflecting paraboloid in defocused instruments is

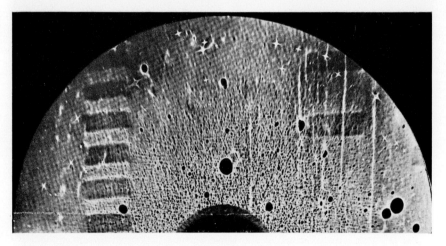

FIG. 1. Mirror projection image of a magnetic recording pattern (made by L. Mayer).

clearly observable from most of the photographic results published. Where electrons strike the specimen surface local negatively charged spots occur. Negative spots give rise to black "bubbles" in the final image. For positively charged spots, caused for instance by a positive ion bombardment on areas where the electrons do not reach the specimen surface, white "stars" emerge on the final screen (Lenz and Krimmel, 1963) (see Fig. 1). When, in a mirror projection microscope, the electrons with the highest energy in the Maxwellian distribution and incident close to the axis are allowed to strike the specimen, the central region on the final screen shows mainly black bubbles. The more off axis electrons reverse their direction before reaching the specimen surface and give mainly white stars for the outer regions (see Fig. 1). On the other hand, the occurrence of some stars in the central region and bubbles in the outer regions is comprehensible owing to the fact that, apart from local charges, the topography of the surface also gives rise to similar effects. In-focus images show hardly any black bubbles and white stars. Near the focusing condition, where the contrast reverses, the bubbles change into stars and vice versa (see Figs. 2a and b).

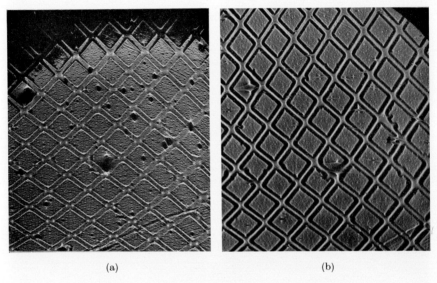

(a)   (b)

FIG. 2. Gold squares, about 20 mm thick, vacuum deposited through a 750 mesh grid on a layer of gold (magnification × 330).

(a) slightly under focus.
(b) slightly over focus.

The main subject of this article is the theory, design and construction of a mirror electron microscope, suitable for focused images. It consists of rotationally symmetric lenses, a magnetic prism, a wide aperture condenser system and a contrast aperture in the back focal plane of the objective lens. In Section II a treatment is given of the formation of contrast in a mirror electron microscope, whereas Section III deals with the design and construction of the instrument. Section IV provides a series of photographic results and a brief discussion about possible applications.

II. CONTRAST FORMATION IN A MIRROR ELECTRON MICROSCOPE WITH FOCUSED IMAGES†

A. *Principle*

When a mono-energetic and axially parallel beam of electrons enters a homogeneous electrostatic retarding field, reflection occurs against a flat equipotential plane normal to the z-axis (see Fig. 3) and all electrons return along the same trajectories.

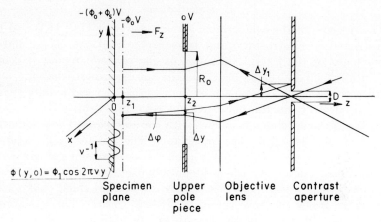

FIG. 3. Retarding field with perturbed specimen and characteristic quantities.

Owing to local deviations from flatness of the reflecting equipotential plane, the electrons which approach these perturbations receive a tangential impulse. These electrons describe a different trajectory after reversal and intersect the aperture plane at a height Δy_1, which depends on the perturbation present. When Δy_1 exceeds $D/2$, the radius of the contrast forming aperture, the electrons impinge on the aperture and are removed from the reflected electron beam. The formation of contrast in a mirror electron microscope resembles the technique (an aperture in the back focal plane of the objective) applied in transmission electron microscopes or the optical "Schlieren" technique. The separation of the tangentially modulated electron pencils from the unperturbed pencils allows for visual observation of perturbations in the reflecting equipotential plane, in terms of current density modulations in the final image. The origin of these perturbations can be twofold, topography of an equipotential specimen surface or electro-

† See p. 206 for a concise list of symbols used in Section II.

static disturbances on a flat specimen surface. In practice, mostly a combination of both is encountered.

The current density modulations in the final image do not provide direct information about the type of perturbation present at the specimen. Perturbations at the specimen of magnetic origin, in comparison with electrostatic and topographic perturbations, hardly affect an axially parallel beam of electrons. Magnetic contrast will be discussed separately in Section II.G.

B. *Calculations*

The purpose of the calculations given in Sections II.C, II.D and II.E —restricted to electrostatic and topographic contrast—is to find the dependence of the lateral shift Δy_1 or the related angular deflection $\Delta\varphi_1$ of a reflected electron pencil on the specimen perturbation.

All three sections assume a mono-energetic $(e\phi_o)$ and axially parallel beam of electrons in the homogeneous retarding field.

In Fig. 3 the specimen coincides with the *xoy*-plane. The calculations are only performed in the *yoz*-plane. The electrostatic retarding potential $(\phi_o + \phi_s)$, a superposition of the accelerating voltage ϕ_o and an additional voltage ϕ_s, is considered homogeneous and unaffected by the bore in the upper pole piece. This is valid provided that z_2 is at least three times the bore radius R_o (Glaser, 1952). This assumption permits the separation of the divergent lens action of the upper pole piece from the contrast formation mechanism near the specimen. Apart from lens defects the divergent lens action does not affect the contrast formation. It only requires a slightly higher excitation of the objective lens to maintain the parallel incidence into the mirror field. The behaviour of electrons in the retarding field can be described either classically by the equations of motion (1a) and (1b) or wave mechanically by the time independent Schrodinger equation (2).

Classically

y-direction

$$-e\,\frac{\partial\Phi(y,\,z)}{\partial y} = -e\,\frac{\partial\phi(y,\,z)}{\partial y} = m\,\frac{d^2y}{dt^2} \tag{1a}$$

z-direction

$$-e\,\frac{\partial\Phi(y,\,z)}{\partial z} = -eF_z - e\,\frac{\partial\phi(y,\,z)}{\partial z} = m\,\frac{d^2z}{dt^2} \tag{1b}$$

Wave mechanically

$$-\frac{\hbar^2}{2m}\left(\frac{\partial^2 u}{\partial y^2} + \frac{\partial^2 u}{\partial z^2}\right) + (V-E)u = 0 \tag{2}$$

where $\Phi(y, z) = \phi_o + \phi_s + \phi(y, z)$ (3)

      e = elementary charge

     m = electron rest mass

     $\phi\ (y, z)$ is the perturbation potential

     F_z is the strength of the retarding field.

For 30 kV across a gap of $3 \cdot 5 \times 10^{-3}$ m, $F_z = 8 \cdot 57 \times 10^6$ V/m.

$$\hbar = \frac{h}{2\pi}, h = \text{Planck's constant}$$

     $u = u(y, z)$ the wave function

     $V = V(z)$ the potential energy in the retarding field

     E is the kinetic energy of the incident beam.

For all calculations following it is assumed that $\phi_s \ll \phi_o$.

Two models A and B, Section II.C and II.D, are based on the equations of motion (1a) and (1b), whereas model C (Section II.E) uses the Schrödinger equation (2).

It would be obvious to describe the formation of contrast in a way comparable with the modulation transfer functions in the light optics. The non-linear character of equations (1a) and (1b), however, does not allow for such a description because no linearity exists in the case of sufficient contrast between the perturbation amplitude at the specimen and the tangentially modulated electron pencils. Since it is wished to provide an analytical description of the contrast mechanism, preferably in a way resembling the modulation transfer functions, equations (1a) and (1b) are linearized by the assumption $\dfrac{\partial \phi(y, z)}{\partial z} \ll F_z$ (model A). This causes a sinusoidal specimen to produce a sinusoidal modulation.

Contrary to the approximated model A, model B provides information about the solution of the exact equations (1a) and (1b). The calculations for this model were performed both on a digital computer (Telefunken TR-4) and an analog computer (Applied Dynamics 40). The TR-4 calculations provide the Δy value and the corresponding coordinates of the point of reversal for different heights of incidence. The omitted index 1 for Δy and $\Delta\varphi$ indicates that these values are measured in the plane $z = z_2$, the upper pole piece of the objective lens.

A comparison of the results obtained with models A and B is shown to lead for certain values of the local slope of the reflecting plane to a matching of both models. This means that the non-linear model B shows a linear behaviour.

As $\phi(y, z)$ fulfils the Laplace equation $\Delta\phi(y, z) = 0$ a sinusoidal perturbation potential

$\phi(y, 0) = \phi_1 \cos 2\pi n\nu y$ causes a potential

$\phi(y, z) = \phi_1 \cos 2\pi n\nu y \cdot \exp(-2\pi n\nu z),$

where ν is the spatial specimen frequency and ϕ_1 the perturbation
amplitude.

Since the assumption $\dfrac{\partial\phi(y, z)}{\partial z} \ll F_z$ involves simultaneously moderate

values $\dfrac{\partial\phi(y, z)}{\partial y}$, it can be expected prior to the calculations following

that models A and B match only for specimens slightly perturbed
(small values of $\phi_1\nu$). Specimens with more contrast are not accessible
to a simple analytical description. In that case numerical calculations
should provide information.

Figures 4a and b, both made on an analog computer, demonstrate

the effect of the assumption $\dfrac{\partial\phi(y, z)}{\partial z} \ll F_z$ on the electron trajectories

near a sinusoidal perturbed specimen. In model A all electrons reflect

from a flat equipotential plane whereas in model B $\dfrac{\partial\phi(y, z)}{\partial z}$ leads to a

variable depth of penetration into the retarding field.

Apart from equations (1a) and (b) a second non-linear effect in the

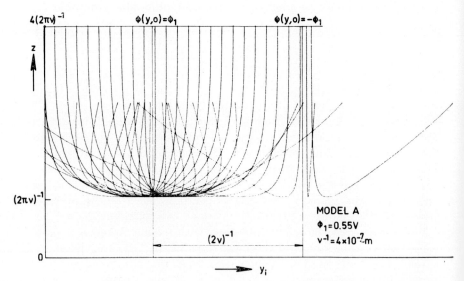

Fig. 4a. Electron trajectories according to model A.

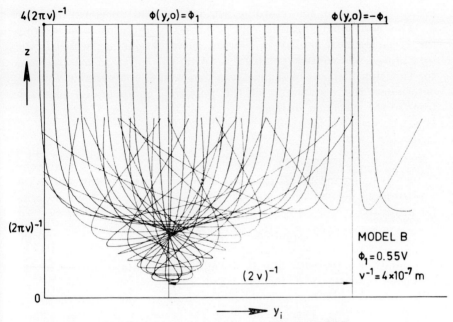

FIG. 4b. Electron trajectories according to model B.

formation of contrast is introduced by the filtering of tangentially modulated electron pencils from the reflected beam with a circular contrast aperture.

Since this effect is independent of the sign of the lateral deflection, the modulated current density distribution in the final image shows the double frequency of a sinusoidal perturbation. This double rectifying effect can be avoided by using a knife edge as aperture or, in the case of two dimensional specimens, two perpendicular edges.

In order to avoid, at this stage, the choice between the non-linearly filtering circular aperture and the linearly operating edge aperture, the lateral displacement Δy and the angular deflection $\Delta\varphi$ are both plotted against the spatial perturbation frequency and amplitude.

C. Model A

a. *Electrostatic Contrast.* The simplified equations of motion are

$$-e\,\frac{\partial \Phi(y, z)}{\partial y} = m\,\frac{d^2 y}{dt^2} \tag{4a}$$

$$-e\,F_z = m\,\frac{d^2 z}{dt^2} \tag{4b}$$

where $\quad \Phi(y, z) = \phi_o + \phi_s + \phi(y, z); \; \phi_s = \phi_{\mathrm{SC}} + \phi_{\mathrm{SV}} \tag{4c}$

Superimposed on the retarding potential ϕ_o is a "specimen" voltage ϕ_s. This additional negative voltage prevents electrons from striking the specimen near positively charged perturbations. ϕ_s, which is the sum of ϕ_{SC}, the contact potential between the specimen material and the tungsten filament in the electron gun, and ϕ_{SV}, a variable voltage, determines the distance z_1 of the reflecting equipotential plane in front of the specimen surface.

If ϕ_{SC} is corrected for then

$$z_1 = \frac{\phi_{SV}}{F_z} \ll z_2$$

In this linearized model it is useful to represent an electrostatic (or topographic) perturbation $\phi(y, 0)$ along the y-axis as a Fourier series (or integral).

$$\phi(y, 0) = \sum_{n=1}^{\infty} \phi_n \cos 2\pi n v y \qquad (5)$$

The omitted term with $n = 0$, an additional voltage on top of the specimen potential, is defined as ϕ_{SA}.

In these calculations no incident electrons are allowed to reach the physical specimen surface, because the electron scattering effects which would occur, destroy the validity of the results obtained. Experiments revealed, that for a specimen biased slightly positive (some tenth of volts) with respect to the accelerating voltage ϕ_o, the image contrast deteriorated considerably due to the electron scattering phenomena at the specimen surface.

The lateral impulse $\Delta m v_y$ given an electron travelling towards and from the specimen amounts to

$$\Delta m v_y = -2 \int_{z_1}^{z_2} e \frac{\partial \phi(y, z)}{\partial y} dt \approx -2 \int_{0}^{\infty} e \frac{\partial \phi(y, z)}{\partial y} dt \qquad (6)$$

Inserting equations (4c) and (5) into (6) and neglecting the lateral displacement during reversal, which is permissible for small values of $\Delta m v_y$, it follows that:

$$\Delta m v_y = e \sum_{n=1}^{\infty} [\phi_n 2\pi n v \cdot \exp(-2\pi n v z_1) \cdot$$

$$\cdot \sin 2\pi n v y] 2 \int_{0}^{\infty} \exp[-2\pi n v(z - z_1)] dt$$

$$\Delta m v_y = \left(\frac{4\pi^2 m e}{F_z}\right)^{\frac{1}{2}} \sum_{n=1}^{\infty} [\phi_n (n v)^{\frac{1}{2}} \sin 2\pi n v y \cdot \exp(-2\pi n v z_1)] \qquad (7)$$

ϕ_{SV} has been introduced by writing $(z - z_1)$ instead of z.

If y_i represents the height of incidence above the z-axis and y_r the corresponding height for the reflected electrons, both measured in the plane $z = z_2$, then

$$y_r - y_i = \Delta y \Big|_{z=z_2} = v_y t$$

$$\Delta y \Big|_{z=z_2} = \frac{2\pi}{F_z} (2_{z_2}\nu)^{\frac{1}{2}} \sum_{n=1}^{\infty} [\phi_n n^{\frac{1}{2}} \sin 2\pi n\nu y \cdot \exp(-2\pi n\nu z_1)] \qquad (8)$$

After interaction with the specimen the reflected electrons follow parabolic trajectories in the retarding field, owing to the lateral momentum received.

The corresponding angular deflection $\Delta\varphi$ is

$$\Delta\varphi = \frac{\Delta y}{2z_2} \qquad (9)$$

For $n = 1$ Δy and $\Delta\varphi$ are determined as a function of ϕ_1 and ν, the spatial frequency of a sinusoidal perturbation at the specimen surface,

$$\Delta y \Big|_{z=z_2} = \Delta y_v = \frac{2\pi}{F_z} (2z_2\nu)^{\frac{1}{2}} \phi_1 \sin 2\pi\nu y \cdot \exp\left(-2\pi \frac{\phi_1}{\phi_o} z_2\nu\right) \cdot$$

$$\exp\left(-2\pi \frac{\phi_{\mathrm{SA}}}{\phi_o} z_2\nu\right) \qquad (10)$$

The index v in Δy_v and $\Delta\varphi_v$ refers to electrostatic contrast in $z = z_2$. $z_1 = \dfrac{\phi_1 + \phi_{\mathrm{SA}}}{\phi_o} z_2$ (with ϕ_{SC} corrected for) has a minimum value $\dfrac{\phi_1}{F_z}$ which prevents electrons in the case of electrostatic contrast from striking the specimen near positively charged perturbations.

The difference $\phi_{\mathrm{SV}} - \phi_1 = \phi_{\mathrm{SA}}$ corresponds to an additional voltage for adjusting the reflecting equipotential plane away from the perturbed specimen surface.

In Fig. 5 the maximum values of Δy_v and $\Delta\varphi_v$, following from equation (10), are plotted against ν with ϕ_1 as curve parameter. For all curves presented it is assumed that $\sin 2\pi\nu y = 1$, $\phi_{\mathrm{SC}} = \phi_{\mathrm{SA}} = 0$. Since the local "slope" $\phi_1\nu$ of the perturbations in the reflecting equipotential plane plays the main role in this contrast mechanism, Δy_v is also plotted against $\phi_1\nu$ (see Fig. 6).

The maxima of the plotted Δy_v and $\Delta\varphi_v$ values for each curve $F(\phi_1\nu)_{\phi_1}$ coincide with the straight line $\dfrac{(\phi_1\nu)_{opt}}{F_z} = 8 \times 10^{-2}$ or in this instrument $(\phi_1\nu)_{opt} = \dfrac{F_z}{4\pi} = 6\cdot8 \times 10^5$ V/m.

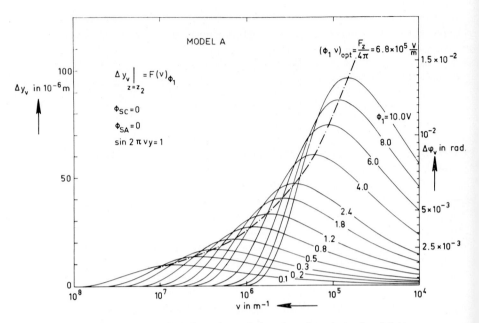

Fig. 5. Modulation functions Δy_v and $\Delta \varphi_v$ plotted against v (model A).

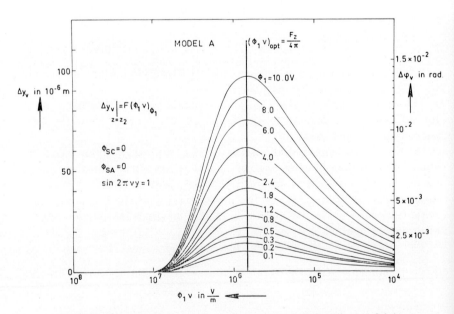

Fig. 6. Modulation functions Δy_v and $\Delta \varphi_v$ plotted against $\phi_1 v$ (model A).

b. *Topographic Contrast.* Provided that $\dfrac{\partial d(y, z)}{\partial y} \ll 1$ the topographic "displacement" Δy_t is found by substituting in equation (8)

$$\phi_n = d_n F_z \text{ and } z_1 = \frac{\phi_{\text{SA}}}{F_z} \tag{11}$$

These relations are only applicable for small perturbations with moderate curvatures $\left(\dfrac{\partial^2 d(y, z)}{\partial y^2} \ll 1\right)$ because then the z component of the field strength near the specimen equals F_z.

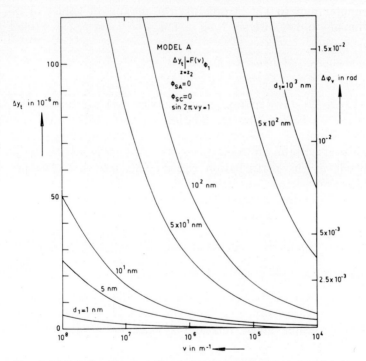

FIG. 7. Modulation functions Δy_t and $\Delta \varphi_t$ plotted against ν (model A).

d_n represents the amplitude of the component harmonics in the specimen topography

$$d(y, 0) = \sum_{n=1}^{\infty} d_n \cos 2\pi n \nu y \tag{12}$$

Contrary to electrostatic contrast all topographic perturbations coincide with one equipotential plane. It follows that the minimum required bias for electrostatic contrast $\left(\text{for } n = 1 \text{ is } z_1 = \dfrac{\phi_1}{F_z}\right)$ can be omitted for topographic contrast.

The electrostatic contrast calculations of Barnett and Nixon (1967a, b) have apparently overlooked this effect. Contrary to the title of their publication, it means that they have in fact calculated topographic contrast.

If $\phi_{SA} = 0$ and $n = 1$ all electrons reach exactly the "topographic specimen" and

$$\Delta y_t \Big|_{z = z_2} = 2\pi (2z_2 \nu)^{\frac{1}{2}} d_1 \sin 2\pi \nu y \tag{13}$$

A similar plot to Fig. 5 presents the maximum values of Δy_t as a function of ν and d_1 with $\sin 2\pi \nu y = 1$, $\phi_{SA} = 0$ and $\phi_{SC} = 0$ (see Fig. 7).

c. *Conclusions and Remarks for Model A*

Conclusions

1. An electrostatic or topographic cosine perturbation at the specimen surface gives a sine modulation on the angle of the reversing electron pencils. For electrostatic contrast the specimen should be at least biased with an additional negative voltage, equal to the positive amplitude of the perturbation signal. This prevents electrons from reaching the specimen surface.

2. It follows directly from the linear character of the equations (4a) and (4b) that the modulation effect of an arbitrary periodic perturbation, either electrostatic or topographic, can be calculated by summing the separate modulation effects of the component harmonics.

3. For topographic contrast with $\dfrac{\partial d(y, 0)}{\partial y} \ll 1$ and $\dfrac{\partial^2 d(y, 0)}{\partial y^2} \ll 1$ a linear relation exists between the perturbation amplitude and the modulation effect (Δy_t or $\Delta \varphi_t$) on the angle of the reversing electron pencils. A similar linear relation is valid for electrostatic contrast provided that $2\pi \phi_1 \nu \ll F_z$.

4. A discrimination between topographic and electrostatic contrast is possible in principle, because for topography the modulation function is independent of the position at the specimen. For electrostatic contrast it varies from place to place owing to the additional damping factor.

5. The fact that $\phi_1 \nu$ possesses a constant value corresponding to the maxima of the plotted Δy_v or $\Delta \varphi_v$ values (Figs. 5 and 6) means that once ϕ_1 is given, the optimum ν value for maximum electrostatic contrast follows immediately.

The conclusions stated above involve the assumptions:

1. the illuminating beam in the retarding field is parallel to the z-axis;

2. $\dfrac{\partial \phi(y, z)}{\partial z} \ll F_z$;

3. $\dfrac{\partial d(y, 0)}{\partial y} \ll 1$ and $\dfrac{\partial^2 d(y, 0)}{\partial y^2} \ll 1$ for topographic contrast;

4. electrons do not reach the specimen surface;

5. the illuminating beam is mono-energetic $(e\phi_o)$.

Remarks

1. The current density distribution in the final image can be determined by projecting the $\varDelta y$ value found onto the contrast aperture plane. For the combination of objective and electrostatic lens (see Fig. 8) the lateral shift $\varDelta y_1$ in the contrast aperture plane amounts to

$$\varDelta y_1 = \left(1 + \frac{x_4}{x_2}\right)\left(1 + \frac{x_5}{x_3}\right)\varDelta y = G_1 \varDelta y$$

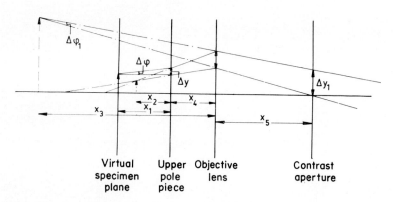

FIG. 8. Relation between $\varDelta y$ on the upper pole piece and $\varDelta y_1$ in the contrast aperture plane.

This expression holds for both the virtual and the real imaging mode of the objective lens.

Every point of the specimen surface inside the field of view is illuminated by an electron pencil incident parallel to the z-axis. When a circular contrast aperture is used and the illuminating pencils fill this aperture, the common area 0_c of the contrast aperture and the cross section of each lateral deflected electron pencil directly indicates the current density in the final image. From Fig. 9 is found

$$0_c = \tfrac{1}{2}D^2 \left\{ \arccos \frac{|\varDelta y_1|}{D} - \frac{|\varDelta y_1|}{D} \left[1 - \left(\frac{\varDelta y_1}{D} \right)^2 \right]^{\frac{1}{2}} \right\} \tag{14}$$

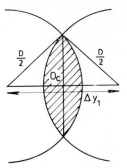

Fig. 9. Common area of contrast aperture and reflected electron pencil.

Plotting $f_c = \dfrac{0_c}{\frac{1}{4}\pi D^2}$ against $\dfrac{|\varDelta y_1|}{D}$ shows an almost proportional relation (see Fig. 10).

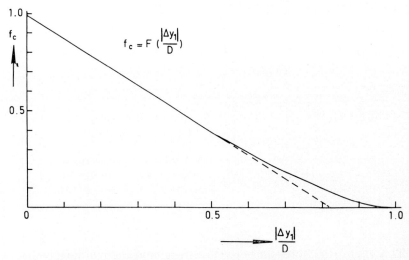

Fig. 10. Filtering characteristic f_c for a "filled" circular contrast aperture.

Starting from a sinusoidal perturbed specimen

$$\phi(y, 0) = \phi_1 \cos 2\pi\nu y \quad \text{for electrostatic contrast}$$

and

$$d(y, 0) = d_1 \cos 2\pi\nu y \quad \text{for topographic contrast,}$$

the corresponding current density distribution $j(y)$, calculated back on the specimen surface, follows from

$$j(y) = j_o f_c \tag{15}$$

where j_o = current density of the illuminating beam without perturbations at the specimen and

$$f_c = \frac{2}{\pi} \left\{ \arccos \frac{G_1 |\varDelta y|}{D} - \frac{G_1 |\varDelta y|}{D} \left[1 - \left(\frac{G_1 \varDelta y}{D} \right)^2 \right]^{\frac{1}{2}} \right\} \quad \phi_{\mathrm{SA}} = \phi_{\mathrm{SC}} = 0$$

and

$$\varDelta y = \varDelta y_v \Big|_{z=z_2} = \frac{2\pi}{F_z} (2z_2 \nu)^{\frac{1}{2}} \phi_1 \sin 2\pi \nu y . \exp\left(- 2\pi \nu \nu_1 \right) \tag{8}$$

for electrostatic contrast

or

$$\varDelta y = \varDelta y_t \Big|_{z=z_2} = 2\pi (2z_2 \nu)^{\frac{1}{2}} d_1 \sin 2\pi \nu y$$

for topographic contrast.

Another filtering characteristic f_c would be

$$f_c = 1 \ \text{for} \ 0 \le G_1 |\varDelta y| \le D/2$$

and

$$f_c = 0 \ \text{for} \ G_1 |\varDelta y| > D/2$$

It assumes a cross-over for the illuminating electron beam small compared with D.

Since D is restricted in practice to a lower limit of 50 μm, the cross-over of the illuminating beam will neither fill the contrast aperture entirely nor has a size which makes it much smaller than D. The shape of the practical f_c curves for a circular aperture resembles a trapezium with the edges rounded off.

For a knife edge aperture and a small cross-over

$$f_c = 1 \ \text{for} \ G_1 \varDelta y < 0$$

$$f_c = 0 \ \text{for} \ G_1 \varDelta y > 0$$

2. The assumption of specimen illumination with an electron beam perfectly parallel to the z-axis requires an infinitely small semi-angular aperture β of the composing pencils.

However, in order to fill the aperture, β needs to have a fairly large value. The value of β, depending on the circular contrast aperture D, follows from Fig. 8 where

$$\varDelta \varphi_1 = \beta \ \text{for} \ \varDelta y_1 = D/2$$

$$\beta = \frac{D}{2x_1} \left[\left(1 + \frac{x_5}{x_3} \right) \left(1 + \frac{x_4}{x_2} \right) \right]^{-1} = \frac{D}{2x_1 G_1} \tag{16}$$

The effect of $\beta \neq 0$ is twofold:

(a) Owing to the spherical aberration of the homogeneous retarding field, electrons reflect against equipotential planes at $z = z_s(\beta)$ instead of $z = 0$ ($\phi_{\mathrm{SA}} = \phi_{\mathrm{SC}} = 0$).

The electron trajectories drawn in Fig. 11 correspond to one electron travelling parallel to the z-axis, having its point of reversal in $z = 0$,

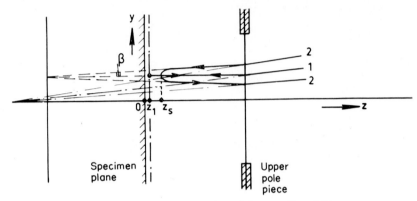

FIG. 11. Spherical aberration of the retarding field.

and a second electron entering the retarding field at an angle β with a point of reversal in $z = z_s$. The difference in depth of penetration amounts to

$$z_s = \beta^2 z_2 \text{ or in terms of voltage } \phi_{\mathrm{SA}} = \beta^2 \phi_o \qquad (17)$$

Here ϕ_{SA} represents the voltage equivalent of z_s.

(b) Not derivable from this simplified contrast model are the high sensitivities of Δy_t and Δy_v for electrons approaching the perturbed reflecting equipotential plane at an angle with the z-axis. This effect, being more predominant than the damping effect $\phi_{\mathrm{SA}} = \beta^2 \phi_o$, is examined both with a digital and an analog computer on the basis of the exact equations (1a) and (1b).

See for further results Fig. 16b and c of Section II.D and Section II.E.

3. As can be learned from equation (8)

$$\Delta y_v \Big|_{z = z_2} = \frac{2\pi}{F_z} (2z_2 v)^{\frac{1}{2}} \sum_{n=1}^{\infty} [\phi_n \sin 2\pi n v y \cdot \exp(-2\pi n v z_1) \cdot n^{\frac{1}{2}}]$$

a strongly increasing damping $n^{\frac{1}{2}} \exp(-2\pi n v z_1)$ occurs in the case of electrostatic contrast for higher order harmonics. Equation (8) corresponds to a low pass filter. It causes structures with higher order harmonics to be hardly distinguishable from a sinusoidal perturbation.

4. The assumption of a mono-energetic beam of electrons is not feasible in practice owing to the Maxwellian energy distribution for

electrons emitted from a heated filament. As a result of the Boersch effect, the retarded electron beam in front of the specimen shows even a larger spread in energy than following from the Maxwellian distribution. The corresponding variations in depth of penetration lead to a decrease in contrast because of the exponential z damping in the contrast modulation function.

The contribution of electrons with an energy between E_1 and $E_1 + \Delta E_1$ (after passing the retarding potential ϕ_0) to the formation of image contrast can be determined by multiplying the exponential z damping factor (curve 2 in Fig. 12).

$$\exp(-2\pi\nu z) = \exp\left[-\frac{2\pi\nu}{eF_z}(e\phi_{\mathrm{SV}} - E_1)\right] \qquad (18)$$

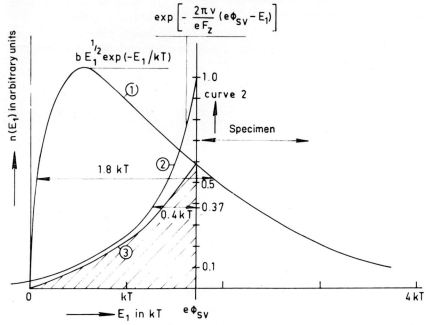

FIG. 12. Effective contribution of electrons out of the Maxwellian energy distribution to the formation of image contrast.

with the energy distribution in front of the specimen. Neglecting the broadening of the Maxwellian energy distribution by the Boersch effect, this distribution is given by

$$\int_0^\infty n(E_1)dE_1 = \int_0^\infty bE_1^{\frac{1}{2}}\exp(-E_1/kT)dE_1 \quad (b = \text{constant}) \qquad (19)$$

The integrand of equation (19) is plotted in Fig. 12 (curve 1). The hatched area covered by

$$\int_0^{e\phi_{sv}} \exp\left(-2\pi v z\right) . n(E_1)dE_1 =$$

$$= \int_0^{e\phi_{sv}} \exp\left[-\frac{2\pi v}{eF_z}(e\phi_{sv} - E_1)\right] . bE^{\frac{1}{2}}_1 . \exp\left(-E_1/kT\right)dE_1 \quad (20)$$

represents the effective contribution of electrons out of the Maxwellian energy distribution to the formation of image contrast. The remaining part only gives a continuous background illumination in the final image. Curve 3 stands for the integrand of equation (20). The choice of $e\phi_{sv}$ as the upper limit in equation (20) involves the assumption that electrons reaching the specimen surface are ignored or in a more pessimistic view only contribute to the background illumination.

Introduction of a visibility criterion for the image contrast leads to a value of v in equation (18) for which just enough lateral contrast is obtained.

Assuming 20% contrast yields a value for v of

$$\frac{2\pi v}{eF_z}(e\phi_{sv} - E_1) = 1 \text{ with } (e\phi_{sv} - E_1) = 0 \cdot 4 \text{ kT}.$$

The lateral resolving power (along the specimen) is then approximated by

$$v^{-1} = \frac{0 \cdot 8 \, \pi kT}{eF_z} = 70 \text{ nm } (\text{T} = 2800°\text{K}) \quad (21)$$

From Fig. 12 it is evident that already a rough filtering of the illuminating beam leads to a substantial improvement in contrast and therefore in lateral resolving power.

Concerning the lateral resolving power a point resolution of 80...100 nm has been reached in Fig. 27b.

For the axial resolving power (step perturbations) the reader is referred to the appendix (Section V.B).

D. Model B (*solution of the exact equations of motion 2.1a and b*)

a. *Electrostatic Contrast Calculations with a Digital Computer.* By means of a digital computer a plot, similar to Fig. 5, for the maximum values of Δy_v and $\Delta\varphi_v$ depending on v and ϕ_1 as curve parameters, is presented in Fig. 13.

The condition $\sin 2\pi v y = 1$ of model A is not tenable for model B because it does not coincide with the maximum values of Δy_v and $\Delta\varphi_v$. For comparison the curves of model A (dashed lines) and model B (full lines) are both pictured in Fig. 13. In this computer program the two second order differential equations were split into four first order equations. Integration of these equations is performed with the Nordsieck procedure, containing a variable steplength combined with a

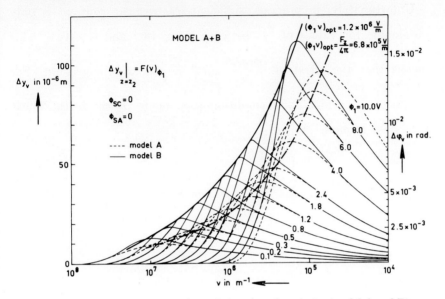

FIG. 13. Modulation functions Δy_v and $\Delta \varphi_v$ plotted against ν (model A and B).

preset desired accuracy for each equidistant interval along the time axis.

As characteristic example of the computer results obtained Δy_v, $(y_{top} - y_i)$ and z_{top} are plotted (Fig. 14) against y_i, the height of incidence, for a sinusoidal perturbation with $\nu^{-1} = 4 \times 10^{-7}$ m and $\phi_1 = 0.5$ V.

The coordinates z_{top} and y_{top} represent the point of reversal. For $y_i = 10^{-7}$ m, thus at the positive maximum of the perturbation, the electrons just reach the specimen surface, whereas for $y_i = 3 \times 10^{-7}$ m z_{top} possesses its maximum.

For both values of y_i $(y_{top} - y_i) = 0$, thus the electron trajectories are straight lines.

Considering Figs. 13 and 14 the following conclusions for model B emerge:

1. For small values of Δy_v and $\Delta \varphi_v$, corresponding to a low contrast, both groups of curves match accurately as expected. Comparison of

the curves $\Delta y_v = F(\phi_1 v)\phi_1$ of model A (Fig. 6) with those of model B leads to a matching within 2% of A and B provided that

$$\frac{\phi_1 v}{F_z} < 1\cdot2 \times 10^{-2} \text{ and } \frac{\phi_1 v}{F_z} > 6 \times 10^{-1} \quad (F_z = 8\cdot57 \times 10^6 \text{ V/m}) \quad (22)$$

Unfortunately, in practice, sufficient electrostatic contrast is only obtainable for values of $\dfrac{\phi_1 v}{F_z}$ between $1\cdot2 \times 10^{-2}$ and 6×10^{-1}. This means that the effect of $\dfrac{\partial \phi(y, z)}{\partial z}$ must be taken into account. For values of $\dfrac{\phi_1 v}{F_z}$ which fulfil condition (22), model B shows the linear behaviour of model A.

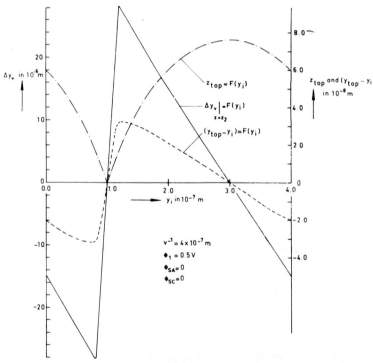

FIG. 14. Δy_v and the coordinates y_{top} and z_{top} of the point of reversal plotted against y_i.

In order to make model A applicable in practice, the circular contrast aperture D should be at least smaller than 10 μm. This value for D, however, is not feasible owing to the centring problems then arising.

Since $\dfrac{\partial \phi(y, z)}{\partial z}$ has the same sign as F_z near positively charged spots, it is comprehensible that the maximum values of Δy_v and $\Delta \varphi_v$ in model B exceed those of model A.

2. Like model A, the maxima of all B curves coincide with a line

$$\frac{(\phi_1 \nu)_{opt}}{F_z} = 1 \cdot 4 \times 10^{-1}.$$

For this value Δy_v and $\Delta \varphi_v$ show almost a sawtooth shape instead of sinusoidal (see Fig. 14). A cosine perturbation thus causes a modulation effect consisting of a spectrum of sines, approximated by a constant

$$\times \sum_{n=1}^{\infty} n^{-1} \sin 2\pi n \nu y.$$

The increasing damping effect (always present for electrostatic contrast) on higher order harmonics, both for model A and B, is demonstrated in Fig. 15, where Δy_v is plotted against y_i for the following perturbations:

1. sine,
2. triangular,
3. trapezium,
4. square wave.

Except for differences in the maximum Δy_v value, the shapes of the curves resemble each other closely. This indicates that although the modulation is by no means sinusoidal for the chosen values of ν and ϕ_1 almost no information about the accurate shape of an electrostatic perturbation can be obtained from the current density distribution in the final image.

b. *Topographic Contrast.* Concerning topographic contrast an exact solution of equations (1a) and (1b) requires a relaxation of the electrostatic field near the specimen. Since these relaxation procedures lead to a fairly long computer time further calculations are omitted.

c. *Electrostatic Contrast Calculations with an Analog Computer.* As a check on the digital computer results and to have a method which provides directly plotted results, equations (1a) and (1b) were also solved with an analog computer. The functions $\dfrac{\partial \phi(y, z)}{\partial z}$ and $\dfrac{\partial \phi(y, z)}{\partial y}$ were partially generated with a combined sine and cosine potentiometer. The exponential damping factor in these derivatives was introduced into the machine with a ten point diode function generator. An increase of the interaction between the specimen perturbation and the reversing electrons is reached by letting the electrons enter the retarding

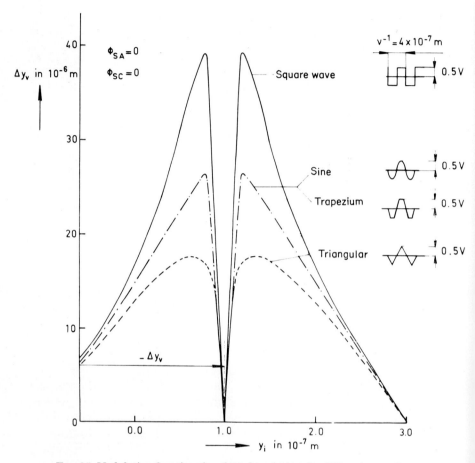

Fig. 15. Modulation function Δy_v plotted against y_i for different wave forms.

field at $z = 4(2\pi\nu)^{-1}$ instead of $z = z_2$. The error involved, owing to $\phi\{y, 4(2\pi\nu)^{-1}\} \neq 0$, is $\exp(-2\pi\nu z) = 1\cdot8 \times 10^{-2}$.

Addition of logic circuitry provided automatic plotting of the electron trajectories for given values of ϕ_o, ϕ_1, ν, δy_i (steplength along the y_i-axis) and α, the angle of illumination.

For three different angles α the electron trajectories are plotted against y_i in Figs. 16a, b and c with

$$\phi_o = 30 \text{ kV} \qquad\qquad \delta y_i = (80\,\nu)^{-1}$$
$$\phi_1 = 0\cdot55 \text{ V} \qquad\qquad \nu^{-1} = 4 \times 10^{-7} \text{ m}$$
$$F_z = 8\cdot57 \times 10^6 \text{ V/m} \qquad \phi_{SA} = \phi_{SC} = 0$$

The positively charged part of the sine wave clearly acts as a concave

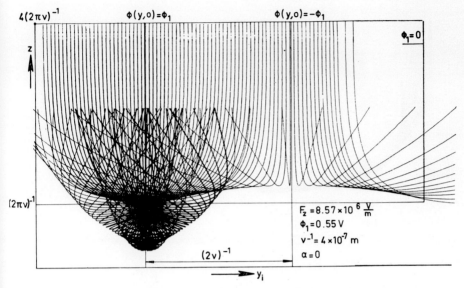

FIG. 16a. Electron trajectories (model B) for $\alpha = 0$.

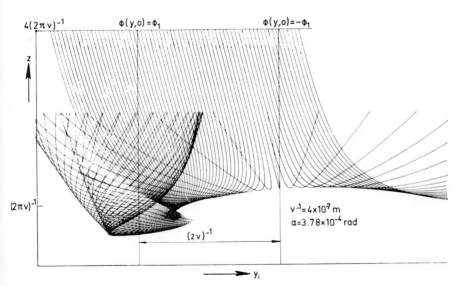

FIG. 16b. Electron trajectories (model B) for $\alpha = 3 \cdot 78 \times 10^{-4}$ rad.

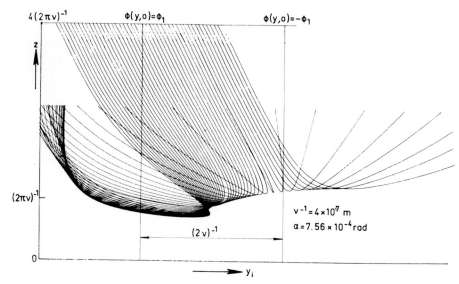

FIG. 16c. Electron trajectories (model B) for $\alpha = 7\cdot56 \times 10^{-4}$ rad.

mirror. When the focus of this "micro mirror" is imaged it appears as a white line (corresponding to the stars in the actual pictures) in the final image. The negatively charged part is then related to a black band (bubbles in the actual pictures).

From the measured Δy_v values relative current density distributions $\dfrac{j(y)}{j_o}$, referred back to the specimen, were plotted. As example $\dfrac{j(y)}{j_o}$ is presented for $a = 0$ (dashed curve in Fig. 17). For this curve the angular aperture of the illuminating pencils is assumed to be infinitely small ($\beta \approx 0$). The filtering characteristic f_c used is

$$f_c = 1 \text{ for } 0 \leq G_1|\Delta y_v| \leq D/2$$
$$f_c = 0 \text{ for } \quad G_1|\Delta y_v| > D/2$$
$$\text{with } D = 100 \ \mu\text{m}.$$

Since $\beta \approx 0$, the cross-over of the illuminating electron beam is small compared with D.

A remarkable and unexpected change in $\dfrac{j(y)}{j_o}$ occurs for a finite value of β or the related cross-over in the aperture plane. For a semi-angular aperture of only $\beta = 10^{-3}$ rad, $\dfrac{j(y)}{j_o}$ (full line in Fig. 17) has been plotted against y_i, extending for a full period $\nu^{-1} = 4 \times 10^{-7}$ m

across the specimen surface. This curve has been constructed graphically
by adding the relative current density distributions with $\beta = 0$ for six
different but equidistant values of a. The cross-over size of the illuminating beam corresponding to $\beta = 10^{-3}$ rad amounts to 20 μm (focal

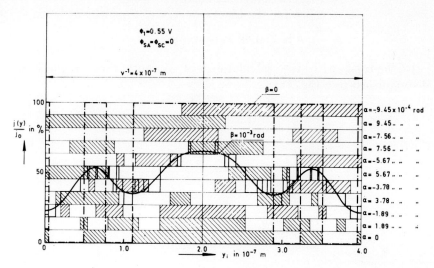

FIG. 17. Relative current density distribution for $\beta = 0$ and $\beta = 10^{-3}$ rad.

length of the combination objective and electrostatic lens is 10 mm)
and is therefore still small compared to $D = 100$ μm. Yet the influence
of the cross-over size turns out to be tremendous.

The current density distribution for $\beta = 10^{-3}$ rad shows, although
not exactly, the double frequency of the sinusoidal perturbation as a
result of the double rectifying effect of the contrast aperture. The non-
linear behaviour of model B for an electrostatic perturbation with
$\phi_1 = 0.55$ V and $\nu^{-1} = 4 \times 10^{-7}$ m gives rise to the occurrence of
rectified higher order harmonics in $\dfrac{j(y)}{j_o}$.

E. *Model C*

Although the application of wave mechanics to a "macroscopic"
problem such as the formation of contrast is not expected to lead to other
results than the classical calculations, it is performed in extension of
the calculations of Wiskott (1956).

This model uses the time independent Schrödinger equation (2)

$$-\frac{\hbar^2}{2m}\left(\frac{\partial^2 u}{\partial y^2} + \frac{\partial^2 u}{\partial z^2}\right) + Vu = Eu \tag{2}$$

where
$u = u(y, z)$, the wave function
$V = V(z)$, the potential energy
$E =$ constant, the kinetic energy of the incident beam
$\hbar = \dfrac{h}{2\pi}$, h is Planck's constant.

In contrast with the calculations of Wiskott three regions are distinguished (see Fig. 18).

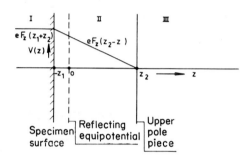

Fɪɢ. 18. Potential energy for region I, II and III.

I. $z \leq -z_1$, where $V = eF_z(z_1 + z_2)$

$$F_z = \frac{E}{ez_2} = \frac{\phi_o}{z_2} = 8 \cdot 57 \times 10^6 \text{ V/m}$$

II. $-z_1 < z < z_2$, where $V = eF_z(z_2 - z)$
III. $z \geq z_2$, where $V = 0$.

The separation into these regions matches the experimental set-up more accurately. Furthermore the retarding potential ϕ_0, although large with respect to ϕ_1, must have a limited value.

The physical specimen surface is located at $z = -z_1$, whereas electron reflection occurs from the equipotential at $z = 0$. This configuration has been chosen here, contrary to model A, to include directly the effect of the specimen voltage ϕ_{SV}, which is related to z_1. Similar to model A, $|z_1| \ll z_2$ and $z_1 \to \dfrac{\phi_1}{F_z}$ provide maximum interaction of the specimen perturbation on the reversing electron beam. This wave mechanical calculation, performed in the yoz-plane, deals firstly with the unperturbed solution of equation (2) for the regions I, II and III.

The complete solution, a continuous wave function extending over the entire yoz-plane, is found by matching regions I, II and III. This is achieved by fulfilling the boundary conditions at $z = -z_1$ and $z = z_2$

$$uu_I \Big|_{z=-z_1} = uu_{II} \Big|_{z=-z_1} \quad \text{and} \quad \frac{duu_I}{dz} \Big|_{z=-z_1} = \frac{duu_{II}}{dz} \Big|_{z=-z_1}$$

$$uu_{II} \Big|_{z=z_2} = uu_{III} \Big|_{z=z_2} \quad \text{and} \quad \frac{duu_{II}}{dz} \Big|_{z=z_2} = \frac{duu_{III}}{dz} \Big|_{z=z_2}$$

(23)

$uu_I(z)$, $uu_{II}(z)$ and $uu_{III}(z)$ represent the unperturbed solutions corresponding to the regions mentioned above.

Since the perturbations on the specimen, in terms of voltage, are assumed to be small with respect to the retarding potential ϕ_0, the usual perturbation calculus can be applied. This allows for the determination of the perturbed wave function $u(z)$ from the unperturbed wave function $u_u(z)$. The calculations are simplified by neglecting second and higher order effects.

1. Determination of the unperturbed wave functions.

Region I:

equation (2) changes into

$$-\frac{\hbar^2}{2m} \frac{d^2 uu_I}{dz^2} + uu_I e F_z(z_1 + z_2) = e F_z z_2 uu_I$$

$$uu_I = C_1 \exp\left[\gamma \left(\frac{z_1}{z_2}\right)^{\frac{1}{2}} z\right] + D_1 \exp\left[-\gamma \left(\frac{z_1}{z_2}\right)^{\frac{1}{2}} z\right]$$

where

$$\gamma^2 = \left(\frac{mv_z}{\hbar}\right)^2 = \left(\frac{2\pi}{\lambda}\right)^2$$

For 30 kV electrons $\lambda = 7.07 \times 10^{-12}$ m and $\gamma = 8.88 \times 10^{11}$ m^{-1}. Obviously the second exponential term in uu_I goes to infinity for large negative values of z, so $D_1 = 0$.

$$uu_I = C_1 \exp(\gamma_1 z)$$

(24)

where

$$\gamma_1 = \gamma \left(\frac{z_1}{z_2}\right)^{\frac{1}{2}} = 1.50 \times 10^{13} z_1^{\frac{1}{2}} \text{ m}$$

$$0 \leq z_1 < 10^{-7} \text{ m and } z_2 = 3.5 \times 10^{-3} \text{ m}$$

Region II:

equation (2) changes into

$$-\frac{\hbar^2}{2m} \frac{d^2 uu_{II}}{dz^2} + uu_{II} e F_z(z_2 - z) = e F_z z_2 uu_{II}$$

Substitution of $z = \zeta l$ yields

$$\frac{d^2 u u_{II}}{d\zeta^2} + u u_{II}\,\zeta = 0, \text{ where } l^3 = \frac{\hbar^2}{2me F_z} = \frac{z_2}{\gamma} = 1\cdot64 \times 10^{-9} \text{ m}$$

$$u u_{II} = C_2 Ai(-z\gamma_2) + D_2 Bi(-z\gamma_2) \tag{25}$$

where $$\gamma_2 = l^{-1} = 6\cdot08 \times 10^8 \text{ m}^{-1}$$

$Ai(z)$ and $Bi(z)$ are Airy functions†.

Region III:

equation (2) changes into

$$-\frac{\hbar^2}{2m}\frac{d^2 u u_{III}}{dz^2} - e F_z z_2 u u_{III} = 0$$

$$u u_{III} = C_3 \exp{(i\gamma z)} + D_3 \exp{(-i\gamma z)}$$

Since the intensity of the illuminating beam is known, $C_3 = 1$

$$u^n{}_{III} = \exp{(i\gamma z)} + D_3 \exp{(-i\gamma z)} \tag{26}$$

2. On the basis of the boundary conditions, equation (23), the constants C_1, C_2, D_2 and D_3 are defined.

$$C_1 = -\frac{2i\gamma \exp{(i\gamma z_2)}\,.\,\exp{(\gamma_1 z_1)}}{\pi\gamma_2(x_4 x_1 - x_2 x_3)}$$

$$C_2 = -\frac{2i\gamma x_2 \exp{(i\gamma z_2)}}{\gamma_2(x_4 x_1 - x_2 x_3)}$$

$$D_2 = \frac{2i\gamma x_1 \exp{(i\gamma z_2)}}{\gamma_2(x_4 x_1 - x_2 x_3)}$$

$$D_3 = -\exp{(2i\gamma z_2)}\frac{x_4 x_1 - x_2 \bar{x}_3}{x_4 x_1 - x_2 x_3}$$

where $$x_1 = \frac{\gamma_1}{\gamma_2} Ai(z_1\gamma_2) + Ai'(z_1\gamma_2)$$

$$x_2 = \frac{\gamma_1}{\gamma_2} Bi(z_1\gamma_2) + Bi'(z_1\gamma_2)$$

$$x_3 = i\frac{\gamma}{\gamma_2} Ai(-z_2\gamma_2) - Ai'(-z_2\gamma_2)$$

$$x_4 = i\frac{\gamma}{\gamma_2} Bi(-z_2\gamma_2) - Bi'(-z_2\gamma_2) \tag{27}$$

The bars indicate complex conjugate quantities.

Since $|D_3| = 1$, the illuminating beam is entirely reflected. As a result of $|D_3| = 1$ it follows that $C_1 = 0$, which is rather obvious because the chosen $V(z)$ distribution does not allow tunnel effects.

† See *Handbook of Mathematical Functions*, p. 440, by Abramowitz, M. and Stegun, I. A. (Dover Publications, New York) 1965.

Calculation of the current density j_{zIII} of the unperturbed beam for region III provides an additional check.

$$j_{zIII} = \varrho_{III} v_z \quad (\varrho = \text{electron density})$$

$$j_{zIII} = \frac{\hbar}{2im} \left[\bar{u}_{u_{III}} \frac{du_{u_{III}}}{dz} - u_{u_{III}} \frac{d\bar{u}_{u_{III}}}{dz} \right] =$$

$$= \frac{\hbar}{2im} (2 - 2D_3^2) i\gamma = v_z(1 - D_3^2)$$

For $|D_3| = 1 \rightarrow j_{zIII} = 0$, the incident beam is fully reflected.

Solution $u_{u_I} = C_1 \exp(\gamma_1 z)$ represents a stationary electron cloud for $z \leq 0$.

3. Superimposed on the equipotential at $z = -z_1$ is a sinusoidal perturbation

$$\phi_t(y, -z_1) = \phi_1 \cos 2\pi v y = \frac{\phi_1}{2} [\exp(2\pi i v y) + \exp(-2\pi i v y)]$$

$$\phi_1 = \text{perturbation amplitude in volts}$$
$$v = \text{spatial specimen frequency}$$

Considering only the first exponential term $\frac{\phi_1}{2} \exp(2\pi i v y)$ in the following calculation, then

$$\phi(y, -z_1) = \phi_1 \exp(2\pi i v y)$$

According to equation (4)

$$\phi(y, z) = \phi_1 \exp(2\pi i v y) \cdot \exp(-|z + z_1| v) \qquad (28)$$

With reference to the superimposed perturbing field $\phi(y, z)$ equation (2) for region II changes into

$$-\frac{\hbar^2}{2m} \left(\frac{\partial^2 u_{II}}{\partial y^2} + \frac{\partial^2 u_{II}}{\partial z^2} \right) - eF_z z u_{II} =$$

$$= -u_{II} \eta \phi_1 \exp(2\pi i v y) \cdot \exp(-|z + z_1| v) \qquad (29)$$

$u_{II} = u_{II}(y, z)$ represents the perturbed wave function for region II. It appears allowable to assume that the general perturbed wave function $u(z)$ consists of the unperturbed electron beam u_u, diminished with ηu_2, and a transverse beam ηu_1 modulated with the same periodicity v as the specimen perturbation.

$$u = u_u + \eta u_1 \exp(2\pi i v y) - \eta u_2 \qquad (30)$$

For instance for region II

$$u_{II} = u_{u_{II}} + \eta u_{1_{II}} \exp(2\pi i v y) - \eta u_{2_{II}}$$

The axially directed beam ηu_2 stands for the decrease of u_u owing to the generated transverse beam $\eta u_1 \exp(2\pi i v y)$. Factor η is an arbitrary small factor, which is introduced to distinguish higher order effects. Since this calculation only intends to describe first order effects, terms with η^2 and higher powers of η are omitted. Inserting equation (30) into (29) and similar equations for regions I and III yields

Region I:

$$\underbrace{-\frac{d^2 u u_I}{dz^2} + \gamma^2 \frac{z_1}{z_2} u u_I}_{0} + \eta \exp(2\pi i \nu y)\left(-\frac{d^2 u 1_I}{dz^2} + \gamma_3^2 u 1_I\right) +$$

$$+ \eta\left(\frac{d^2 u 2_I}{dz^2} - \gamma^2 \frac{z_1}{z_2} u 2_I\right) = - u u_I \eta \phi_1 \frac{2m}{\hbar^2} \exp(2\pi i \nu y).$$

$$. \exp\left(-|z+z_1|\,\nu\right)$$

Multiplying the remaining part with $\dfrac{}{\eta} \exp(-2\pi i \nu y)$ and integrating between

$$y_1 = -\frac{\pi}{\nu}$$

and

$$y_2 = \frac{\pi}{\nu}$$

results in

$$-\frac{d^2 u 1_I}{dz^2} + \gamma_3^2 u 1_I = f_1$$

where

$$\gamma_3^2 = \gamma^2 \frac{z_1}{z^2} - 4\pi^2 \nu^2$$

and

$$f_1 = - u u_I \phi_1 \frac{2m}{\hbar^2} \exp\left(-|z + z_1|\,\nu\right)$$

Variation of constants provides

$$u 1_I = C 1_I \exp(\gamma_3 z) + D 1_I \exp(-\gamma_3 z) +$$

$$+ \frac{1}{2\gamma_3}\left[\exp(\gamma_3 z) \int_{-\infty}^{-z} \exp(-\gamma_3 z) \cdot f_I \cdot dz - \right.$$

$$\left. - \exp(-\gamma_3 z) \int_{-\infty}^{-z} \exp(\gamma_3 z) \cdot f_I \cdot dz\right] \qquad (31)$$

Again $D 1_I = 0$ to avoid $u 1_I$ going to infinity for large negative values of z.

Region II: a similar procedure leads to

$$u_{1_{II}} = C_{1_{II}} Ai(-z\gamma_4) + D_{1_{II}} Bi(-z\gamma_4) -$$

$$- \pi \left[Ai(-z\gamma_4) \int_{-z_1}^{z} Bi(-z\gamma_4) \cdot f_{II} \cdot dz - \right.$$

$$\left. - Bi(-z\gamma_4) \int_{-z_1}^{z} Ai(-z\gamma_4) \cdot f_{II} \cdot dz \right] \quad (32)$$

where
$$\gamma_4 = \left(\frac{\gamma^2}{z_2} - 4\pi^2\nu^2 \right)^{\frac{1}{3}}$$

and
$$f_{II} = - u_{u_{II}} \phi_1 \frac{2m}{\hbar^2} \exp\left(- |z + z_1| \nu\right)$$

Region III:

$$u_{1_{III}} = C_{1_{III}} \exp(i\gamma_5 z) + D_{1_{III}} \exp(-i\gamma_5 z) +$$

$$+ \frac{1}{2i\gamma_5} \left[\exp(i\gamma_5 z) \int_{z_2}^{z} \exp(-i\gamma_5 z) \cdot f_{III} \cdot dz - \right.$$

$$\left. - \exp(-i\gamma_5 z) \int_{z_2}^{z} \exp(i\gamma_5 z) \cdot f_{III} \cdot dz \right]$$

where
$$\gamma_5 = (\gamma^2 - 4\pi^2\nu^2)^{\frac{1}{2}}$$

and
$$f_{III} = - u_{u_{III}} \phi_1 \frac{2m}{\hbar^2} \exp\left(- |z + z_1| \nu\right)$$

Since the perturbations at the specimen are assumed to be small $f_{III} = 0$. The illumination of the specimen with an electron beam parallel to the z-axis makes $C_{1_{III}}$, representing skew illumination, equal to zero.

$$u_{1_{III}} = D_{1_{III}} \exp(-i\gamma_5 z) \quad (33)$$

Summarizing

$$\lambda = 7 \cdot 07 \times 10^{-12} \text{ m}$$

$$\gamma = 8 \cdot 88 \times 10^{11} \text{ m}^{-1}$$

$$\gamma_1 = \gamma \left(\frac{z_1}{z_2} \right)^{\frac{1}{2}} = 1 \cdot 50 \times 10^{13} z_1^{\frac{1}{2}} \text{ m}$$

$$\gamma_2 = \left(\frac{\gamma^2}{z_2} \right)^{\frac{1}{3}} = 6 \cdot 08 \times 10^8 \text{ m}^{-1}$$

$$z_2\gamma_2 = 2 \cdot 13 \times 10^6$$

$$0 \leq z_1 < 10^{-7} \text{ m and } z_1 \geq \frac{\phi_1}{F_z} = 1 \cdot 17 \times 10^{-7} \phi_1 \text{ m}$$

$$\frac{\gamma}{\gamma_2} = 1{\cdot}46 \times 10^3$$

$$\gamma_3 = \left(\gamma^2 \frac{z_1}{z_2} - 4\pi^2\nu^2\right)^{\frac{1}{2}}$$

$$\gamma_4 = \left(\frac{\gamma^2}{z^2} - 4\pi^2\nu^2\right)^{\frac{1}{3}}$$

$$\gamma_5 = (\gamma^2 - 4\pi^2\nu^2) \approx \gamma$$

$$z_2 = 3{\cdot}5 \times 10^{-3} \text{ m}$$

4. Similar boundary conditions as equation (23) make the values of the perturbed wave functions u_{1_I}, $u_{1_{II}}$, $u_{1_{III}}$ and its first derivatives in z again coincident at $z = -z_1$ and $z = z_2$. For $z = -z_1$ and $z = z_2$ it follows that

$$u_{1_I} \Big|_{z=-z_1} = C_{1_I} \exp\left(-\gamma_3 z_1\right) +$$
$$+ E_{1_I} \exp\left(-\gamma_3 z_1\right) + F_{1_I} \exp\left(\gamma_3 z_1\right)$$

where
$$E_{1_I} = \frac{1}{2\gamma_3} \int_{-\infty}^{-z_1} \exp\left(-\gamma_3 z\right) . f_I . dz$$

and
$$F_{1_I} = -\frac{1}{2\gamma_3} \int_{-\infty}^{-z_1} \exp\left(\gamma_3 z\right) . f_I . dz$$

$$u_{1_{II}} \Big|_{z=-z_1} = C_{1_{II}} Ai(z_1\gamma_4) + D_{1_{II}} Bi(z_1\gamma_4)$$

$$u_{1_{II}} \Big|_{z=z_2} = (C_{1_{II}} + E_{1_{II}}) Ai(-z_2\gamma_4) + (D_{1_{II}} + F_{1_{II}}) Bi(-z_2\gamma_4)$$

where
$$E_{1_{II}} = -\pi \int_{-z_1}^{z_2} Bi(-z\gamma_4) . f_{II} . dz$$

and
$$F_{1_{II}} = \pi \int_{-z_1}^{z_2} Ai(-z\gamma_4) . f_{II} . dz$$

and
$$u_{1_{III}} \Big|_{z=z_2} = D_{1_{III}} \exp\left(-i\gamma_5 z_2\right)$$

The boundary conditions for u_{1_I}, $u_{1_{II}}$ and $u_{1_{III}}$ lead to expressions for the constants C_{1_I}, $C_{1_{II}}$, $D_{1_{II}}$ and $D_{1_{III}}$.

$$C_{1_I} = [C_{1_{II}} Ai(z_1\gamma_4) + D_{1_{II}} Bi(z_1\gamma_4)] \exp\left(\gamma_3 z_1\right) -$$
$$- E_{1_I} - F_{1_I} \exp\left(2\gamma_3 z_1\right)$$

$$C_{1_{II}} = \frac{1}{x_6 x_7 - x_5 x_8} [2\gamma_3 x_8 F_{1_I} \exp\left(-\gamma_3 z_1\right) + E_{1_{II}} x_6 x_7 + F_{1_{II}} x_6 x_8]$$

$$D_{1_{II}} = \frac{1}{x_6 x_7 - x_5 x_8} [2\gamma_3 x_7 F_{1_I} \exp\left(\gamma_3 z_1\right) + E_{1_{II}} x_5 x_7 + F_{1_{II}} x_5 x_8]$$

$$D_{1_{III}} = \frac{2i}{\gamma_5} \exp\left(i\gamma_5 z_2\right)[(C_{1_{II}} + E_{1_{II}})\bar{x}_7 + (D_{1_{II}} + F_{1_{II}})\bar{x}_8] \tag{34}$$

where
$$x_5 = \gamma_3 Ai(z_1\gamma_4) + \gamma_4 Ai'(z_1\gamma_4)$$
$$x_6 = \gamma_3 Bi(z_1\gamma_4) + \gamma_4 Bi'(z_1\gamma_4)$$
$$x_7 = i\gamma_5 Ai(-z_2\gamma_4) - \gamma_4 Ai'(-z_2\gamma_4)$$
$$x_8 = i\gamma_5 Bi(-z_2\gamma_4) - \gamma_4 Bi'(-z_2\gamma_4)$$

The thus determined general perturbed wave function contains as most interesting part u_{III_r}, representing the reflected beam in region III. Including also the negative exponential term

$$\frac{\phi_1}{2} \exp(-2\pi i v y)$$

of the sinusoidal specimen perturbation $\phi_t(y, -z_1)$, it is found that

$$u_{III_r} = D_3 \exp(-i\gamma z) +$$
$$+ \tfrac{1}{2} D_{1_{III}} \exp(-i\gamma_5 z) \left[\exp(2\pi i v y) + \exp(-2\pi i v y)\right]$$
$$u_{III_r} \approx \exp(-i\gamma z)\{D_3 + \tfrac{1}{2} D_{1_{III}} \left[\exp(2\pi i v y) + \exp(-2\pi i v y)\right]\} \quad (35)$$

In this equation $D_3 \exp(-i\gamma z)$ represents the unperturbed reflected beam (zero order) and the remaining part both transversely modulated beams as first order diffracted beams.

The results obtained resemble closely the diffraction pattern of an optical grating illuminated with a parallel beam of coherent light of wave length λ_l. On the basis of Abbe's diffraction formula $\Theta_l = \lambda_l v$, applied for a sinusoidal grating with a periodicity of v, the "effective" electron wave length λ_e in the perturbed retarding field can be compared with λ_l.

The diffraction angle Θ_e for the transverse beam can be derived from equation (35).

$$\Theta_e = \frac{\bar{u}_m \dfrac{\partial u_m}{\partial y} - u_m \dfrac{\partial \bar{u}_m}{\partial y}}{\bar{u}_m \dfrac{\partial u_m}{\partial z} - u_m \dfrac{\partial \bar{u}_m}{\partial z}} \qquad (36)$$

where
$$u_m = \tfrac{1}{2} D_{1_{III}} \exp(-i\gamma_5 z) \cdot \exp(2\pi i v y)$$

$$\Theta_e = \frac{2\pi v}{\gamma_5} \approx \frac{2\pi v}{\gamma} = \lambda_e v$$

Setting $\Theta_e = \Theta_l$ provides $\lambda_e = \lambda_l = \lambda$ \qquad (37)

Conclusion

The retarding voltage with the superimposed specimen perturbation $\phi(y, -z_1)$ behaves in first order, regardless of the strongly increasing electron wave length near the specimen, as an optical grating with an identical spatial frequency and irradiated with "light" of wave length λ_l equal to the electron wave length of the incident beam.

This conclusion, which involves that the modulation effect of the perturbations in the electrostatic field only occur near the specimen surface, leads directly to a criterion for the best obtainable lateral resolving power on the basis of Heisenberg's uncertainty principle

$$\Delta p_y \, . \, \Delta y_s \geq h.$$

In this expression Δp_y stands for the tangential impulse determined by the product of objective lens angular aperture a_A and the axially directed impulse mv_z. The uncertainty in the location in the y-direction of an electron near the specimen surface then amounts to Δy_s

where
$$\Delta y_s \geq \frac{h}{a_A m v_z} = \lambda \frac{2f}{D} \tag{38}$$

For 30 kV electrons, $D = 10^{-1}$ mm and $f = 10$ mm (focal length of combination objective and electrostatic lens); it follows that

$$\Delta y_s \geq 1 \cdot 6 \text{ nm.}$$

This figure is valid for coherent illumination and the absence of instrumental shortcomings such as spherical aberration of the retarding field and lens defects. In the case of incoherent illumination Δy_s increases by roughly a factor of two.

When $D_{1_{III}}$ is worked out, information can be obtained whether the perturbed specimen can be considered as a phase specimen or an amplitude specimen. Since no absorption of electrons occurs, it is obvious to state without further calculation of $D_{1_{III}}$ that the perturbed specimen must act as a phase specimen. A second argument in favour of the phase character of the specimen was found experimentally when a focused image was formed. Nearly no phase contrast appeared for values of D in the order of 500 μm and more. Only D values smaller than 200 μm gave sufficient contrast in the final image (see Fig. 26).

The current density distribution can be calculated from the double Fourier transform of the specimen perturbation.

F. *Comparison with the Image Contrast for a Mirror Projection Microscope*

An attempt will now be made to compare the contrast in a mirror microscope with focused images and in a mirror projection microscope. In the first instance only electrons incident at small angles with the z-axis are considered. For the mirror projection microscope an electron probe is formed at $z = z_p$ in front of the specimen. Although all electron trajectories through this electron probe are parabolas, having as

envelope a paraboloid of revolution, electrons incident close to the
z-axis can be considered to reflect from a flat equipotential plane at
$z = z_1$, where $z_1 = \dfrac{\phi_{\mathrm{SV}}}{F_z}$ and $z_p \gg z_1$. In first order approximation the
angular deflection $\varDelta \varphi_p$ for these electrons, as result of a sinusoidal
specimen perturbation, follows from equation (10) belonging to model
A. The index p refers to the projection method. The following calculations are only meant as a qualitative comparison between both types
of contrast.

In the mirror projection microscope the formation of contrast
emerges from local variations of the current density in a plane normal

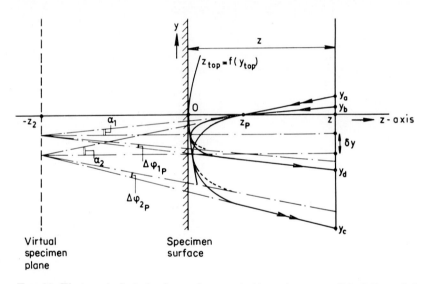

Fig. 19. Electron trajectories in a mirror projection microscope. Calculation of the
current density $j(y)_p$.

to the z-axis. It is shown that these variations are caused by the
curvature of the perturbed specimen. In the projection microscope the
plane $z = z_p$ is conjugate with the final screen. The relative current
density distribution $\dfrac{j(y)_p}{j_p}$ in $z = z_p$ can be compared with $\dfrac{j(y)}{j_o}$. On the
basis of Fig. 19 a relation between $\dfrac{j(y)_p}{j_p}$ and the specimen perturbation
is derived. Both trajectories drawn represent electrons entering the
retarding field at small angles α_1 and α_2.

Local perturbations cause angular deflections $\Delta\varphi_{1p}$ and $\Delta\varphi_{2p}$. Since all electrons inside the current tube, bounded by both trajectories, are reflected it follows that

$$j_p(y_a - y_b) = j(y)_p(y_d - y_c) \qquad (39)$$

j_p is the current density of a reflected beam that is not modulated and $j(y)_p$ the current density of a modulated reflected beam; j_p and $j(y)_p$ are both measured in $z = z_p$. From Fig. 19 it follows that

$$y_a = y_a$$
$$y_b = y_a - \delta y + 2z(a_1 - a_2)$$
$$y_c = y_a - 4za_2 - 2z\Delta\varphi_{2p}$$
$$y_d = y_b - 4za_1 - 2z\Delta\varphi_{1p}$$
$$a = \frac{v_y}{v_z} = \frac{y}{4z}; \; a_2 - a_1 = \frac{\delta y}{4z}$$

Substitution of y_a, y_b, y_c and y_d into equation (39) leads to

$$z = z_p \to j_p = j(y)_p \left[1 - 4z_p \frac{\partial \Delta\varphi_p}{\partial y} \right]$$

Provided that $4z_p \dfrac{\partial \Delta\varphi_p}{\partial y} \ll 1$, which involves a slowly varying specimen perturbation, the relative current density distribution in $z = z_p$ is

$$\frac{j(y)_p}{j_p} = 1 + 4z_p \frac{\partial \Delta\varphi_p}{\partial y} \propto \frac{d^2\phi(y, 0)}{dy^2} \qquad (40)$$

The current density $j(y)_p$ is proportional to the curvature of the specimen for small values of $4z_p \dfrac{\partial \Delta\varphi_p}{\partial y}$.

Combination of equation (10) for $\Delta\varphi_p$ and equation (40) yields

$$\frac{j(y)_p}{j_p} = 1 + \frac{8\pi^2}{F_z} (2v^3 z_p)^{\frac{1}{2}} \phi_1 \cos 2\pi vy \, . \, \exp\left(- 2\pi v z_1\right) \qquad (41)$$

An approximation of equation (19) is chosen for the calculation of the relative current density distribution $\dfrac{j(y)}{j_o}$ of the imaging mirror microscope

$$f_c = 1 - \frac{G_1 |\Delta y|}{D}$$

$$\frac{j(y)}{j_o} = 1 - \frac{2\pi}{F_z} \frac{G_1}{D} (2v z_2)^{\frac{1}{2}} \phi_1 \sin 2\pi vy \, . \, \exp\left(- 2\pi v z_1\right) \qquad (42)$$

For a 50% modulation in both relative current density distributions, which means

$$\frac{j(y)_p}{j_p} = \frac{j(y)}{j_o} = 0.5,$$

the crucial factors in the modulation terms of equations (41) and (42) can be compared. It leads to the conclusion that the contrast for a mirror projection microscope is superior to that of a mirror microscope with focused images when

$$\nu > \frac{G_1}{4\pi D}\left(\frac{z_2}{z_p}\right)^{\frac{1}{2}} \tag{43}$$

Substitution of practical values

$$z_2 = 3.5 \text{ mm}; \ z_p = 1 \text{ mm}; \ G_1 = 1.5$$

leads to

$$\nu > \frac{2.2 \times 10^{-1}}{D} \tag{43a}$$

Application of a circular contrast aperture, for instance $D = 100 \ \mu$m, yields

$$\nu > 2.2 \times 10^3 \text{ m}^{-1}.$$

When the circular contrast aperture is replaced by a knife edge aperture, D stands for the cross-over of the illuminating beam. For magnifications at the final screen larger than $1500 \times$, D is in the order of 1 μm so

$$\nu > 2.2 \times 10^5 \text{ m}^{-1} \tag{43b}$$

The size of the cross-over of the illuminating beam for magnifications larger than $1500 \times$ is mainly determined by the optical reduction of the electron source in the condenser system.

The conclusion that the contrast of a mirror projection microscope exceeds the contrast of a mirror microscope with focused images, in case equation (43a) or (43b) is fulfilled, is rather trivial. Since the mirror electron microscope deals with phase specimens a defocusing leads to an increase of contrast at the expense of resolving power. The through-focal series of photographs (Fig. 27) clearly shows this effect for the out of focus images. Near the in focus condition image contrast can be regained by either decreasing the diameter D of the contrast aperture or reduction of the illuminating cross-over for the knife edge technique.

Not considered for the projection method is the blur due to the out of focus Fresnel diffraction. This type of diffraction results from the interference of reflected electrons from different parts of the specimen.

Since the retarding field macroscopically acts as a flat mirror, the specimen can be considered as being illuminated in transmission by the virtual electron probe at $3z_p$ behind the specimen (see Fig. 20). The

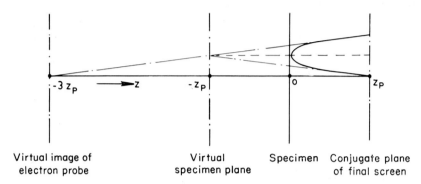

FIG. 20. Calculation of the Fresnel diffraction at the specimen.

distance d_F, calculated back at the specimen, between the geometrical projection and the maximum of the first fringe of the diffraction pattern amounts to

$$d_F \Big|_{z=0} = (z_p \lambda_p)^{\frac{1}{2}}$$

where λ_p is the electron wave length near $z = z_p$.

The conclusion of Section II.E (p. 198) justifies the use of this equation as a criterion for the theoretical resolving power in a mirror projection microscope

$$d_F \Big|_{z=0} = (z_p \lambda_p)^{\frac{1}{2}} = \left(z_p \frac{\lambda_e^2 \phi_0}{F_z} \right)^{\frac{1}{4}} = (z_p z_2 \lambda_e^2)^{\frac{1}{4}} \qquad (44)$$

For 30 kV electrons the wave length λ_e of the incident beam amounts to 7×10^{-12} m. For $F_z = 8.6 \times 10^6$ V/m and $z_p = 10^{-3}$ m

$$d_F \Big|_{z=0} = 1.2 \times 10^{-7} \text{ m} \quad (\nu_F = 8.3 \times 10^6 \text{ m}^{-1}).$$

Because of the fourth root, the value of d_F is hardly affected by the distance z_p of the electron probe in front of the specimen. z_p must be sufficiently large to obtain an adequate field of view. In view of this theoretical resolving power, a resolution of 10^{-7} m (along the specimen surface) with the mirror projection microscope claimed by Litton Systems Inc. (1968) seems rather doubtful and is certainly not shown.

For the off axis electrons in a mirror projection microscope, the z coordinate of the point of reversal moves away from the specimen surface and causes, in first order, an additional exponential damping on the contrast. This damping effect depends on the locus of $y_{top}=f(z_{top})$, where y_{top} and z_{top} represent the coordinates of the point of reversal.

The locus is calculated for an unperturbed specimen. Perturbations present only give rise to slight local deviations.

According to equation (17) (see Fig. 19)

$$z_{top} = \left(\frac{v_y}{v_z}\right)_p^2 z_p = \frac{y_{top}^2}{4z_p} = \frac{y_p^2}{16z_p} \tag{45}$$

$\left(\frac{v_y}{v_z}\right)_p$ and y_p refer to the plane $z = z_p$.

This equation represents a parabola, having its apex at $z = 0$ or at $z = z_1$ for $\phi_{\mathrm{SV}} \neq 0$ and a curvature $\varrho_c \approx \dfrac{1}{8z_p}$.

For small values of z_p, ϱ_c tends to infinity. It means that an electron probe is formed at the specimen surface. Although in a mirror projection microscope without beam separation a focused image of the specimen is then formed, the field of view is very limited. Scanning the electron probe across the specimen surface provides a focused image with sufficient field of view. A scanning mirror electron microscope has been described by Garrood and Nixon (1968). Use of large values of z_p nullifies ϱ_c, or the specimen is illuminated with an electron beam parallel to the z-axis.

The additional distance z_{top} from the reflecting equipotential ($z = z_1$) can be introduced into equation (8) by

$$dt = - \left(\frac{m}{2eF_z}\right)^{\frac{1}{2}} (z - z_1 - z_{top})^{\frac{1}{2}} dz$$

Neglecting again the additional modulation damping during travelling of the reversing electrons parallel to the specimen surface,

$$\Delta\varphi_p = \frac{\pi}{F_z} \left(\frac{2v}{z_2}\right)^{\frac{1}{2}} \phi_1 \sin 2\pi v y \, . \, \exp\left\{- 2\pi v \left(z_1 + \frac{y_p^2}{16z_p}\right)\right\}$$

Since

$$\frac{y_p}{8z_p} = \frac{1}{2}\left(\frac{v_y}{v_z}\right)_p \ll 1$$

$$\frac{j(y)_p}{j_p} \approx 1 + \frac{8\pi^2}{F_z} (2v^3 z_p)^{\frac{1}{2}} \phi_1 \cos 2\pi v y \, . \, \exp\left\{- 2\pi v \left(z_1 + \frac{y_p^2}{16z_p}\right)\right\}$$

This equation shows that the modulation factor $\dfrac{j(y)_p}{j_p}$ for a mirror projection microscope falls off exponentially for off axis areas of the specimen. This effect is observable in most mirror projection pictures (see Fig. 21a). When the specimen is biased slightly positive so that

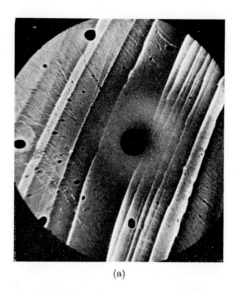

(a)

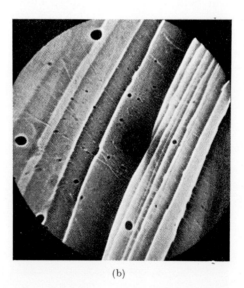

(b)

Fig. 21. Mirror projection pictures of a cleaved surface of rock salt covered with an evaporated gold layer. Magnification about 100 ×. (These pictures were made by M. E. Barnett and W. C. Nixon, and are published with their kind permission.)

    a. The specimen is negatively biased with − 0·2 V.

    b. The specimen is positively biased with + 1·8 V.

$- z_1 =: \dfrac{y_p^2}{16 z_p}$ the exponential damping effect decreases at the expense of electrons striking the specimen surface near the axis. Then the image quality near the axis deteriorates owing to electron scattering at the surface, whereas the off axis areas reveal more detail (see Fig. 21b).

G. *Magnetic Contrast*

The modulation effect of a magnetic perturbation on the approaching electrons is in first order approximation equal but oppositely directed to the modulation on the reflected electrons. This means that in the case of normal incidence on the reflecting equipotential plane practically no magnetic contrast can be obtained.

The Lorentz force $\bar{K}$, caused by the magnetic perturbation, shows that for skew incidence magnetic contrast can be obtained.

$$\bar{K} = - ({}^v \times \bar{B})$$
$$K_x = - (ev_z B_y + ev_y B_z)$$
$$K_y = - (ev_x B_z + ev_z B_x) \tag{46}$$
$$K_z = - (ev_y B_x + ev_x B_y)$$
$$\bar{B} = \text{magnetic induction}$$

K_z can be neglected because this component only modulates the depth of penetration into the retarding field. Since v_z changes sign at the point of reversal all deflection terms with v_z compensate one another if model A, with no lateral displacement during reversal, is used.

The remaining effect $K_r = (K_x^2 + K_y^2)^{\frac{1}{2}} = ev_t B_z$ shows that magnetic contrast only originates (in model A) from the interaction of the normal component B_z of the magnetic perturbing field and the tangential electron velocity v_t. Therefore only skew illumination reveals magnetic contrast. In projection images magnetic contrast increases with the distance from the axis. This distance is proportional to v_t. This effect is known from Kranz and Bialas (1961). It is expected that the calculations according to model A are an even coarser approximation than those for electrostatic contrast.

According to a suggestion of Bethge (1964) sufficient magnetic contrast for in focus images can be achieved by converting the magnetic perturbing effect into an electrostatic perturbation. Therefore a magnetic perturbed specimen is covered with an insulating film and a layer which possesses a large Hall effect.

The emerging potential distribution, which occurs for current flow through this Hall layer, can be imaged.

H. *Concise List of Symbols Used in Section II*

x, y, z	Cartesian coordinates; x and y coincide with the specimen surface; z along the electron optical axis
v_y, v_z	electron velocity in the y- and z-direction
z_1	coordinate of the reflecting equipotential plane
z_2	coordinate of the upper pole piece of the objective lens
z_p	coordinate of the electron probe (projection method)
λ_2	electron wavelength at $z = z_2$
v_2	electron velocity at $z = z_2$
ϕ_o, F_z	retarding potential and retarding field strength
$\Phi(y, z)$	potential distribution between the perturbed specimen and upper pole piece of objective lens
$\phi(y, z)$	perturbation potential
ϕ_s	specimen voltage
ϕ_{SC}	contact potential between specimen and tungsten filament
ϕ_{SA}	additional negative specimen bias
ϕ_n	perturbation amplitude of the n-order harmonic
ν	spatial specimen frequency
$d(y, 0)$	topography of the specimen
y_i	height of incidence at $z = z_2$
y_r	intersection of reflected electron with $z = z_2$
y_{top}, z_{top}	coordinates of the point of reversal
$\Delta y_v, \Delta \varphi_v$	lateral displacement and angular deflection, both at $z = z_2$ for electrostatic contrast
$\Delta y_t, \Delta \varphi_t$	lateral displacement and angular deflection, both at $z = z_2$ for topographic contrast
$\Delta y_1, \Delta \varphi_1$	lateral displacement and angular deflection in the contrast aperture plane
D	diameter of contrast aperture
$D_o = 2R_o$	diameter of the upper pole piece of the objective lens
f_c	filtering characteristic
j_o	current density of the parallel illuminating beam (in the retarding field)
$j(y)$	modulated current density, referred to the specimen
j_p	current density of reflected beam, being not modulated, at $z = z_p$
$j(y)_p$	modulated current density of reflected beam at $z = z_p$
M_1	magnification between specimen and object plane of the projector lens

β semi angular aperture of the electron pencils at $z = z_2$

α angle of illumination at $z = z_2$

$u = u(y, z)$ wave function

$V = V(z)$ potential energy in the retarding field

E kinetic energy of the incident beam

E_1 arbitrary energy in the one-dimensional Maxwellian energy distribution

uu_I, uu_{II}, uu_{III} unperturbed wave functions for region I, II and III

u_I, u_{II}, u_{III} perturbed wave functions for region I, II and III

η arbitrary small factor

$\left(\dfrac{v_y}{v_z}\right)_p$, y_p quantities measured in $z = z_p$

III. Description and Design of a Mirror Electron Microscope with Focused Images

A. Description

Summary

Apart from the requirement of separated axes for the illuminating and imaging system, a continuously variable magnification of 250....4000 × at the final fluorescent screen was desired. On the basis of Fig. 22 the instrument will be described by following the accelerated electron beam in the illuminating system towards the specimen surface (the mirror electrode) and after reflection towards the final fluorescent screen. Fig. 23 shows an overall picture of the microscope.

a. *The Illuminating System.* When a magnetic prism is used, it is favourable to minimize the deflection angle with regard to errors caused by the deflection field. These effects are proportional to the second and higher powers of the angle of deflection.

In order to create sufficient clearance for the miniaturized illuminating system, the angle between this system and the vertical main axis is fixed at 30°. The main axis represents the centreline of the specimen and the projector lens. The deflection angle of the magnetic prism is further reduced to 15° by mounting on top of the second condenser lens an additional deflector, which matches the axis of the illuminating system with the proper direction of incidence for the prism.

A 30 kV electron beam is produced by a conventional triode electron gun. The first condenser lens with iron pole pieces demagnifies the electron source 10....40 ×. The second condenser lens, a miniature magnetic lens without iron circuit (Le Poole, 1964a), images the demagnified electron source through the injector-deflector, prism and intermediate lens into the contrast aperture.

An adjustable holder for three apertures is mounted between the first and second condenser lens in order to obtain a fixed angular aperture of the illuminating electron beam.

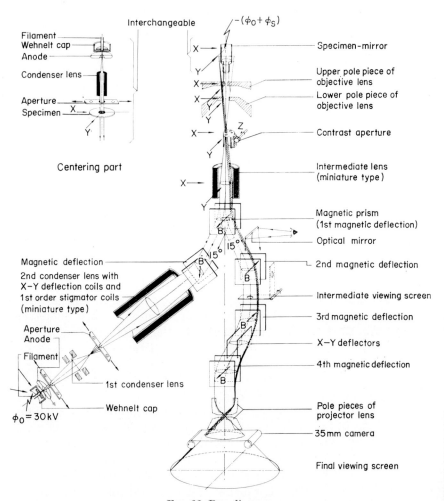

FIG. 22. Ray diagram.

Each deflector (the injector-deflector, the prism and the later discussed additional "bridge-deflectors") consists of pairs of circular air coils. The inner sides of the coils are covered with thin sheets of transformer iron. The resulting magnetic field is quite homogeneous and has an almost rectangular boundary (Fig. 24).

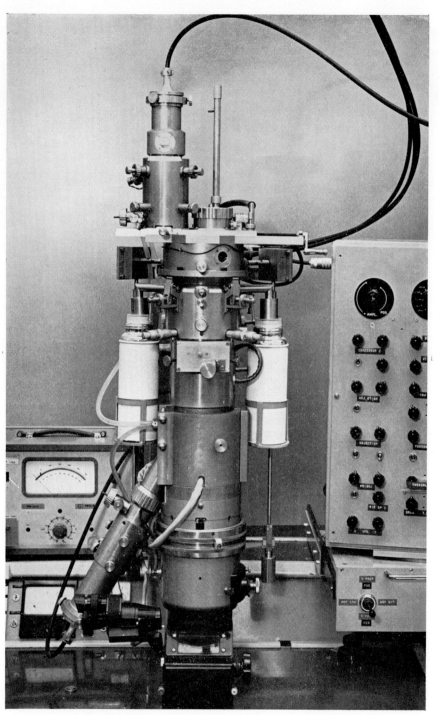

Fig. 23. Mirror electron microscope for focused images.

Experiments revealed that, for electron beams with a diameter nearly equal to the separation of the iron sheets, the image distortion still remains within admissible limits. Without iron sheets the usable cross-section of the deflection field, with respect to image distortion, is too restricted for application in this instrument. The small amount of iron in the sheets leads to a negligible non-linearity.

The second condenser lens is provided with additional pairs of x- and y-deflectors for centring the lens and two quadrupoles for correcting astigmatism. Iron tubing screens the illuminating system against stray

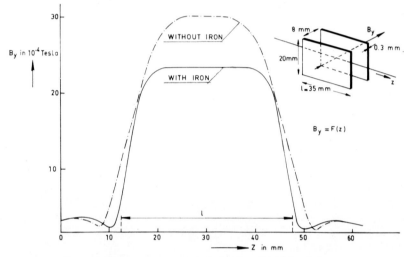

Fig. 24. Transverse magnetic induction $B(y)$ of a deflector.

magnetic fields. Near the second condenser lens the iron tubing is connected with a rectangular iron plate, covering a hole in the vertical main column housing.

Care has been taken to separate the magnetic fluxes in the illuminating system from the fluxes in the main column. Especially, interaction of magnetic fluxes generated in different parts of the instrument gives rise to problems in the centring of the electron optics. In order to minimize this coupling effect, additional concentric iron cylinders are used near the joining of the illuminating system and the main column, and around the intermediate lens.

b. *Imaging System.* The imaging system consists of the objective lens with contrast aperture and the intermediate lens. The combination of objective, intermediate and projector lens allows a continuously variable magnification of 250 .... 4000 $\times$ at the final fluorescent screen.

The imaging system and the prism form a group of electron optical components which are passed both by the illuminating and the reflected beam. This feature sets high requirements on the centring accuracy of both lenses. The axes of both lenses should coincide perfectly with the main axis of the vertical column, because each residual inclination or decentring produces a transverse magnetic field which acts as a prism. In order to make the lens axes, in practice always inclined and de-centred, coincident with the main axis, these lenses are centred by combining current reversing with pole piece centring, as proposed by Haine (1947).

Limitation of the field of view can be avoided by positioning the contrast aperture, as mentioned in Section II, in the back focal plane of the combination objective and electrostatic lens. The negative lens action of the upper objective lens pole piece, which forms the earthed boundary of the retarding field, necessitates a slightly higher excitation of the objective lens to assure normal incidence onto the mirror plane. Since for changes in the magnification the objective lens excitation has to be varied, the contrast aperture is, apart from the x- and y-centring, also adjustable along the main axis (z-direction). To each setting of the objective lens current there corresponds an optimum z-position of the contrast aperture providing maximum field of view. Both the objective and the intermediate lens are provided with two crossed quadrupoles for correcting astigmatism.

In order to prevent magnetic flux interaction of the objective and the intermediate lens, the objective lens has been magnetically insulated from the surrounding iron tubing by positioning the iron circuit between two brass plates.

c. *The Specimen Stage.* The separation between the upper pole piece of the objective lens and the specimen plane amounts to 3·5 mm. For a retarding voltage of 30 kV the corresponding strength F_z of the electrostatic field measures $8·57 \times 10^6$ V/m. Although a higher F_z value would improve the image quality, this value proved to be a proper choice to minimize electrical breakdowns. Once an electrical breakdown occurs the specimen will usually be damaged by a spark-over or heavy ion bombardment (see Fig. 26d). The specimen carrier is a massive metal cylinder with a diameter of 10 mm.

The high sensitivity of this microscope for differences in topography requires for most applications a polished specimen surface. Therefore most of the specimens pictured in Section IV consist of vacuum deposited layers onto an accurately polished glass disk, fitting into the specimen carrier.

The electrical conductivity of the specimen surface must be adequate to maintain the electrical potential. So when insulating materials have to be examined with this microscope, a conductive coating of the specimen is required. Since the retarded electrons possess rather low velocities near the specimen, the effective cross-section for ionization of the gas molecules present increases considerably. The positive ions thus generated bombard the negatively biased specimen and cause specimen damage. Moreover, a contamination layer of cracked organic molecules will be deposited onto the specimen surface.

This makes an ultra high vacuum (pressures lower than 10^{-6} Nm^{-2}) desirable. A further reduction of specimen contamination during observation is obtained by heating the specimen up to 500°C.

The specimen stage, consisting of two concentric stainless steel cylinders, of which the exterior cylinder is fixed and the interior one movable, is supported on a glass plate and insulated from earth potential. To minimize high electrical field concentrations the specimen surface is shielded (except for a central hole of 6 mm) with a rounded and polished stainless steel electrode.

In order to avoid an additional lens action due to this screening cap and to maximize the electrical field at the specimen surface, the specimen is mounted directly behind the electrode. The thickness of this electrode near the specimen surface measures 0·3 mm.

By means of three glass insulators the specimen can be translated within a circle of 6 mm in diameter.

Positioning of the specimen surface perpendicular to the main axis for normal mirror microscopy, or tilting the specimen with regard to the main axis over a controllable angle for dark field mirror microscopy, is performed by moving the stage across a sphere with its centre at the intersection of the main axis and the specimen surface. Within mechanical tolerances this construction permits tilting of the specimen without translation.

The operation of this instrument is facilitated by an air lock system for changing specimens. Both the air lock system and a second electron gun, which is used for alignment of the microscope (see Section III.Ag), are adjustable normal to the main axis. At will, either the air lock system or the second electron gun can be centred on the main axis.

Moreover, since the specimen stage protrudes into the cup-shaped upper pole piece of the objective lens, sufficient shielding near the specimen surface against stray magnetic fields is assured.

d. *The Deflection Bridge and the Projector Lens with Camera.* In previous mirror electron microscopes, equipped with a magnetic prism (Mayer *et al.*, 1962; Bartz, *et al.*, 1956; Schwartze, 1967) the reflected electron beam (after passing the magnetic prism) is observed by a skew projection system.

In this microscope the reflected beam, after passing the prism, is made to coincide again with the main axis by means of three additional deflectors. The magnetic prism with the three following deflectors form the deflection bridge.

The advantages of making the reflected beam again coincident with the main axis are:

1. The effective deflection of the bridge is zero. Therefore this system shows an achromatic behaviour for high voltage fluctuations. In addition, when the deflectors of the bridge are energized in series, correction against fluctuations in the series current is established.

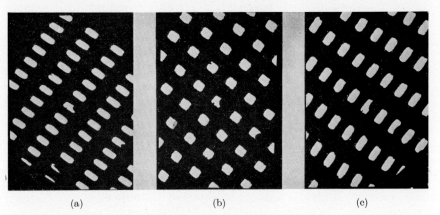

(a)   (b)   (c)

Fig. 25. Images of a grid, illuminated in transmission, for different excitations (i_{defl}) of the deflector bridge. Magnification $120 \times$.
(a) $i_{defl} = 0.38\ A$; (b) $i_{defl} = 0.46\ A$ (corrected); (c) $i_{defl} = 0.52\ A$.

Without this compensation the required current stability for the prism amounts to a few parts per million. The use of the deflection bridge lowers the stability required for achieving identical quality in the final image at least by two orders of magnitude;

2. except for the illuminating system, the main column can be erected vertically, which makes it easier to achieve the high requirements for mechanical stability;

3. a considerable facilitation for the alignment of the electron beam through the microscope is achieved (Section III.Ag).

As an aid for alignment of the reflected beam through the bridge an intermediate fluorescent screen can be inserted between the second and third deflector. This screen can be observed through a glass window and an obliquely positioned glass mirror.

Thanks to the insensitivity of the deflection bridge for variations of the series current over a wide range (0·3 .... 0·6 A) a current setting has been selected for which the astigmatism, which is a by-product of the deflection fields and results in different magnifications in perpendicular directions, is minimized. Residual astigmatism is corrected with a stigmator mounted around the fourth deflector. Figs. 25a, b and c clearly show the effect of the bridge excitation on the image distortion at the final screen.

Between the third and fourth deflector an additional set of x- and y-deflectors is mounted for precise alignment of the electron beam. The projector lens with camera housing and final fluorescent screen is of standard Philips design (EM 75) and described by Le Poole (1964b).

A continuously variable projector lens magnification (35 .... 110 ×) is available by varying the pole piece separation. Image registration is performed either on 35 mm film, mounted directly below the projector pole pieces, or on a 60×60 mm^2 plate camera below the final fluorescent screen. A 7 × binocular viewer permits accurate focusing.

e. *Electrical Supplies*. The high voltage generator produces a continuously variable accelerating voltage between 0 and 30 kV with an overall stability of better than 20 parts per million over a period of 15 min. The filaments in the electron guns are heated with a 1% stable d.c. current. Emission control is obtained by adjusting the Wehnelt resistor.

A variable d.c.-power supply, insulated against 30 kV, heats the specimen furnace. The specimen temperature can be measured with a thermo-couple. The tungsten furnace coil is wound bifilarly to minimize stray magnetic fields near the specimen surface.

The stabilized voltage ϕ_s, variable between $-$ 12 V and $+$ 12 V, is superimposed on top of the retarding potential to prevent electrons from reaching the specimen surface and to correct for the contact potential ϕ_{SC}.

Electrical instabilities in the current supplies for the second condenser, objective, intermediate lens and all deflectors are stable within 5 parts per million over 15 min.

Most lens stigmators and x- and y-deflectors are energized in series with the lens currents. All electrical supplies are remote controlled in order to reduce stray magnetic fields in the vicinity of the column.

f. *Vacuum System.* In this microscope two oil diffusion pumps (baffled pumping speed about 100 l/s) provide a vacuum in the order of 5×10^{-3} Nm^{-2} in the main column and 5×10^{-4} Nm^{-2} near the specimen. Both pumps are backed by a mercury booster pump in combination with a rotary pump for roughing. The booster pump with additional vacuum vessel allows the switching off of the rotary pump after prevacuum pumping of the column.

An air lock system is incorporated for reloading the plate camera without breaking the vacuum in the main column and the specimen chamber.

g. *Alignment Procedures.* Owing to the large number of electron optical components, the alignment of the electron beam through this microscope should preferably be performed in steps. In this microscope the alignment is achieved on the basis of four procedures. The first procedure starts with the centring of the second electron gun, on top of the specimen stage, with respect to the projector lens.

The other lenses and deflectors are not energized, and the specimen stage is kept earthed. The second gun, centred on the projector lens, defines the main axis. Next, the axis of the objective lens is made coincident with this main axis by means of current reversing and pole piece centring.

After following a similar centring method for the intermediate lens, the contrast aperture is inserted in the electron beam.

The second procedure involves the electrical adjustment of the deflection bridge. Proper excitation of the bridge and the accessory stigmators provides a distortion free image on the final screen.

For the third procedure the second electron gun is earthed. The illuminating system then focuses the first electron source onto the filament tip of this gun. When the resulting current flow from the filament tip to earth potential only shows minor changes for current reversal of the objective and intermediate lens, the adjustment of the illuminating system is complete. The fourth and final alignment procedure involves biasing the specimen stage to the proper retarding voltage and positioning the specimen plane normal to the main axis.

All current settings of the critical electron optical components are quickly readable from a four decade digital voltmeter. This control feature reduces considerably the time required for accomplishing the four alignment procedures.

For a calculation of the electron optical parameters and the ray diagrams for both the real and the virtual imaging mode of the objective lens the reader is referred to the author's thesis (Bok, 1968).

A. B. BOK *ET AL.*

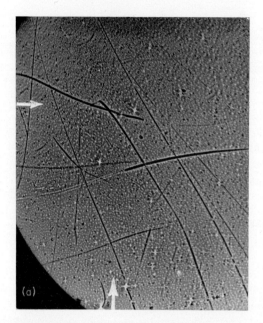

FIG. 26a. Under-focus (distance off focus −50 μm).

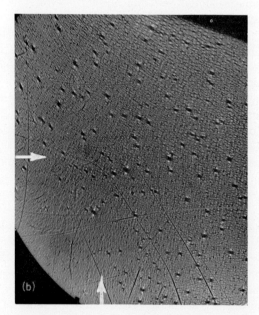

FIG. 26b. In-focus. Magnification 600 ×.

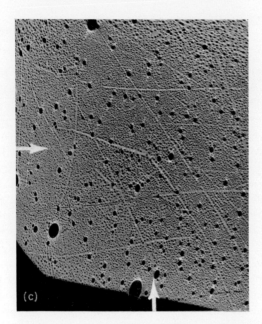

FIG. 26c. Over-focus (distance off focus + 50 μm).

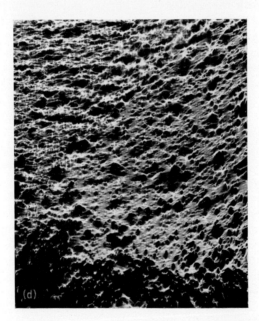

FIG. 26d. Same specimen area after a spark-over to the specimen surface.

IV. Results and Applications

A. *Results*

Since interpretation of mirror microscope images is rather complicated, only test specimens are selected having a known composition and topography. Most of the pictures presented in this chapter are therefore vacuum deposited layers on accurately polished glass disks. A first conductive layer makes the glass surface coincident with an equipotential plane, whereas additional layers, mostly evaporated through a grid of known dimensions, provide a regular pattern. The advantage of using a regular pattern is found in the easy determination of image distortion and magnification. Although in principle discrimination between electrostatic and topographic contrast is possible (conclusion 4 of Section II.Cc), no clear practical evidence is found yet in the images obtained. Due to the high sensitivity for slight differences in height, the possibility of preparing a specimen with purely electrostatic contrast appears to be rather doubtful. In view of this difficulty it was decided to concentrate primarily most of our efforts on the focused imaging of specimens with topographic contrast. The photographic results presented are only meant to demonstrate the remarkable improvement in image quality of the focused mirror microscope in comparison with the results from mirror projection microscopes. The authors are aware that this series of photographs only provides a limited outlook at the large, but hardly explored, field of possible applications. The achieved improvement in image quality and resolving power of this type of mirror electron microscope, the main purpose of this instrument, makes it worthwhile to initiate a more systematic research for widening the scope of useful applications.

Figures 26a, b and c represent a through focal series of a vacuum deposited gold layer (30 nm thick) on a polished glass disk. The magnification amounts to about 600 × and the diameter of the contrast aperture was 200 μm. All white stars in the under focus image clearly change into black bubbles in the over focus image (reverse of contrast). The point of intersection of the arrows indicates the same specimen spot. The occurrence of these local perturbations is mainly due to a continuous ion bombardment of the specimen surface. Uncovered areas of the glass surface, having a poor surface conductivity, cause numerous local charges. Slight differences in the surface conductivity, for example by the bombardment of ions, then lead to a fluctuating potential distribution across the specimen. This phenomenon gives at the final image the impression of the specimen being "alive". The mechanical scratches in the thin gold film fade away for the

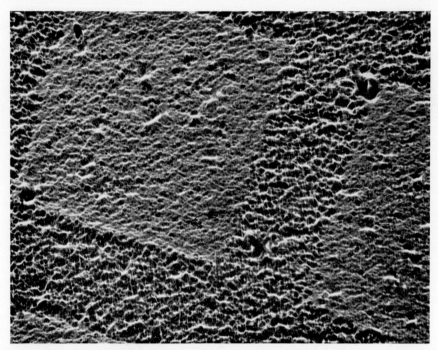

FIG. 27a. Under-focus (distance off focus −5 μm).

FIG. 27b. In-focus. Magnification 2700×.

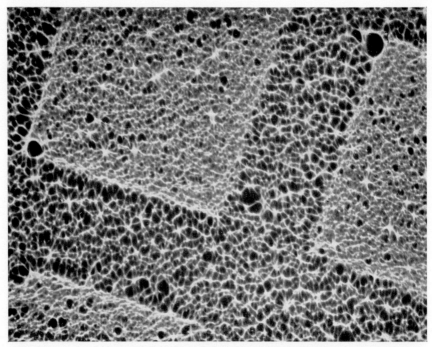

Fig. 27c. Over-focus (distance off focus + 5 μm).

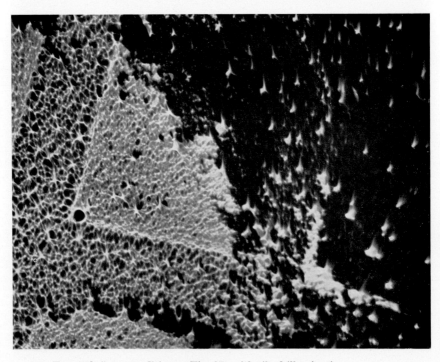

Fig. 27d. Same condition as Fig. 27c with tilted illumination.

in-focus condition (Fig. 26b). This effect illustrates the statement, made in Section II.E, that the perturbed mirror electrode can be considered as a phase specimen.

Figures 28a, b and c demonstrate the effect of an increasing tilt of the illuminating beam. Accurate positioning of the contrast aperture in the back focal plane of the combination objective and electrostatic lens gives a knife edge filtering for the entire illuminated area at the

FIG. 28a. Slightly tilted illumination magnification 900 ×.

specimen. The specimen pictured is the same as in Fig. 27. The magnification is 1000 ×. The inclined incidence of the illuminating beam causes a narrowing of the bars (actual width about 8 μm) between the squares. The increasing concave mirror action of the bars for tilted illumination provides sharp line foci in the final image.

Figure 29 is the same specimen as pictured in Fig. 27 but at a magnification of 250 ×. In spite of the low magnification hardly any image

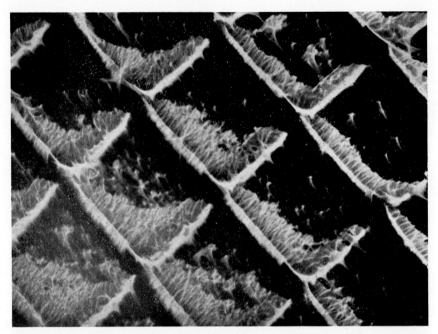

Fig. 28b. Tilted illumination (same specimen area as Fig. 28a).

Fig. 28c. Grazing incidence of the illuminating beam (same specimen area as Figs. 28a and b).

FIG. 29. Minimum magnification (250×).

distortion is present. This image is again equally sharp across the
entire field of view (about 600 μm).

The in-focus specimen pictured (Fig. 30) is an accurately polished
stainless steel surface with copper squares evaporated through a
400 mesh grid (the squares are about 33×33 μm²). The thickness of
the copper squares amounts to (10 ± 3) nm. The magnification, which
is not equal in all directions owing to astigmatism of the deflection
bridge, is $3000 \times \pm 15\%$. The traverse lines are scratches from
polishing. The sharply imaged "groove" around the copper square
shows a more gradual change in contrast than found for purely topographic specimens (Figs 27 and 28).

Figures 31a and b show droplets of aquadag on a gold evaporated
glass surface (a shows the in-focus image and b the slightly under-focus
condition). The magnification is 1700 ×. Much detail is revealed
within the droplets. A slight amount of astigmatism of the objective

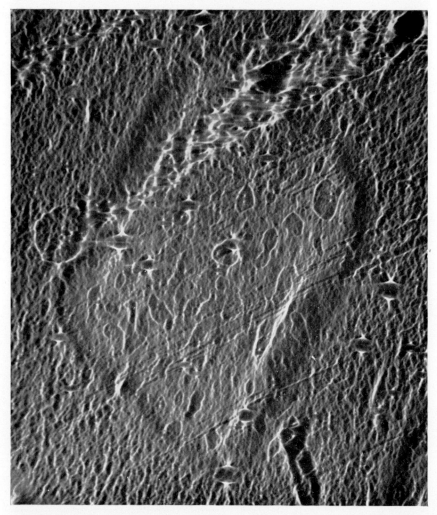

FIG. 30. Electrostatic and topographic contrast of a stainless steel surface covered with a copper square (magnification 2500× ± 15%).

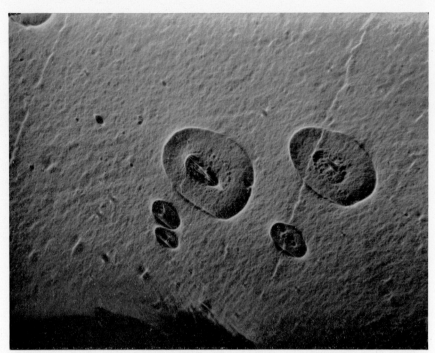

FIG. 31a. In-focus (1500 ×).

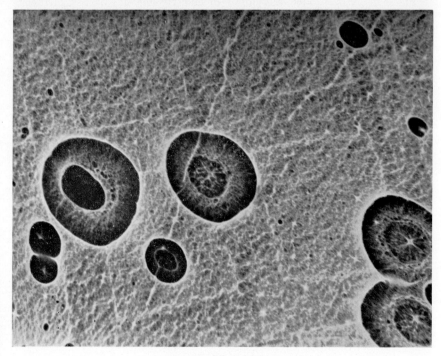

FIG. 31b. Under-focus.

lens or caused by the specimen itself is visible in the lower left region of 31a. It is often difficult from observation of the fluorescent screen to correct entirely for residual astigmatism of the objective lens.

B. *Applications*

The applications mentioned in brief hereafter are not meant to provide the reader with a complete survey about all possibilities of the mirror electron microscope. It only presents a number of applications which might be of interest to physicists investigating surface phenomena at a microscopic scale.

a. *The Investigation of Semi-conductor Electronics (Micro-circuits).* Apart from the surface topography, electric properties as potential distributions across resistors, condensers, etc., and current flow in separate components can be observed. Especially the visualization of the dynamic behaviour of micro-circuits allows for determination of interruptions and breakdowns in the circuitry (Igras and Warminski, 1965, 1966, 1967; Maffit and Deeter, 1966; Ivanov and Abalmazova, 1968).

b. *The Investigation of Surface Conductivity, Diffusion of Metals* (Igras and Warminski, 1967) *and Ferro-electric Domain Patterns* (Igras *et al.*, 1959). The movement of electric charges across surfaces, having a poor surface conductivity, can be studied dynamically. Measuring the propagation velocity of electric charges provides information about the surface conductivity (Mayer, 1967a). The storage of charges on photo-sensitive layers (image intensifiers) can be visualized at a high magnification. When the mirror electrode is replaced by a photo-sensitive layer, an image intensifier with a high resolving power could perhaps be realized. Changes in the local work function, resulting from diffusion of metals or doping effects, lead to current density modulations in the final image.

c. *The Investigation of Thin Films.* The high sensitivity for topography and local charges offers the possibility to test the quality of evaporated layers. Contaminations and impurities can be easily detected.

d. *The Dynamic Observation of Magnetic Domain Patterns* (Spivak *et al.*, 1955; Mayer, 1959a, b). This is, for instance, the imaging of patterns recorded on magnetic tape (Mayer, 1958; Spivak *et al.*, 1963) and magnetic stray fields on grain boundaries (Mayer, 1959a). Some experimenters have successfully reported on observations of ferro-magnetic phenomena (Spivak *et al.*, 1955).

e. *The Investigation of the Local Work Function, as Already Performed in the Emission Electron Microscope.* In addition to the mirror images, secondary emission images are obtainable by bombarding the specimen with low energy electrons (in the order of tens of electron volts). This feature provides the possibility to obtain two different types of image from the same specimen area. A stable 100 V source on top of the accelerating voltage is under construction. In how far low energy electron diffraction (LEED) is possible in this instrument remains to be seen.

f. *An Improvement of the Vacuum near the Mirror Electrode and a Rough Filtering of the Illuminating Beam Yield Interesting Perspectives for Investigation of Physi- and Chemi-sorption Phenomena.* Contrast will be obtained in this case by changes in the local work function resulting from adsorption.

Apart from paying more attention to the practical application of our mirror electron microscope, the following instrumental improvements are planned for the near future:

1. filtering of the illuminating beam by means of a Wien filter,
2. improvement of the vacuum near the specimen.

V. APPENDIX

A. *A Scanning Mirror Electron Microscope with Magnetic Quadrupoles* (Bok et al., 1964)

Le Poole's basic idea of using two crossed quadrupoles for a scanning mirror electron microscope originated as a side line during the process of designing a transmission electron microscope with magnetic quadrupoles (Le Poole, 1964b). From the theory of quadrupoles it is known that a single quadrupole has small coefficients of spherical aberration and, moreover, that it offers more possibilities for correction than rotationally symmetric lenses. However, since the formation of a real image requires at least two quadrupoles arranged in such a way that they counteract each other to a large extent, the actual errors for such a doublet or triplet are of the same magnitude or larger than those of the rotationally symmetric lenses. The new idea was, that when anamorphotic imaging (different magnifications in perpendicular directions) is accepted, a quadrupole doublet can be made which possesses a highly reduced spherical aberration. Since calculations revealed that the formation of a focal line of 0·1 nm in width should be possible, this idea might involve a new method to improve the resolving power for a transmission electron microscope. Obviously good resolution is only required in the direction of the line since no electrons hit the specimen outside the line focus of the illuminating system.

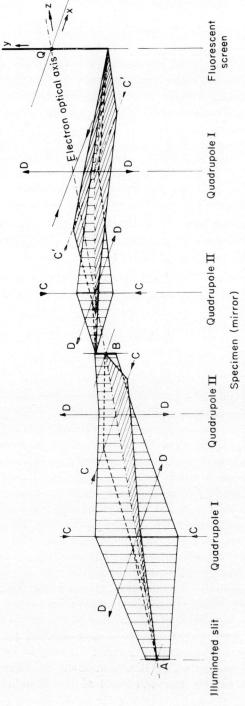

Fɪɢ. 32. Ray diagram (the specimen is at *B*).

Sufficient field of view is obtained by scanning the line focus across the specimen.

As objective lens an identical quadrupole doublet can be used on the same axis but rotated over 90 degrees. If the specimen is replaced by a mirror both doublets are combined in one. The lens action of a magnetic quadrupole changes over 90 degrees for reversed directions of the electron beam.

The irreversible lens action of these lenses thus provides an elegant method for designing a mirror electron microscope for focused images.

In the summer of 1963 the design of a scanning mirror electron microscope was started in collaboration with J. Kramer.

In Fig. 32 the ray diagram for the transmission system is presented. For the scanning mirror instrument, part BQ of the z-axis should be rotated around B (the specimen) till Q and A coincide. The convergent and divergent lens action of the quadrupoles are marked with C and D.

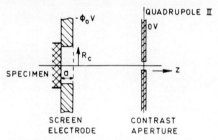

FIG. 33. Mirror electrode configuration.

The anamorphotic imaging on the specimen (mirror) of an illuminated aperture slit at A gives a reduction of roughly $100 \times$ in width and $15 \times$ in length. For the reflected beam this effect reverses, so the length is multiplied by $100 \times$ and the width by $15 \times$. In this case too, sufficient field of view is obtained by sweeping the focal line across the specimen by means of two deflection systems. The small amplitude of this movement at the specimen is again magnified at the screen. Proper excitation of the deflection systems provides equal magnification in perpendicular directions at the screen. One double deflector wobbles the illuminating beam, without lateral displacement, in the aperture slit at A. A second deflector, mounted around quadrupole I, directs the deflected beam backwards through an aperture located in quadrupole II. In order to avoid, for a larger field of view, vignetting of this contrast aperture for reflected electron pencils, the retarding field is shaped to act like a concave mirror. The contrast aperture, which is adjustable along the z-axis, is made coincident with

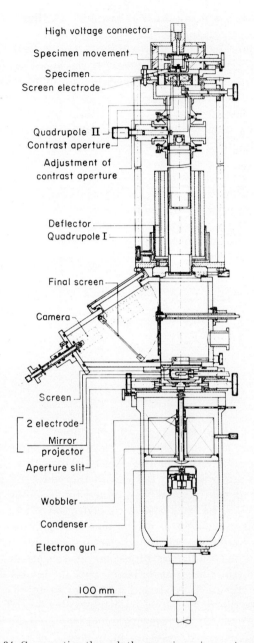

High voltage connector

Specimen movement

Specimen

Screen electrode

Quadrupole II

Contrast aperture

Adjustment of
contrast aperture

Deflector

Quadrupole I

Final screen

Camera

Screen

2 electrode

Mirror
projector

Aperture slit

Wobbler

Condenser

Electron gun

100 mm

FIG. 34. Cross-section through the scanning mirror microscope.

the centre of curvature of the concave mirror. The concave mirror action of the retarding field is obtained by using a cup-shaped screen electrode (see Fig. 33). The focal length of this mirror changes for variations of the ratio a/R_c. A slightly off axis two electrode mirror (Le Rütte, 1952), which is mounted below the fluorescent screen

FIG. 35. Scanning mirror microscope with quadrupoles.

projects part of the image with an additional magnification of $20 \times$ (total $2000 \times$) onto the obliquely positioned final screen. Figure 34 gives a cross-section through the instrument, Fig. 35 an overall photograph and Figs 36a and b the best results obtained.

The specimen pictured is a well polished brass surface covered by an aluminium layer vacuum deposited through a 750 mesh (33 μm)

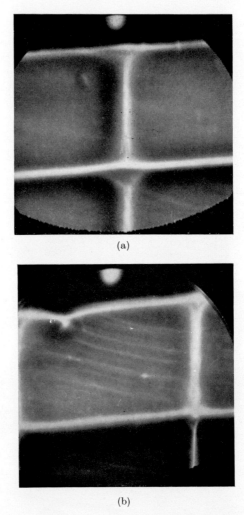

(a)

(b)

FIG. 36. Brass surface with vacuum deposited aluminium squares. (a) focused on the brass "bars", (b) focused on the specimen surface (topography). Magnification 1200 ×.

grid. The magnification, although not perfectly equal in perpendicular directions, is in the order of 1200 ×. Since a further improvement of the image quality with this instrument could only have been achieved with an entirely redesigned apparatus (better shielding against stray magnetic fields, better mechanical and electrical stability), it was then decided (summer, 1964) to leave the idea of magnetic quadrupoles and to change over to a new mirror electron microscope with rotationally symmetric lenses and a magnetic prism.

B. *Calculation of the Influence of the Specimen Perturbation on the Phase of the Reflected Electron Beam in a Mirror Electron Microscope*

H. DE LANG

Philips Research Laboratories, Waalre, Netherlands

The influence of the specimen perturbation on the phase of the reflected wave will be calculated making use of the concept of phase length along optical rays. Since the axial component of the electron velocity goes through zero near the plane of the mirror (specimen), the wave length of an axial beam becomes infinite there. It might seem doubtful, therefore, if the simple concept of ray-optical phase length would make any sense here. We will show, however, that such an approximation is justified.

a. *Justification of the Use of the Ray-optical Phase Length in Relation to Calculations Near the Specimen in a Mirror Electron Microscope.* Suppose a plane monochromatic wave directed along the z-axis (normal to the specimen) enters the homogeneous retarding mirror field. The reflected wave front shows a phase ripple with an amplitude equal to the modulation amplitude of the phase length, provided that the interaction area near the specimen is sufficiently "thin". By two different arguments it will be demonstrated that this condition is fulfilled for this specific problem. In the first case use is made of the uncertainty principle, in the second the radius of the first Fresnel zone is compared with the perturbation period.

1. *The Uncertainty Principle.* The following assumptions are involved:

the specimen possesses a sinusoidal perturbation with spatial frequency ν and amplitude Δz;
the axial thickness of the effective interaction area amounts to $(\pi\nu)^{-1}$;
the lateral uncertainty in location for an electron entering the interaction area is a fraction of ν^{-1} (say $(8\nu)^{-1}$).

We are allowed to use the ray-optical phase length if the lateral wandering, as a result of the lateral electron velocity, is also within a fraction of ν^{-1} (say again $(8\nu)^{-1}$).

From the uncertainty principle for the electrons entering the retarding mirror field

$$\Delta y \cdot \Delta v_y \approx h/m \qquad (47)$$

where

$$\Delta y = (8\nu)^{-1}$$

It follows that

$$\Delta v_y \approx 8 \frac{h \cdot \nu}{m} \qquad (48)$$

Multiplying Δv_y with the time the electron travels through the interaction area, the lateral electron wandering then found must not exceed $(8\nu)^{-1}$ which yields

$$4 \left(\frac{z_2}{\pi\nu}\right)^{\frac{1}{2}} v_2^{-1} 8 \frac{h\nu}{m} < (8\nu)^{-1}$$

or

$$\nu^{-1} > 26{\cdot}5(z_2\lambda_2^2)^{\frac{1}{3}} \tag{49}$$

Taking $z_2 = 3 \times 10^{-3}$m and $\lambda_2 = 7 \times 10^{-12}$m (30 kV) we find

$$\nu^{-1} > 140 \text{ nm} \tag{49a}$$

which means that ν does not need to be much smaller than the practical resolving power in the lateral direction.

2. *The Radius of the First Fresnel Zone Compared with the Perturbation Period.* After travelling through the interaction area the radius of the first Fresnel zone has to remain within a fraction of ν^{-1} (say $(8\nu)^{-1}$).

The wave length $\lambda = \lambda(z)$ is

$$\lambda = \frac{h}{m} (v_y^2 + v_z^2)^{-\frac{1}{2}}$$

Subtracting the unperturbed phase-length:

$$2\pi dz h^{-1} m v_z$$

from the perturbed one:

$$2\pi dz h^{-1} m (v_y^2 + v_z^2)^{\frac{1}{2}} \cos^{-1} a = 2\pi dz h^{-1} m (v_y^2 + v_z^2) v_z^{-1}$$

leads to:

$$2\pi v_y^2 h^{-1} m v_z^{-1} dz = 2\pi v_y^2 h^{-1} m \left(\frac{z_2}{z}\right)^{\frac{1}{2}} v_2^{-1} dz.$$

Integration of this expression over the interaction path must result in π. Thus we obtain

$$8\pi^{\frac{1}{2}} v_y^2 h^{-1} m z_2^{\frac{1}{2}} v_2^{-1} \nu^{-\frac{1}{2}} = \pi \tag{50}$$

The requirement that the radius of the first Fresnel zone (equal to the lateral velocity times electron travelling time) be smaller than $(8\nu)^{-1}$ gives

$$v_y < \pi^{\frac{1}{2}} \nu^{-\frac{1}{2}} v_2 z_2^{-\frac{1}{2}} / 32 \tag{51}$$

Combining equations (50) and (51) we obtain

$$\nu^{-1} > 17{\cdot}3(z_2\lambda_2^2)^{\frac{1}{3}} \tag{52}$$

Thus we may conclude that consideration of the radius of the first Fresnel zone (eqn. 52) as well as the application of the uncertainty principle (eqn. 49) both justify the use of a simple ray-optical phase-length treatment for ν-values occurring in practice. This greatly facilitates the calculation of the phase modulation in the reflected wave.

b. *Calculation of the Phase Modulation in the Reflected Wave as a Function of the Amplitude Δz and the Spatial Frequency ν of a Sinusoidal Specimen Perturbation.* The unperturbed wavelength as a function of z is

$$\lambda(z) = \left(\frac{z_2}{z}\right)^{\frac{1}{2}} \lambda_2$$

where $\left(\dfrac{z_2}{z}\right)^{\frac{1}{2}}$ is the "refractive index".

In order to take into account the perturbation by the specimen we select two rays, i.e. one to a " peak" of the specimen and another to a "valley".

Along the rays to the " peaks" the perturbed wavelength λ' is

$$\lambda'(z) = \lambda(z - \Delta z \cdot e^{-2\pi\nu z}) = \lambda_2 z_2^{\frac{1}{2}} (z - \Delta z \cdot e^{-2\pi\nu z})^{-\frac{1}{2}} \tag{53}$$

Along the rays to the "valleys" the perturbed wavelength λ'' is

$$\lambda''(z) = \lambda(z + \Delta z \cdot e^{-2\pi\nu z}) = \lambda_2 z_2^{\frac{1}{2}} (z + \Delta z \cdot e^{-2\pi\nu z})^{-\frac{1}{2}} \tag{54}$$

The amplitude R_ζ of the phase modulation is the reflected wave at the plane $z = \zeta$ is

$$R_\zeta = 2\pi\left\{ \int_{-\Delta z}^{\zeta} (\lambda'')^{-1} dz - \int_{\Delta z}^{\zeta} (\lambda')^{-1} dz \right\} \tag{55}$$

With equations (53) and (54), equation (55) transforms into

$$R_\zeta = 2\pi\lambda_2^{-1}z_2^{-\frac{1}{2}}\left\{ \int_{-\Delta z}^{\zeta} (z + \Delta z \cdot e^{-2\pi\nu z})^{\frac{1}{2}} dz - \int_{\Delta z}^{\zeta} (z - \Delta z \cdot e^{-2\pi\nu z})^{\frac{1}{2}} dz \right\} \tag{56}$$

In order to calculate the final value R of the amplitude of the phase modulation we have to take $\zeta = \infty$ in (56). Assuming further that $\Delta z \ll (\nu)^{-1}$ we can easily evaluate the integrals in (56). In doing so we obtain for the final amplitude R of the phase modulation in the reflected wave

$$R = 2^{\frac{1}{2}}\pi\lambda_2^{-1}z_2^{-\frac{1}{2}}\nu^{-\frac{1}{2}}\Delta z \tag{57}$$

Remarks: For a distance $z = (\pi\nu)^{-1}$ from the specimen the reflected wave has already received 95% of its total modulation. Thus the supposed thickness $(\pi\nu)^{-1}$ for the interaction area is in fact reasonable. A similar calculation holds for an electrostatically perturbed specimen. In that case Δz in equation (57) has to be replaced by $\Delta\phi/F_z$.

c. *Visibility of Small Differences in Specimen Relief (Axial Resolving Power)*. For a weakly modulated specimen (sinusoidal perturbation) the reflected wave consists of a direct beam (zero order) and both diffracted beams of the first order. The relative amplitude of the diffracted beam with respect to the direct beam amounts to (see equation 57):

$$\tfrac{1}{2}R = 2^{-\frac{1}{2}}\pi\lambda_2^{-1}z_2^{-\frac{1}{2}}\nu^{-\frac{1}{2}}\Delta z \tag{58}$$

If with a knife edge aperture one of the first order beams is intercepted, the remaining first order beam and the direct beam give in the imaging plane an interference pattern with a sinusoidal intensity modulation. The modulation depth is

$$(I_{max} - I_{min})/(I_{max} + I_{min}) = R = 2^{\frac{1}{2}}\pi\lambda_2^{-1}z_2^{-\frac{1}{2}}\nu^{-\frac{1}{2}}\Delta z \tag{59}$$

Taking a modulation depth of 0·1 as the threshold value, the corresponding minimum specimen ripple amplitude then amounts to

$$(\Delta z)_{min} = 2 \cdot 2 \times 10^{-2}\lambda_2 z_2^{\frac{1}{2}}\nu^{\frac{1}{2}} \tag{60}$$

Taking
$$\lambda_2 = 7 \times 10^{-12}\text{m} \ (30 \text{ kV})$$
$$z_2 = 3 \times 10^{-3}\text{m}$$
$$\nu^{-1} = 3 \times 10^{-7}\text{m}$$

the threshold value of specimen relief $(\Delta z)_{min}$ is found to be 16×10^{-12}m.

d. *Influence of the Energy Spread of the Incident Electron Beam on the Axial Resolving Power*. The results obtained in the previous sections are only valid for a mono-energetic electron beam. To approximate the influence of an energy spread a rectangular energy distribution (width $\Delta\phi$) is supposed to be present in the incident beam.

The contribution to the image intensity of a narrow interval $d\phi$ out of the energy width $\Delta\phi$ is represented by:

$$dI = d\phi(A + B \cdot e^{-2\pi (d\phi/\phi_2) \nu z_2} \cdot \sin 2\pi\nu y),$$

where A, B are constants.   (61)

Integrating equation (61) from zero up to $\Delta\phi$ we find

$$I = A\Delta\phi + B\Delta\phi \frac{\phi_2}{2\pi\nu z_2\Delta\phi} (1 - e^{-2\pi (\Delta\phi/\phi_2) \nu z_2}) \sin 2\pi\nu y$$

For the factor f by which the modulation depth of the image intensity is reduced as a result of a rectangular energy spread $\Delta\phi$ we find

$$f = \frac{\phi_2}{2\pi\nu z_2\Delta\phi} (1 - e^{-2\pi (\Delta\phi/\phi_2) \nu z_2}) \tag{62}$$

Taking

$$\left. \begin{aligned} \phi_2 &= 3 \cdot 10^4 \, V \\ z_2 &= 3 \cdot 10^{-3} \text{m} \\ \nu^{-1} &= 3 \cdot 10^{-7} \text{m} \\ \varDelta\phi &= 0 \cdot 5 \, V \end{aligned} \right\} \quad \text{we find } f = 0 \cdot 55$$

Conclusion: The minimum detectable specimen relief as expressed by (60) and (62) is three orders of magnitude smaller than the "ad hoc" value $\dfrac{\varDelta\phi}{F_z}$. In fact the influence of the energy spread $\varDelta\phi$ on the axial resolving power is not severe as long as the energy spread does not exceed the thickness (in terms of energy) of the interaction layer. It might seem paradoxical that such a favourable axial resolving power is possible with non-monochromatic radiation. It must be pointed out, however, that with interference microscopes using broadband visible light, observation of depth differences several orders of magnitude smaller than the wave length is quite common.

C. *Practice of Mirror Electron Microscopy*

H. BETHGE AND J. HEYDENREICH

Institut für Festkörperphysik und Elektronenmikroskopie, Halle, DDR

a. Since the first investigations in mirror electron microscopy almost 35 years ago, the philosophy of design of the devices and the application of this method have changed considerably. The first mirror electron microscopes (Hottenroth, 1937; Orthuber, 1948; Bartz *et al.*, 1952) were devices with a simple magnetic prism, some of which were glass devices. In order to avoid the difficulties connected with the astigmatism of the magnetic deflection field, most of the later mirror electron microscopes were straightforward devices without a magnetic prism, that means with the same beam axis for the incoming as well as the reflected electrons (see e.g. Mayer, 1955; Bethge *et al.*, 1960; Spivak *et al.*, 1961).

Devices of this type have proved very good and are also nowadays successful in operation. Nevertheless, they have some disadvantages, the most essential of which are: 1. The centre of the image cannot be observed, because the aperture for the passage of the primary beam is in the way. 2. The manipulation of the beam of reflected electrons, e.g. a further magnification by a projector lens, is not possible without influencing of the primary electron beam. 3. With the aid of such a straightforward device, only shadow images can be obtained and not focused images (see e.g. Schwartze, 1965). Since possibilities exist for the production of deflection fields without noticeable astigmatism (see e.g. Archard and Mulvey, 1958), an increasing number of mirror electron microscopes are being produced with a magnetic prism (see e.g. Mayer et al., 1962; Schwartze, 1966; Bok, 1968).

b. An example of a straightforward electron optical device with the same beam axis for the primary electrons and the reflected electrons, is given in Fig. 37, which is a sectional view of a mirror electron microscope. This is described more in detail by Bethge et al., (1960). Viewing from bottom to top, the microscope column consists of the following main parts: electron source, viewing chamber, projection tube and specimen chamber.

The electron gun is a simple triode system (T), and possibilities for mechanical alignment (A) of the electron beam are included. The primary beam, after having passed two apertures in the viewing chamber, reaches the specimen (O), where it is reflected. The actual electron mirror system consists of the specimen (O), mounted on an insulated table and aperture (H) placed in front of it. This system acts as a diverging mirror. The resulting "magnification" on the screen (S), is affected firstly by the size of the aperture (H) and secondly by the distance between the specimen and the aperture. The specimen holder can be shifted horizontally under vacuum by the aid of the screw S_1; the distance specimen–aperture is also changeable under vacuum (screw S_2). A further possibility for changing the magnification is given by the lens action of two cylindrical electrodes (C). The screen (S) can be observed either through the small viewing hole (V) at the side or through a window (W) by the aid of a mirror (M) inclined at 45 degrees. Photographs are taken from outside by the aid of a standard camera placed in front of this window. For changing the specimen, the upper part of the specimen chamber is removed. The beam voltage can be varied between 4 and 15 kV; the resulting maximum magnification is about $\times$ 300.

c. As an example of a modern mirror electron microscope with a magnetic prism, a device described in detail by Heydenreich (1970) is shown in a sectional view in Fig. 38. In the device a deflection angle of 20 degrees is used. The main parts of the microscope (magnetic prism, electron source and specimen chamber) are mounted on top of a recording chamber (containing screen and plates) as used in commercial electron microscopes.[†] The electron source is fitted at the side of the magnetic prism column. Starting from the electron gun (G) the electron beam is aligned with the aid of combined electric and magnetic adjustment equipment (A) and is eventu-

† In this case: Electron optical Plant EF (VEB Carl Zeiss Jena).

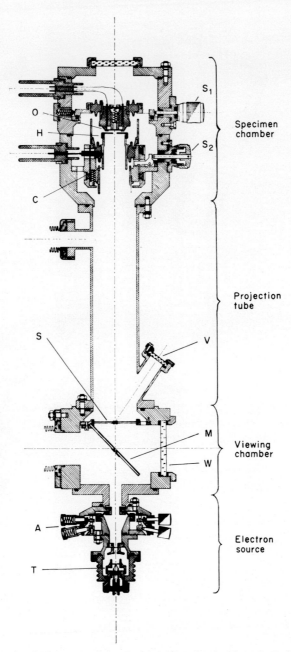

FIG. 37. Sectional view of a mirror electron microscope without beam deflection.

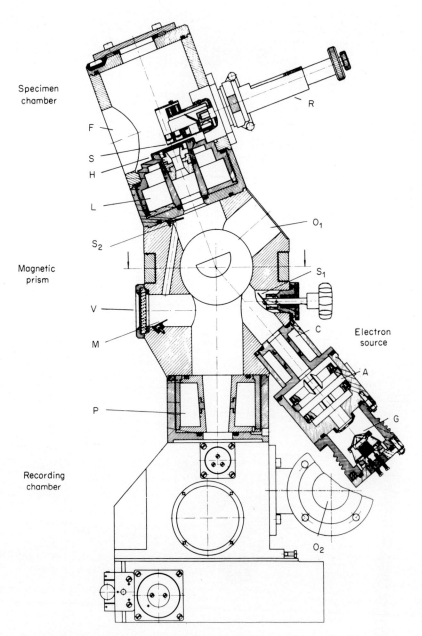

Specimen chamber

F

S

H

L

S₂

Magnetic prism

V

M

P

Recording chamber

R

O₁

S₁

C

Electron source

A

G

O₂

Fig. 38. Sectional view of a mirror electron microscope with a magnetic prism.

ally focused by a condensor lens (C). Adjustment aids are a removable screen (S_1) and a further auxiliary screen (S_2). Both the screens are observed through a special viewing window (V), the screen (S_2) via a small inclined mirror (M). The coils of the magnetic prism are above and below the plane of the drawing; in the sectional view the form of the pole pieces can be seen. The specimen (S), which is insulated with respect to the microscope column, is mounted on a rod (R), which allows the necessary horizontal movement of the specimen during the electron optical investigation to take place. Specimen change is effected through the front opening (F) of the specimen chamber, which is closed with a sealing plate during routine operation. This sealing plate can be replaced by a special attachment for specimen treatments (also not drawn in the sectional view). After having brought the specimen into place in this attachment by shifting the specimen rod, the specimen to be investigated can be exposed to an ion beam for cleaning the surface. Furthermore, this attachment contains evaporation sources for the formation of thin layers to be investigated in the mirror electron microscope. The specimen chamber has been so designed that it will readily accept equipment for cooling, heating, magnetizing or straining the specimen. The aperture (H), necessary for producing a shadow image, is mounted on the objective lens (L), which is important for producing focused images. With the aid of the projector lens (P), shown in the sectional view without pole pieces, a considerable increase in the end magnification of the image is given. The microscope column is pumped out to a vacuum of about 10^{-5} torr through two openings (O_1, O_2).

The stability of the beam voltage and the currents to be used (including that of the magnetic prism) is about 10^{-5}. The beam voltage can be varied over a wide range; it proved favourable to use voltages between 10 and 20 kV. For the shadow image technique, which is used in the majority of cases, the resolving power is about 1000 Å; the maximum attainable magnification lies in the region of $3000 \times$.

As is well known the mirror electron microscope is suitable for imaging the relief as well as electric or magnetic inhomogeneities of a specimen surface.

d. The imaging of *geometric inhomogeneities* of a specimen surface, of which impressive examples in the early days of this technique have been given by Bartz, Weissenberg and Wiskott (1954) on metal surfaces, suffers from the considerably restricted resolving power of the mirror electron microscope used in shadow image technique (e.g. Schwartze, 1965, 1966) which for practical reasons is of the order of 1000 Å. Nevertheless, because of the special type of image formation in the mirror electron microscope, the imaging of the surface relief is of some interest. Since the information about the specimen surface in the imaging electron beam is given by the potential field in front of the object, which is especially sensitive to different "heights" of surface inhomogeneities, a kind of "spatial" detection of the specimen surface is carried out. In this way one gets pronounced shifting effects of image structures, which are larger, the rougher a surface irregularity is. In many cases this effect is regarded as a disturbance of the image and leads to the requirement for smooth specimen surfaces in mirror electron microscopy. On the other hand, this shifting effect of the image structure gives additional information about the specimen surface, in so far as the shifting effect is detectable by comparing different pictures of the same specimen region.

An example of this is given in Fig. 39, which shows a series of pictures of the same region of a NaCl cleavage face (lightly coated with evaporated Ag). In (a) is shown the image taken with the mirror electron microscope, in (b) the corresponding conventional microscope image and in (c) the electron mirror image of the "matched" cleavage face. Assuming the conventional microscope image (b) is a "true" image of the object, we see cleavage steps partly joining, thereby forming tip-like regions (B, C). Because of the special type of contrast formation in the mirror electron microscope (shadow image technique), each step is imaged as a bright-dark double line, the bright edge of which marks the lower plateau adjoining the step. For this reason the electron optical image (a) shows that the regions A and C are depressions and the region B is a protrusion. As briefly mentioned in the introduction (see also Heydenreich, 1966) a further consequence of the mechanism of contrast formation is that, especially at rough specimen surfaces, the double lines marking the steps are not found at the position corresponding to the object according to a conventional microscopic imaging process, but the double lines are shifted in the "image plane" in the direction of the lower plateau. In the example, this effect is easily recognizable from the fact that the lower regions A and C seem to be much smaller in the electron optical image than in the conventional microscope image, and that the plateau region B seems to be correspondingly larger. As can be seen from Fig. 39c the effect is reversed in the "matched" cleavage face. According to interferometric measurements, the height of the steps amounts to about 700 Å. Conclusions about the roughness of the specimen surface are possible from the shift of the image structures, pre-supposing that adequate calibration has taken place, in which careful attention must be paid to specimen potential.

e. The pictorial representation of *electrical surface inhomogeneities*

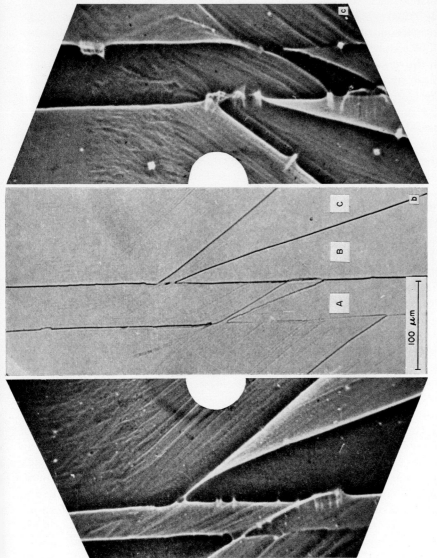

Fɪɢ. 39. Rough cleavage steps on a NaCl cleavage face. (a) and (c) Electron optical micrographs (cleavage face and opposite cleavage face), (b) Conventional microscope image.

with the aid of the mirror electron microscope is of special importance. Naturally, the scanning electron microscope, which is now well established as a commercial device, can also be used for the imaging of electrical specimen inhomogeneities. However, in this case one has to take into account the fact that in the formation of the image a great number of electrons is shot into the specimen surface. This is especially critical in the investigation of semiconductor surfaces, which are the specimens most frequently used for such investigations. In this case the mirror electron microscope, in which there is no direct interaction between the imaging electron beam and the specimen surface, proves a very useful instrument because of the absence of possible specimen modification. Investigations of electrical surface inhomogeneities by the aid of the mirror electron microscope have already been carried out by Orthuber (1948), who was interested in potential distributions in metallic and semi-conducting surfaces. As special examples of the imaging of electric structures can be cited the imaging of p-n junctions, carried out first in 1957 by Bartz and Weissenberg, and the observation of ferro-electric domains (barium titanate) by Spivak *et al.* (1959a). Of special interest is the pictorial representation of electric conductivity distributions which is possible in the mirror electron microscope by producing a corresponding current flow in the specimen surface (e.g. Mayer, 1957a). As an example of this technique, Fig. 40 shows the electron optical image of a thick germanium layer (900 Å) which is covered by a very small quantity of evaporated indium with an average thickness of about 40 Å. Without any current flow (a) the only image contrasts which are seen are those which are related to the rough surface of the evaporated layer. When a voltage is applied between the left and the right-hand sides of the specimen surface a current flow results and one sees contrast lines of increasing clearness in the images (b), (c) and (d) (respectively: $+2$, $+10$ $+20$ volt) which show that strong locally acting electric fields are present. These strong fields are related to the boundaries of the flat indium islands, which are more or less insulated from each other. When current flows through the specimen the more negatively charged areas appear darker in the image of the border region than the more positive areas. In this example the image contrast corresponding to electrical inhomogeneities at large current flows is much stronger than that corresponding to surface roughness which is simultaneously present. By a suitable adjustment of the specimen potential one always succeeds in suppressing the weaker contrast (in the present case, geometrical). Fig. 41 shows that, for the specimen under discussion, the specimen potential necessary for this suppression is between -30 and -40 volt.

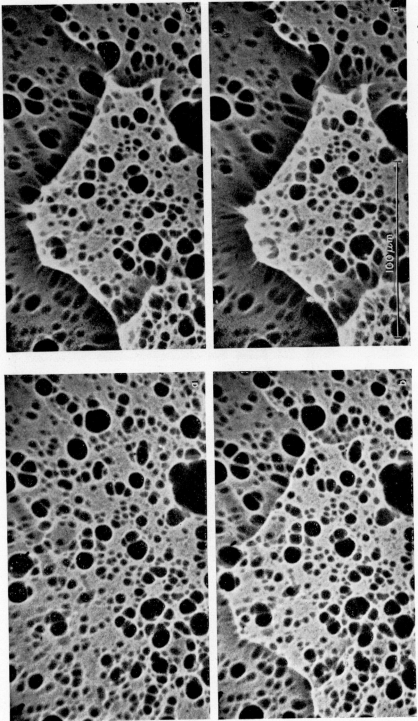

Fig. 40. Rough germanium layer, coated with a thin indium layer. Electron optical micrographs with different currents flowing in the layer, produced by different applied voltages: (a) 0 V, (b) +2 V, (c) +10 V, (d) +20 V.

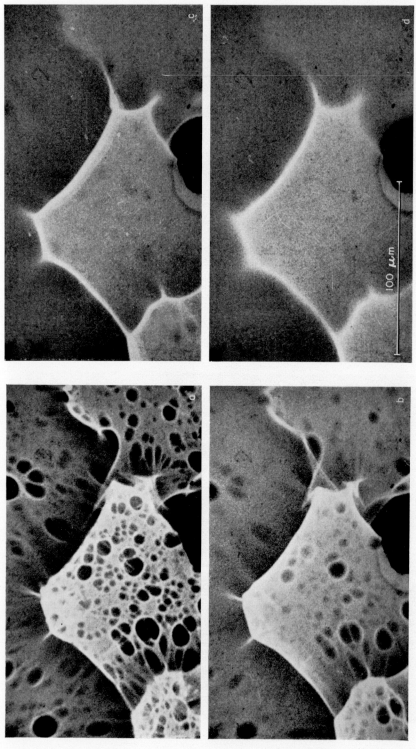

FIG. 41. See Fig. 40. Applied voltage for current flow in the layer: +50 V. Specimen potential: (a) −10 V, (b) −20 V, (c) −30 V, (d) −40 V.

f. In the mirror electron microscopy of *magnetic inhomogeneities*, the first investigations of which were carried out by Mayer (1957b) and by Spivak *et al.* (1955), the action of the normal component of the magnetic field at the specimen surface on the radial component of the electron velocity in the reflection plane, is responsible for the formation of contrast. Since, in the usual shadow image technique, the radial components of the electron velocity in the region of the intersection of the beam axis with the specimen surface are zero (or very small) and only increase with increasing distance from the axis, one can only image contrasts from magnetic surface irregularities at a distance from the centre of the image. The possibility thus exists of distinguishing between image contrasts which are related to geometric or electric surface inhomogeneities and those which are related to magnetic inhomogeneities. The main field of application of the mirror electron microscope for studying magnetic structures is the pictorial representation of magnetic domains, which was successfully achieved by this technique for the first time by Mayer (1959a,b).

A demonstration of the special types of contrast formation in the imaging magnetic structures is given in Fig. 42 which shows a special magnetic tape with recordings (rectangular pulses). At the centre of the image (which, in this case, does not lie on the screen centre and which is marked by a diffuse spot of electron impact) the magnetic contrast is less pronounced than towards the edge of the screen. Contrast reversal occurs between the upper and lower regions of the specimen resulting from a reversal in the radial component. To the right and left of the image there is no magnetic contrast since, because of the direction of the radial components of the electron velocity, the magnetic force acts in the same direction as the stripe structures in the image. The difference between the widths of the stripes ("recordings") on the left- and right-hand sides of the image, shows that there is also a shifting effect of the magnetic contrast lines which is more pronounced the stronger the magnetic fields. These shifts are in opposite directions on opposite sides of the centre of the image and give the impression of different widths of the recordings (rectangular pulses). The true width of the recordings can be obtained by calculating the average.

g. Based on the three main possibilities of the mirror electron microscope (detection of geometric, electric and magnetic surface irregularities), the device is being used more and more for special problems of solid state physics. Some examples of these applications will now be briefly reviewed. Attempts to image dislocations in germanium emerging at the surface by the detection of the space charge region around the dislocations were carried out by Igras (1962). With

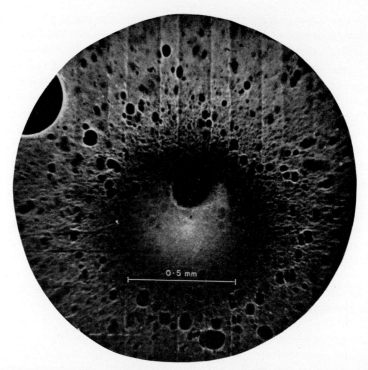

F<small>IG</small>. 42. Demonstration of the imaging of magnetic structures: magnetic tape with recordings on it (rectangular pulses).

certain restrictions, the local detection of electric microfields in semi-conducting surfaces can, as Igras has also shown, be used for example in the investigation of the segregation of impurity atoms, or for the detection of surface drifts. Using a suitable helium cooling stage in the specimen chamber Bostanjoglo and Siegel (1967) were able to image the local distribution of the magnetic flux of superconductors in the intermediate state. Mainly nobium specimens were used; these were kept at a temperature of $3 \cdot 5°K$ during the electron mirror observation. According to a brief report by Wang *et al.* (1966), the Abrikosov-Vortex structure of some superconducting materials of the type II have also been studied in the mirror electron microscope. In the future solid state circuits and electronic devices in thin film techniques will probably also be studied by mirror electron microscopy.

h. Summarizing, it can be said that, in spite of its restricted resolving power, the high sensitivity of the mirror electron microscope in the shadow image technique for electric or magnetic fields on the specimen surface (remembering that the geometry of the surface also influences this field distribution) gives a great deal of information about these properties of the specimen surface. The shifting effect of image structures discussed above gives, on the one hand, additional information about the strength of fields present, as far as it can be detected exactly, but on the other hand this effect causes an uncertainty in the localization of the image structure of these fields. In mirror electron microscopy with focused images, the resolving power is much better than in the shadow image technique and the shifting effects mentioned above are not present, so that there is no problem in image localization. However, this method does not have the high sensitivity to field inhomogeneities of the shadow image method. Magnetic structures, for instance, cannot be imaged by this technique because there are no radial components of the electron velocity. The possibility presents itself of using both techniques in combination so that in the same region of the specimen the localization of the inhomogeneities can be precisely detected in the focused image and the surface fields present are thereafter detected with sufficient differentiation using the shadow image.

In the field of design and development of apparatus, the tendency is towards reliable instruments equipped with a magnetic prism. Special developments such as, for example, the stroboscopic mirror electron microscope (Lukjanow and Spivak, 1966) or the scanning mirror microscope (Garrood and Nixon, 1968) are very useful. For the future it is desirable to develop devices with universal possibilities for specimen treatment. In order to take full advantage of the high sensitivity of the mirror electron microscope for surface potential distributions, it is necessary to aspire to the investigation of clean surfaces (free of adsorption layers). For this reason ultrahigh vacuum devices are becoming of increasing importance in the field of mirror electron microscopy.

D. Shadow Projection Mirror Electron Microscopy in "Straight" Systems

M. E. BARNETT

Imperial College, London, England

Work on mirror microscopy in England was begun in 1962–63 at the Engineering Department of Cambridge University. More recently, the subject has been studied in London University, at University College and Imperial College. Work has been largely confined to the shadow projection mode, using "straight" systems with magnetic lenses. Figure 43 shows the basic design of system (Szentesi and Barnett, to be published), and is in fact a schematic scale diagram of the instrument in use at present at the Electrical Engineering department of University College. This is a small, low magnification system,

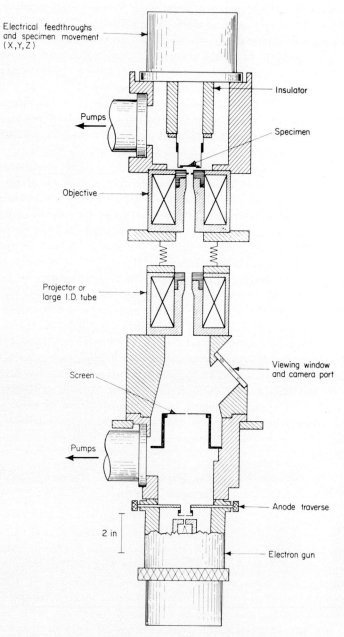

Electrical feedthroughs
and specimen movement
(X,Y,Z)

Insulator

Pumps

Specimen

Objective

Projector or
large I.D. tube

Viewing window
and camera port

Screen

Pumps

Anode traverse

2 in

Electron gun

FIG. 43. A simple "straight" design of a mirror electron microscope.

similar in layout to the column originally used in Cambridge in 1963. Other columns have been variations on this basic design. A condenser lens can be placed between gun and screen (Barnett and Nixon, 1964) to improve the image brightness at magnifications greater than a few hundred. This is necessary since a short focal length projector lens greatly reduces the intensity of the upward beam. In a more elaborate

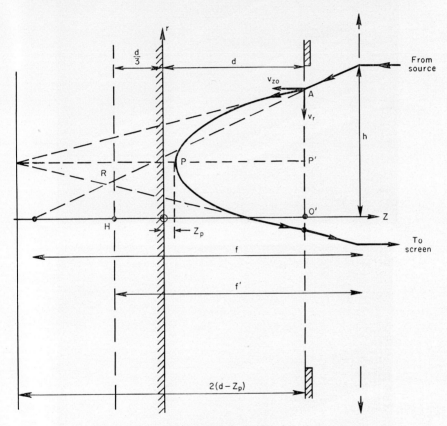

FIG. 44. Geometrical ray paths in the objective stage.

system (Barnett and Nixon, 1967a), designed for magnifications up to 2500 ×, a two lens projector system, as well as a condenser system was incorporated. In this column, a 45° mirror was located under the transparent viewing screen, so that photography normal to the screen was possible (cf. the micrographs of Fig. 45). In simpler systems the screen is photographed at an oblique angle, hence the foreshortening in most micrographs.

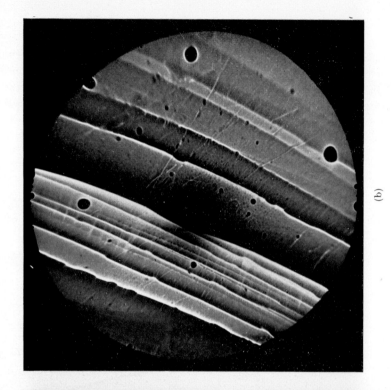

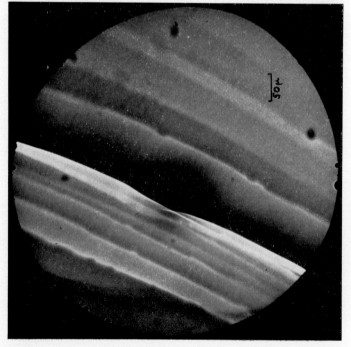

(b)

(a)

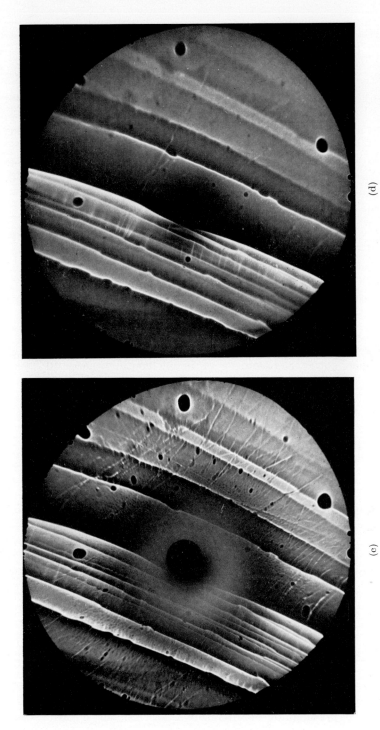

(c)

(d)

Fig. 45. Cleaved surface of rock salt crystal (Barnett and Nixon 1967b), with changing specimen bias: (a) $V_b = +10$ volts, (b) $V_b = -4.3$ volts, (c) $V_b = -0.2$ volts, (d) $V_b = +1.8$ volts.

a. *The Objective Stage.* Since the image contrast results from the modification, by the surface microfields, of the radial velocities of the electrons, it is useful to know what the radial velocities at the specimen would be in the absence of microfields, i.e. if the specimen were a non-magnetic plane equipotential surface. Fig. 44 shows geometrical ray paths in the objective stage assuming a plane specimen (at $z = 0$) and a uniform retarding field between the top surface of the objective lens (at $z = d$) and the specimen. It is easily shown (Barnett and Nixon, 1967b) that the radial velocity of the electrons at their point of reversal is given by

$$v_r = \frac{r}{6d} \left(\frac{2eV}{m}\right)^{\frac{1}{2}} \left\{1 - \frac{8d}{3(f - f')}\right\} \qquad (63)$$

where V is the original beam energy, f is the focal length of the objective lens and f^1 is the distance between the principal plane of the objective lens and the principal plane HR ($z = -d/3$) of the mirror. This expression shows that v_r changes sign at $f = f' + \dfrac{8d}{3}$ and again at $f = f'$ (the latter value corresponding to the formation of a focused probe at the specimen surface). Reversals of image contrast occur at these values of f.

The existence of radial velocity components in the beam can be shown to cause the beam to be reflected at a paraboloid of revolution rather than along a plane. If the specimen is at a small bias voltage V_b negative with respect to the cathode the locus of reversal points is

$$z = \frac{V_b}{E} + \frac{r^2}{36d} \left\{1 - \frac{8d}{3(f - f')}\right\}^2 \qquad (64)$$

E being the field strength at the surface.

The curvature of the reversal surface gives rise to a falling off of sensitivity with increasing distance from the axis. This effect is clearly evident in the series of micrographs in Figs 45a–d. The specimen is a cleaved rock-salt crystal rendered conducting by the evaporation of a thin layer of gold. The series shows the effect on image contrast of changing the specimen bias voltage; the more positive the specimen, the finer the topographical surface detail revealed. In the final micrograph, the specimen is positive with respect to the cathode, and the central region of the image consists of reflected electrons due to beam impact.

b. *Image Contrast.* An attractive feature of mirror microscopy is its ability to detect electrical potential variations directly. There is, however, the problem of distinguishing between the electrical and

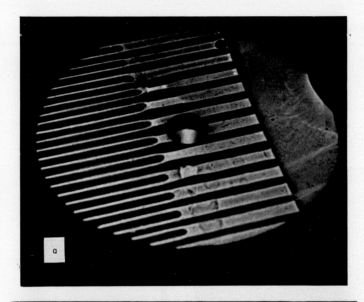

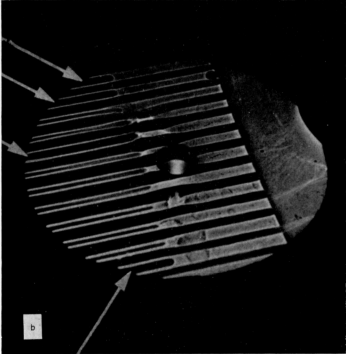

Fig. 46. Interdigital electrode structure (Szentesi and Barnett). (a) Topographical contrast, (b) mixed contrast: right-hand comb 0.75 volt negative with respect to left-hand comb.

topographical origins of image contrast. This can be investigated using specimens of known form (Szentesi and Barnett, 1969). Figure 46a shows a portion of an 800 Å thick copper electrode pattern consisting of two interlocking comb structures; the regions between the (dark) copper bars are of high resistance. The periodicity of the structure on the left-hand side is 100 μm and on the right it is 200 μm. In Fig. 46a there is no voltage between the two combs and hence the contrast is entirely of topographical origin. In Fig. 46b alternate bars (i.e. those

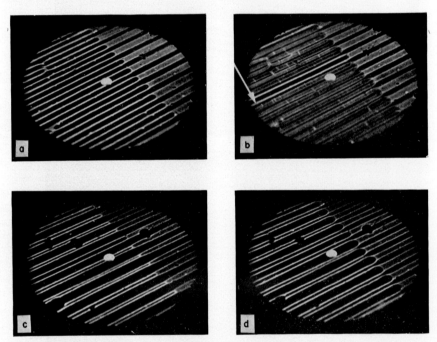

FIG. 47. 66 MHz stroboscopic images of an interdigital structure (Szentesi and Ash). (a) Topographical picture, (b) a.c. potential difference viewed from continuous beam, (c) and (d) a.c. potential variation viewed stroboscopically.

of the right-hand comb) are 0.75 volt negative with respect to the remaining bars. The bars marked by arrows are disconnected and are used as references. It is seen that the effect of the electrical potential difference is to cause the more negative bars to expand at the expense of the more positive.

Figure 46a shows the characteristic image distortion which is a serious defect of shadow projection mirror microscopy. On the left-hand side of the micrograph, the bars appear to be much wider than the intervening spaces, even though in the actual physical structure, the bars (35 μm wide), are considerably narrower than the spaces. The theory of this unavoidable distortion, which exaggerates the area of topographical hills or regions of negative potential, has been worked out for periodic structures by Barnett and England (1968).

The problem of obtaining test specimens in which the contrast is purely of electrical origin can be largely overcome by using evaporated photoconductors on to which a suitable light pattern is focused. A

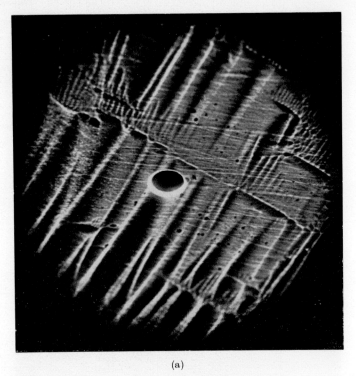

(a)

FIG. 48. Prism face of cobalt crystal (Barnett and Nixon, 1964). (a) Dagger domains along a grain boundary, (b) a cluster of dagger domains (detail), (c) 180° domain walls.

system of this type has been realized at Imperial College by Barnett, Bates and England (1969). If vitreous bismuth-selenium layers are used as specimens, the system operates as an infra-red image converter.

Alternating electrical potentials can be observed using the electron mirror, provided the beam is pulsed at a repetition frequency equal to the frequency of the surface potential variation. Figure 47 shows an example of this stroboscopic technique (Szentesi, 1970); the specimen is an interdigital structure similar to that of Fig. 46. Figure 47a is taken with a continuous beam, with no voltage between the combs. Figure 47b shows a continuous beam image of a 66 MHz variation, in which the amplitude of the alternating potential difference between the combs is about 1 volt. Figures 47c and d are stroboscopic images of this variation. Figure 47d shows the instant at which the left-hand comb is fully negative with respect to the right-hand comb, and Fig. 47c is half a cycle earlier. It is hoped to use this technique for the recording of ultrasonic holograms, using a piezoelectric crystal as the electron mirror.

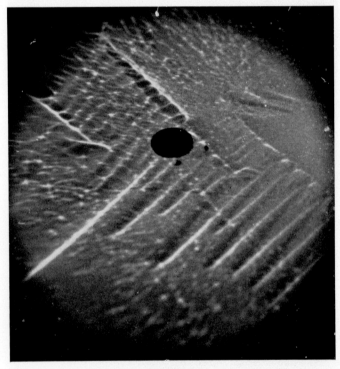

FIG. 48 (b)

Finally, some micrographs are shown which illustrate magnetic contrast in the mirror microscope. The specimen is the polished prism face of a cobalt crystal and the instrument an early three lens column (Barnett and Nixon, 1964). Magnetic contrast is largely due to the interaction between the radial velocity v_r of the electron and the normal component of the surface magnetic field. Equation (63) implies that the most favourable conditions for magnetic contrast will be when f is as close to f' as is consistent with obtaining a sufficiently large field of view at the image. Furthermore, the contrast must disappear in a central region. This is seen in Fig. 48a where contrast due to topographical scratches is continuous through the region above and just to the right of the screen hole, while the magnetic features have faded out. Figure 48a shows a row of dagger domains along a grain boundary; b shows details of a cluster of such dagger domains; c shows a system of parallel 180° domain walls.

FIG. 48 (c)

REFERENCES

Archard, G. D. and Mulvey, T. (1958). *J. Sci. Instrm.* **35**, 279.

Barnett, M. E. and England, L. (1968). *Optik* **27**, 341.

Barnett, M. E. and Nixon, W. C. (1964). *Proc. 3rd. Eur. Conf. Elec. Microsc. Prague* **1**, 37.

Barnett, M. E. and Nixon, W. C. (1967a). *J. Sci. Instrm.* **44**, 893.

Barnett, M. E. and Nixon, W. C. (1967b). *Optik* **26**, 310.

Barnett, M. E., Bates, C. W. and England, L. (1969). "Advances in Electronics and Electron Physics", Vol. 28, p. 545. Academic Press, New York.

Bartz G. and Weissenberg, G. (1957). *Naturwissenschaften* **44**, 299.

Bartz, G., Weissenberg, G. and Wiskott, D. (1954). *Proc. 4th Int. Conf. Elec. Microsc. London*, 1954, p. 395.

Bartz, G., Weissenberg, G. and Wiskott, D. (1952). *Phys. Verh.* **3**, 108.

Bartz, G., Weissenberg, G. and Wiskott, D. (1956). *Radex-Rundschau* **163**.

Bethge, H., Hellgardt, J. and Heydenreich, J. (1960). *Exp. Tech. Phys.* **8**, 49.

Bok, A. B. (1968). Thesis. Delft.

Bok, A. B., Kramer, J. and Le Poole, J. B. (1964). *3rd Eur. Conf. Elec. Microsc. Prague*, A9.

Bostanjoglo, O. and Siegel, G. (1967). *Cryogenics* **7**, 157.

Forst, G. and Wende, B. (1964). *Z. angew Phys.* **17**, 479.

Garrood, J. R. and Nixon, W. C. (1968). *Proc. 4 Eur. Conf. Elec. Microsc. Rome*, 1968, **1**, p. 95.

Glaser, W. (1952). *Grundlagen der Elektronenoptik* 321.

Haine, M. E. (1947). *J. Sci. Instrm.* **24**, 61.

Henneberg, W. and Recknagel, A. (1935). *Z. techn. Physik*, **16**, 621.

Heydenreich, J. (1966). *Proc. 6 Int. Congr. Elec. Microsc. Kyoto*, 1966, **1**, p. 233.

Heydenreich, J. (1970). *Proc. 7th. Int. Conf. Elec. Microsc. Grenoble*, **2**, p. 31.

Hopp, H. (1960). Thesis. Berlin.

Hottenroth, G. (1937). *Ann. Phys. Lpz.* **30**, 689.

Igras, E. (1961). *Bull. Acad. Polon. Sci. Ser. Phys.* **9**, 403.

Igras, E. (1962). *Proc. Int. Conf. Semicond. Exeter*, 1962, p. 832.

Igras, E. and Warminski, T. (1965). *Phys. Stat. Sol.* **9**, 79.

Igras, E. and Warminski, T. (1966). *Phys. Stat. Sol.* **13**, 169.

Igras, E. and Warminski, T. (1967). *Phys. Stat. Sol.* 20, K5.

Igras, E., Spivak, G. V. and Zheludev, I. S. (1959). *Sov. Phys. Cryst.* **4**, 111.

Ivanov, R. D. and Abalmazova, M. G. (1968). *Sov. Phys. Tech. Phys.* **12**, 982.

Knoll, M. Z. (1935). *Tech. Phys.* **16**, 767.

Kranz, J. and Bialas, H. (1961). *Optik* **18**, 178.

Le Poole, J. B. (1964a). *Proc. 3rd. Eur. Conf. Elec. Microsc. Prague* A6.

Le Poole, J. B. (1964b). *Proc. 3rd. Eur. Conf. Elec. Microsc. Prague* A8.

Le Poole, J. B. (1964c). *Discussions on the Conf. Non-conventional Elec. Microsc*, Cambridge.

Le Rütte, W. S. (1952). Thesis. Delft.

Lenz, F. and Krimmel, E. (1963). *Z. Phys.* **175**, 235.

Lukjanow, A. E. and Spivak, G. V. (1966). *Proc. 6th Int. Congr. Elec. Microsc. Kyoto*, 1966, p. 611.

Maffit, K. N. and Deeter, C. R. (1966). Symp. "The Mirror Electron Microscope for Semi-conductors", p. 9.

Mayer, L. (1955). *J. appl. Phys.* **26**, 1228.

Mayer, L. (1957a). *J. appl. Phys.* **28**, 259.

Mayer, L. (1957b). *J. appl. Phys.* **28**, 975.

Mayer, L. (1958). *J. appl. Phys.* **29**, 658.

Mayer, L. (1959a). *J. appl. Phys.* **30**, 1101.

Mayer, L. (1959b). *J. appl. Phys.* **30**, 252s.

Mayer, L. (1960). *J. appl. Phys.* **31**, 346.

Mayer, L., Rickett, R. and Stenemann, H. (1962). *Proc. 5th Int. Congr. Elec. Microsc.* Philadelphia, 1962, D-10.

Orthuber, R. (1948). *Z. Angew. Phys.* **1**, 79.

Recknagel, A. (1936). *Z. Phys.* **104**, 381.

Recknagel, A. and Hanneberg, W. (1935). *Techn. Phys.* **16**, 621.

Ruska, E. Z. (1933). *Phys.* **83,** 492.

Schwartze, W. (1965). *Naturwissenschaften* **52,** 448.

Schwartze, W. (1965–66). *Optik* **23,** 614.

Schwartze, W. (1966). *Exp. Tech. Phys.* **14,** 293.

Schwartze, W. (1967). *Optik* **25,** 260.

Spivak, G. V., Prilegaeva, I. N. and Azovcev, V. K. (1955). *Dokl. Akad. Nauk. U.S.S.R.* **105,** 706, 965.

Spivak, G. V., Igras, E., Pryamkova, I. A. and Zheludev, I. S. (1959a). *Sov. Phys. Cryst.* **4,** 115.

Spivak, G. V. (1959). *Kristallografiya* **4,** 123.

Spivak, G. V., Igras, E., Pryamkova, I. A. and Zheludev, I. S. (1959b). *Izv. Akad. Nauk. U.S.S.R. Ser. Fiz.* **23,** 729.

Spivak, G. V., Pryamkova, I. A., Fetisov, D. V., Kabanov, A. N., Lazareva, L. V. and Silina, A. I. (1961). *Izv. Akad. U.S.S.R. Ser. Fiz.* **25,** 683.

Spivak, G. V., Saparin, G. V. and Pereversev, N. A. (1962). *Izv. Akad. Nauk. U.S.S.R.* **26,** 13332.

Spivak, G. V., Ivanov, R. D., Pavluchenko, O. P., Sedov, N. M. and Shvets, V. F. (1963). *Izv. Akad. Nauk. U.S.S.R. Ser. Fiz.* **27,** 1210.

Szentesi, O. I. and Barnett, M. E. *J. Sci. Instrum.* **2,** 855.

Szentesi, O. I. (1970). Ph.d. Thesis, London.

von Borries, B. and Jansen, S. (1941). *Z. Verein. Dtsch. Ingen.* **85,** 207.

Wang, S. T., Challis, L. J. and Little, W. A. (1966). *Proc. 10 Int. Conf. Low Temp. Phys. Moscow,* 1966, p. 150.

Wiskott, D. (1956). *Optik* **13,** 481.

Energy Analysing and Energy Selecting Electron Microscopes

A. J. F. METHERELL

Cavendish Laboratory,
Cambridge, England

I. Introduction	. . . .	263
II. Energy Analysing and Selecting Devices	. . . . .	266
A. The cylindrical electrostatic analyser	. . . .	267
B. The cylindrical magnetic analyser	. . . .	287
C. Mirror prism device	. . . . .	302
III. Energy Analysing Electron Microscopes	. . . . .	318
A. Instrumentation	. . . . . .	318
B. Applications	. . . .	327
IV. Energy Selecting Electron Microscopes	. . . . .	340
A. Instrumentation	. . . . . .	340
B. Applications	. . . .	345
Acknowledgements	. . . .	358
References	. . . .	359

I. Introduction

CONTRAST is observed in an electron microscope image as a result of the fact that electrons in the incident beam are scattered by the specimen under examination. The scattering processes can be conveniently divided into two classes; elastic scattering in which the incident electron suffers no energy change, and inelastic scattering in which energy is lost to the specimen. For crystalline materials there is only one mechanism responsible for elastic scattering and that is diffraction or Bragg scattering of the incident electrons. The types of inelastic scattering which can occur are, however, numerous. It is possible for the fast electrons to lose energy by the excitation of a crystal quasi-particle such as a phonon, an exciton, a plasmon, etc. or by atomic ionization, by the scattering of atomic electrons from the valence band into the conduction band, by the scattering of atomic electrons from a given state in the conduction band to an empty state in the same band, etc. All inelastic scattering processes excite the specimen from its ground state energy level into higher energy states, and in principle it is possible to study the complicated energy level structure of a solid by measuring the energy loss spectrum of the incident fast electrons. Suppose that a specimen is being examined whose composition varies with position across the field of view. The composition changes are expected to be accompanied by changes in the energy level structure of the solid and hence electrons scattered from different regions of the specimen will have loss spectra characteristic of those regions. Measurement of the loss spectra of electrons arriving at different points in an electron microscope image should therefore provide a powerful method of microanalysis.

The conventional electron micrograph does not distinguish between electrons of different energies, and is therefore the electron optical analogue of a black and white photograph. It is pertinent to consider the possibility of constructing the electron optical analogue of a colour photograph, because a micrograph which is sensitive to the electron energy utilizes all the physical information contained in the final image. The energy range of the scattered electrons is obviously an important factor which must be considered. The use of a "stop" or objective aperture in an electron microscope restricts the angles of scatter of the inelastically scattered electrons, which contribute directly to the image intensity, to values of the order of a Bragg angle (typically $\sim 10^{-2}$ rad. for 100 keV electrons). This restriction reduces the number of inelastic scattering processes which have to be considered, since the majority of the observed energy loss electrons will be produced by loss mechanisms which possess angular distributions sharply peaked in the direction of the beam chosen to form the image. Furthermore only those loss processes with mean free paths of the order of the specimen thickness need be considered. Mean free paths for atomic ionization for example are typically of the order of millimetres, whereas mean free paths for plasmon excitation are typically of the order of thousands of Ångstrom units. Although both processes produce angular distributions of the scattered electrons sharply peaked in the forward direction, the proportion of electrons which suffer energy losses through atomic ionization are negligible in comparison with those which lose energy by plasmon excitation. The most predominant energy loss processes which contribute directly to the inelastic scattering observed in an electron microscope image are (a) plasmon excitations, possibly modified strongly by interband transitions, (b) single electron interactions corresponding to interband transitions and (c) phonon excitations. A typical example of the loss spectrum of electrons contributing to the image of a specimen of aluminium is shown in Fig. 1. The broken line shows the loss profile due to single electron scattering, superimposed on which are the well defined plasmon losses P_1, P_2, etc. Clearly most of

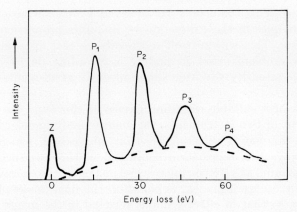

FIG. 1. A typical example of the energy loss spectrum of electrons contributing to an electron microscope image of aluminium of thickness ~ 2000 Å and with an incident beam potential of 80 kV.

the energy losses suffered by the incident electrons lie within several hundred electron volts of the incident beam potential, which is typically of the order of hundreds of kilovolts, and the fractional change in the electron energy is therefore very small.

The difficulties involved in constructing the electron optical analogue of a colour photograph stem mainly from the relatively small differences in the energies of the electrons forming the image of the specimen. The idealized energy analysing electron microscope would record simultaneously the loss spectra from all points x, y (Fig. 2) lying within the

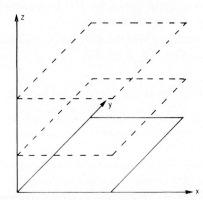

FIG. 2. In an idealized electron optical analogue of a colour photograph, electrons of different energies contributing to the image formed in the field of view xy would be displaced in the z direction and the intensity variation in the volume xyz would then be recorded.

field of view. This could only be achieved by displacing electrons of different energies in the z direction by amounts proportional to their energy losses. This within itself represents no real instrumental difficulty. The problem arises in finding a recording device which can measure the intensity variation simultaneously at all points within the volume xyz.

The lack of a suitable recording device of this type has led to the development of two instruments, the energy analysing electron microscope and the energy selecting electron microscope, which utilize in different ways the information contained in the loss spectra of the electrons forming an electron microscope image. In the energy analysing electron microscope a fine slit is placed in the final image plane of the microscope, so that in effect a line is selected in the image for energy analysis. The slit is made the entrance aperture of an electron spectrometer and electrons of various energies arriving at the image along this line are displaced in a direction perpendicular to the long dimension of the slit by amounts determined by the dispersion of the analyser. A simple two dimensional recording device, such as a photographic plate, can then be used to record simultaneously the spectra from points lying along any line previously selected in the image. In the energy selecting microscope the electrons forming the image are initially dispersed into their various energy loss components by means of an analyser, and by the use of an aperture and subsequent focusing or scanning system, one of the loss components is used to reform the image of the specimen. This image is composed of electrons with a mean energy corresponding to the selected loss component and a spectral width, usually of the order of one or two electron volts, given by the product of the dispersion of the analyser and the width of the selecting aperture. The energy selecting microscope therefore produces the electron optical analogue of a monochrome photograph.

II. Energy Analysing and Selecting Devices

Since Wien (1897) first recognized that an energy analyser could be designed by employing crossed electric and magnetic fields and Leithäuser (1904) constructed the first practical electron spectrometer, based on the principle of dispersion by a magnetic field, a large number and variety of electron energy analysers have been reported in the literature (for a review see Klemperer, 1965). It is not the purpose of this article to present a comprehensive review of the instrumentation and electron optical properties of energy analysers, but rather to consider only those analysers currently employed in energy analysing and energy selecting microscopes. The main restrictions on the choice of analyser used are that the energy resolution should be of the order of 1 eV or less, and that the analyser acceptance aperture should conform to the requirements of the imaging system of the electron microscope. The devices which meet these requirements and which have been incorporated in existing energy analysing and energy selecting electron microscopes are:

(a) the cylindrical electrostic analyser;
(b) the cylindrical magnetic analyser;
(c) the electrostatic mirror-magnetic prism analyser.

Of these, the first has been successfully employed in both energy analysing and energy selecting microscopes, operating at beam potentials in the range 40 to 100 kV. For high voltage electron microscopes, however, the electrostatic system presents considerable high voltage insulation problems and for this reason the cylindrical magnetic analyser is a more suitable device to use. The third system has been used in an energy selecting electron microscope operating in the beam potential range 40 kV to 100 kV. A description of the electron optical properties of these analysers now follows.

A. *The Cylindrical Electrostatic Analyser*

Möllenstedt (1949, 1952) first demonstrated that an electrostatic saddle-field lens of two dimensional symmetry can be used as a high resolution electron velocity analyser. His work has stimulated a number of theoretical papers on the electron optical properties of this system (Laudet, 1953; Lenz, 1953; Archard, 1954; Septier, 1954; Lippert, 1955; Waters, 1956; Dietrich, 1958; Metherell and Whelan, 1965, 1966; Metherell, 1967a; Metherell and Cook, 1970).

The energy resolution ΔE_r of a dispersive system, where ΔE_r is the energy difference between two just resolvable losses, is in general a function of the beam potential E of the incident electrons. For this reason it is important, when comparing the resolving powers of different devices, to compare their *specific* resolving powers, defined as $E/\Delta E_r$.

Metherell and Cook (1970) have shown that under optimum operating conditions a specific energy resolution $(\Delta E/E_r)$ of the order 10^{-7} can theoretically be achieved with the Möllenstedt system. It is doubtful that any other analyser approaches this value for the specific resolution. Furthermore if a cylindrically symmetric system is employed, as opposed to the rotationally symmetric lens originally used by Möllenstedt (1949), the analyser entrance aperture can be made in the form of

a fine slit. For these reasons the cylindrical electrostatic lens is emi-
nently suitable for performing energy analyses in the electron microscope.

A cross-sectional view of the cylindrical lens is shown in Fig. 3. A fine
slit S, of width of the order of a few microns and length of the order of a
centimetre, apertures the incoming electron beam. The slit S is aligned
with its long dimension parallel to the axes of the cylindrical electrodes

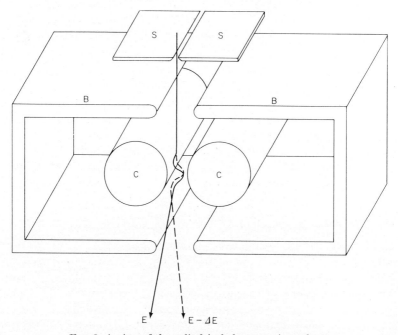

FIG. 3. A view of the cylindrical electrostatic analyser.

C. The apertured electrons pass into a box-shaped electrode system B
at anode (usually earth) potential and then through regions of high
chromatic aberration near the two cylindrical electrodes C which are
biased at cathode potential. The potential distribution in the region of
the electrodes C is shown schematically in Fig. 4 and the point S in this
diagram is the saddle point of the system of equipotential surfaces.
Provided the slit S (Fig. 3) is placed in a suitable off-axis position (see
Fig. 5) electrons of different energies are dispersed, as shown schematically in Figs. 3 and 5, and the energy loss spectrum can then be recorded by placing a photographic plate, or some other suitable recording
device, on the exit side of the lens.

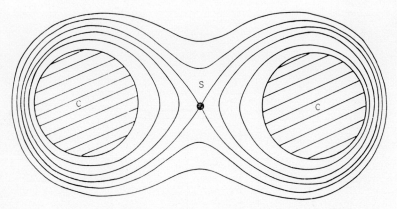

FIG. 4. A cross-section of the cylindrical electrodes showing the potential surfaces in the region of the saddle point S.

A study of the electron optical properties of this system requires a knowledge of the potential distribution in the region between the electrodes C and B (Fig. 3). It is not possible to obtain an analytical expression for the potential distribution in a practical system with an electrode geometry such as that shown in Fig. 3. Various authors have derived expressions for the potential function in idealized electrode systems which approximate to those of practical interest. A particularly simple system, called the line-charge model of the analyser, has been devised by Metherell and Whelan (1966). The model consists of two parallel and infinitely long line charges of charge density q per unit length (Fig. 6) placed equidistant from two earthed parallel plates AB and $A'B'$ both of infinite extent. The potential function $V(x, y)$ for this system is given by

$$V(x, y) = - q\left\{\ln\left[\frac{\cosh\ (\pi(x - D)/d) - \sin\ (\pi y/d)}{\cosh\ (\pi(x - D)/d) + \sin\ (\pi y/d)}\right]\right.$$
$$\left. + \ln\left[\frac{\cosh\ (\pi(x + D)/d) - \sin\ (\pi y/d)}{\cosh\ (\pi(x + D)/d) + \sin\ (\pi y/d)}\right]\right\}$$

Suitable equipotential surfaces can be chosen to simulate the cylindrical electrodes of the analyser shown in Fig. 3. Although the line-charge model does not possess equipotential surfaces which are exactly circular in cross-section and also neglects field penetration through the entrance and exit slots of the earthed electrode B of Fig. 3, Metherell and Whelan (1966) and Metherell (1967a) have shown that it represents to a good approximation the practical electrode system of Fig. 3.

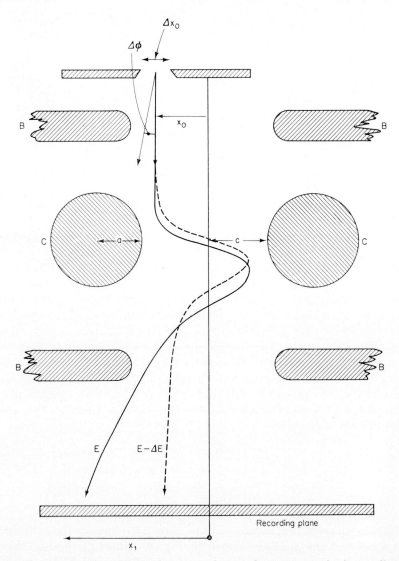

FIG. 5. A cross-section of the analyser. x_0 and x_1 are the entrance and exit co-ordinates of an electron beam passing through an entrance slit of width Δx_0.

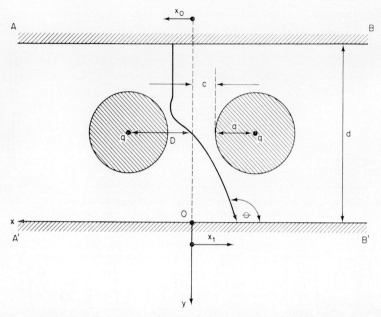

FIG. 6. The line charge model of the analyser. AB and $A'B'$ are parallel earthed plates of infinite extent in the x and y directions. Line charges of charge density q per unit length intersect the plane of the diagram at the points $x = \pm D$ and $y = -0.5d$.

The potential function of the line charge model can be used to calculate the cardinal points, aberration coefficients, etc., of the system. Our interest however lies in the dispersive properties of the lens, rather than its imaging properties and the behaviour of the system as an electron velocity analyser can be best understood by examining the dispersion and resolution in terms of a plot of the entrance and exit co-ordinates x_0 and x_1 (Fig. 5) of a monoenergetic electron beam passing through the lens. An example of the x_0, x_1 curve obtained from the line-charge model is shown in Fig. 7. This curve was obtained by numerical integration of the equations of motion of an electron which was assumed to enter the system parallel to the y-axis (Fig. 6) with a beam potential of 100 kV. The units of x_0 and x_1 refer to $d = 1.0$, x_1 being measured in the plane $y = 0$, and the value of c/a assumed was 0.07. The curve has been drawn schematically after the second maximum (point 3 in Fig. 7) due to the rapidity with which x_1 varies after this point. The trajectories of the electrons arriving at points x_1 corresponding to the maxima and minima 1, 2, 3 and 4 of the x_0, x_1 curve are shown in Figs. 8a, b, c and d. The extrema 1, 2, 3, etc. (Fig. 7) correspond to caustic envelopes of the system and the trajectories of Fig. 8 are tangential to these caustics

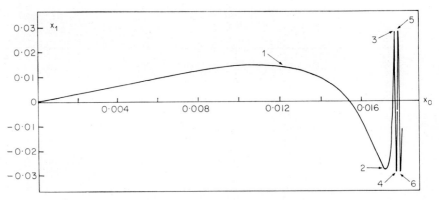

FIG. 7. The x_0, x_1 curve calculated from the line charge model with $c/a = 0 \cdot 07$ and $d = 1 \cdot 0$. The curve has been drawn schematically after the extremum 3 due to the rapidity with which x_1 varies with x_0 after this point.

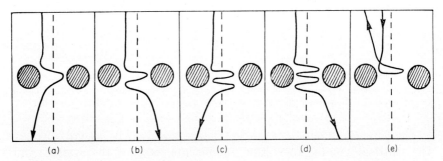

FIG. 8. Trajectories of electrons arriving at points x_1 corresponding to the extrema 1, 2, 3 and 4 of Fig. 7 are shown schematically in (a), (b), (c) and (d) respectively. The trajectory of an electron reflected by the system is shown in (e).

at $y = 0$. In the absence of space charge effects, stray fields etc., an infinite number of extrema occur in the x_0, x_1 curve. The extremum $n = \infty$ corresponds to the case in which the electron enters the region between the inner electrodes and passes through the saddle point in a direction perpendicular to the axes of the two cylindrical electrodes. In this situation the electron becomes trapped by the lens and oscillates backwards and forwards through the saddle point. If x_0 is increased beyond this point the system behaves as a mirror and the trajectory of an electron reflected back towards the entrance side of the analyser shown in Fig. 8e.

The most important design factor controlling the behaviour of the cylindrical lens is the value of the parameter c/a (Figs. 5 and 6). The general form of the x_0, x_1 curve depends on c/a and it is found that four different types of x_0, x_1 curve are produced if c/a is varied. It is therefore convenient to divide the analyser into four classes, depending on the sign of x_1 and $\cos \theta$ (Fig. 6). It is assumed that the electron is incident at the plane $y = -1.0$ at the positive value of x_0 corresponding to the first maximum or minimum value of x_1, measured in the plane $y = 0$. θ is the angle which the trajectory tangential to the first caustic envelope at $y = 0$ makes with the plane $y = 0$. The definitions of the four lens classes are given in Table I, together with the approximate range of

TABLE I

Definitions of the Four Lens Classes

Class	Sign of x_1	Sign of $\cos \theta$	Approx. range of c/a
I	$-$	$-$	$\infty \rightarrow 0.32$
II	$-$	$+$	$0.32 \rightarrow 0.19$
III	$+$	$+$	$0.19 \rightarrow 0.049$
IV	$+$	$-$	$0.049 \rightarrow 0.034$

c/a values for each class. The range of c/a from 0 to ~ 0.03 has been omitted from this table as it is convenient to postpone the discussion about this range until values of c/a > 0.03 have been considered. The electron trajectories tangential to the first four pairs of caustic envelopes at $y = 0$ in the different lens classes are shown in Fig. 9. These trajectories have been drawn schematically for the sake of clarity. Examples of the x_0, x_1 curves for the four classes of lens are shown in Fig. 10 and for each lens class three sets of curves are given corresponding to different positions of the plane in which x_1 is measured. It is worth noting that for lens classes II and IV the effect of projection below the plane $y = 0$ causes the first extremum of the x_0, x_1 curve to disappear. If the maximum and minimum values of x_1 in the x_0, x_1 curve are plotted as a function of y, the curves so obtained will give the caustic envelopes of the system, and these are shown schematically in Figs. 11(a) and (b) for the four different lens classes. It is clear from Fig. 11(b) that the disappearance of the first extremum in the x_0, x_1 curves for lens classes II and IV corresponds to a disappearance of the first pair of caustic envelopes with projection in these classes.

If the gap between the two cylindrical electrodes of the analyser is gradually reduced Table I shows that the system passes from class I to class II to class III and then to class IV. If the value of c/a is further reduced the lens returns to class I and then passes again to classes II,

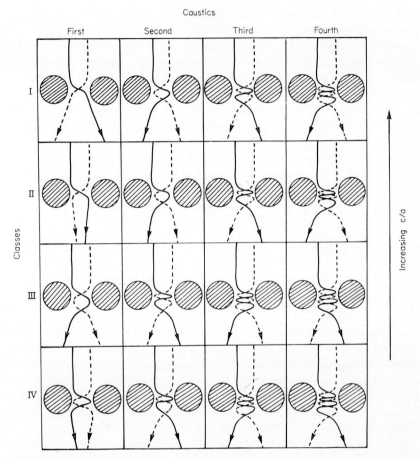

FIG. 9. Electron trajectories, drawn schematically for the sake of clarity, tangential to the first four pairs of caustic envelopes at $y = 0$ (Fig. 6) in the four different lens classes.

III and IV (Fig. 12). Still further reductions in c/a cause the system to cycle from one class to another in the order just given. The range of c/a values for which a particular lens class is operative decreases rapidly after the first few cycles as can be seen from Fig. 12. The reason for this cycling is indicated in Fig. 13 which shows the trajectories tangential to the first caustic envelope at $y = 0$ when the cycle of classes is repeated for the first time. It is apparent that the oscillations of the electrons in the region of the saddle point responsible for the formation of higher order caustics is also responsible for the cycling of the lens from class to class as c/a is decreased. As there is no theoretical limit to

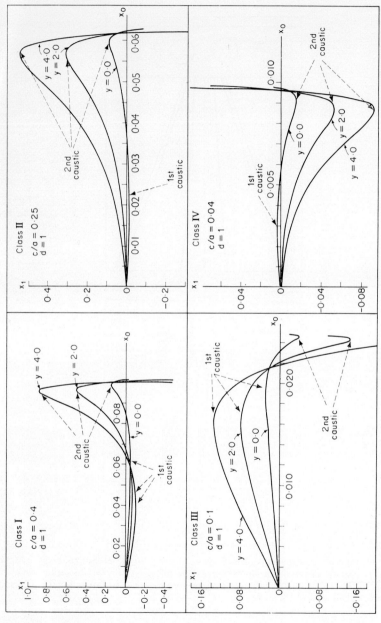

Fig. 10. The x_0, x_1 curves for the different lens classes. The projection distance y, at which x_1 is measured, is indicated for each curve.

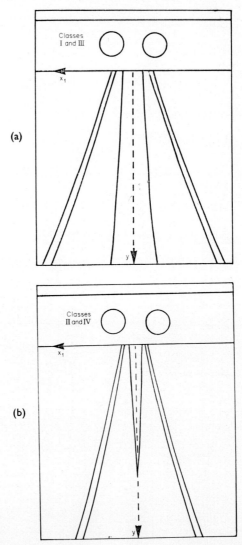

FIG. 11. Behaviour of the caustic envelopes produced by the cylindrical lens. In (a) the innermost caustic envelope produced by a lens of class I or III does not disappear whereas in (b) the innermost caustic envelope disappears. This latter behaviour occurs with lens classes II and IV only.

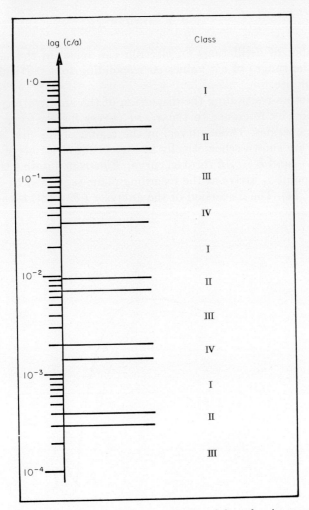

FIG. 12. The range of c/a values over which each lens class is operative.

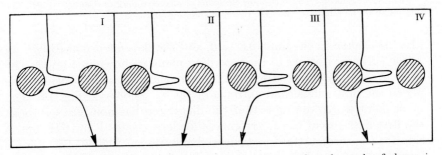

FIG. 13. Caustic trajectories in the four classes of lens when the cycle of classes is repeated for the first time.

the number of oscillations an electron can make in the saddle point region, the ranges of c/a values corresponding to a particular class of lens is infinite.

The factors controlling the dispersion of the analyser can be readily understood by reference to the x_0, x_1 curves for electrons of slightly different energies. These curves, in the region of the first and second extrema are shown schematically in Fig. 14 for electrons of energies E (full curve) and $E - \Delta E$ (broken curve). Electrons entering the analyser at some point x_0 arrive at the recording plane separated by an amount Δx_1 (Fig. 14). The dispersion of the analyser ($|\partial x_1/\partial E|$) is the quantity

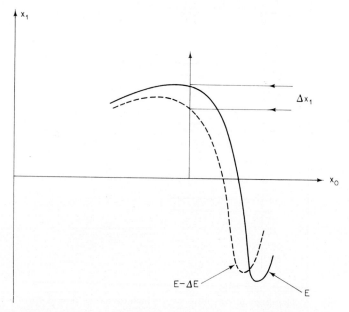

FIG. 14. Schematic x_0, x_1 curves for electrons of discrete energies E and $E - \Delta E$. The dispersion is the quantity $|\Delta x_1/\Delta E|$ taken in the limit $\Delta E \to 0$.

$|\Delta x_1/\Delta E|$ taken in the limit $\Delta E \to 0$, and the dependence of $\partial x_1/\partial E$ on x_0 for various positions of the recording plane is shown in Fig. 15. The vertical arrows numbered 1 and 2 indicate the values of x_0 corresponding to the first and second extrema of the x_0, x_1 curves (Fig. 10). The main point worth noting is that the dispersion has non-zero values at the first extrema of the x_0, x_1 curves for lens classes I and III, but is zero for the second, third and successive extrema. This is also true for lens classes II and IV provided that the recording plane intersects the two innermost caustic envelopes produced by these systems. If, how-

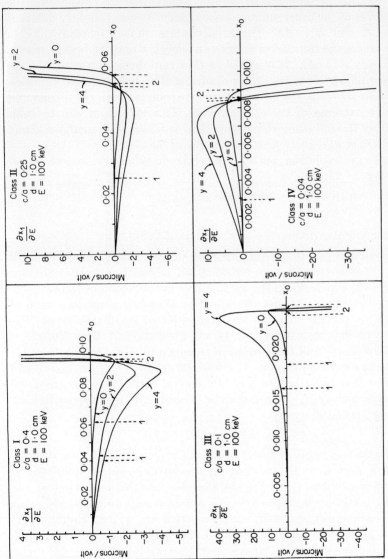

FIG. 15. The behaviour of the dispersion $(\partial x_1/\partial E)$ as a function of x_0 for different projection distances y. The curves in this figure were calculated from the line-charge model by assuming $d = 1$ cm (Fig. 6). The vertical arrows numbered 1 and 2 indicate the first and second caustic positions.

ever, the recording plane lies below the points of intersection of the two innermost caustic envelopes of lens classes II and IV (Fig. 11) the dispersion at all extrema appearing on the recording plane is zero. The reason for this behaviour can best be understood by considering the x_0, x_1 curves of the different lens classes for electrons of slightly different energies E and $E - \Delta E$. These curves are shown schematically in Fig. 16, where the full curves refer to an energy E and the broken curves to an energy $E - \Delta E$. It is apparent that zero dispersion occurs at the cross-over points of the curves for electrons of slightly different energies. In the limit $\Delta E \to 0$, the cross-over points coincide with the second and successive extrema of the x_0, x_1 curves. This behaviour can be summarized by the following statement: *the dispersion of an analyser of any class is zero at all but the first caustic edge of the system.*

The energy resolution, for a given electrode geometry, is determined mainly by

(a) the width of the entrance slit Δx_0 (Fig. 5)

and

(b) the angular divergence of the beam $\Delta \phi$ (Fig. 5).

Consider first a parallel beam ($\Delta \phi = 0$) of monoenergetic electrons of energy E passing through an entrance slit of width Δx_0. It is obvious from Fig. 17, which shows the x_0, x_1 curve in the region of the first two extrema, that the electrons arrive at the recording plane in a strip of width $\delta x_1 = (\partial x_1/\partial x_0)\,\Delta x_0$ to the first order of small quantities. The variation of $\partial x_1/\partial x_0$ with x_0 can be obtained directly from the x_0, x_1 curves of Fig. 10, and this variation is shown in Fig. 18 for a class III lens. At the extrema of the x_0, x_1 curve the quantity $\partial x_1/\partial x_0$ is zero and δx_1 to the first order is also zero. In general a parallel monoenergetic beam passing through an entrance slit of width Δx_0 will appear to have an energy spread given by

$$(\Delta E)_s = (\partial E/\partial x_1)\,(\partial x_1/\partial x_0)\,\Delta x_0 \qquad (1)$$

Now consider a parallel beam containing electrons of two discrete energies E and $E - \Delta E$ incident on an entrance slit of infinitesimal width. It is obvious from Fig. 14 that the electrons arriving at the recording plane are separated by an amount Δx_1, and that electrons with an energy difference ΔE are resolved if $\delta x_1 \leqslant \Delta x_1$. Equation (1) therefore gives the energy resolution limited by slit width.

A similar consideration shows that the resolution limited by the angular divergence $\Delta \phi$ of the beam is given by

$$(\Delta E)_\phi = (\partial E/\partial x_1)\,(\partial x_1/\partial \phi)\,\Delta \phi \qquad (2)$$

Metherell and Whelan (1966) have shown that $(\Delta E)_\phi$ and $(\Delta E)_s$ are of

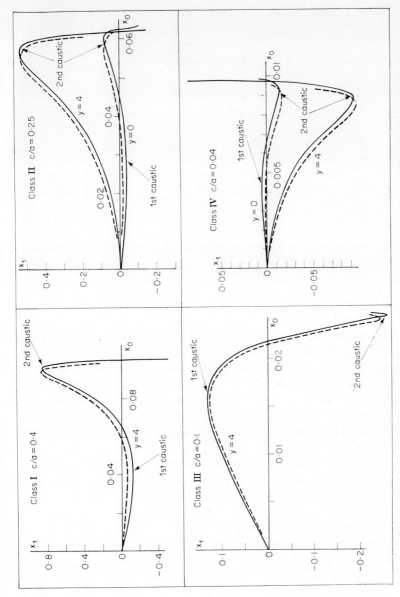

Fig. 16. The x_0, x_1 curves for the four classes of lens for electrons of energy E and $E - \Delta E$. The full curves were calculated from the line-charge model with $d = 1$ (Fig. 6) and the broken curves have been drawn schematically for the sake of clarity.

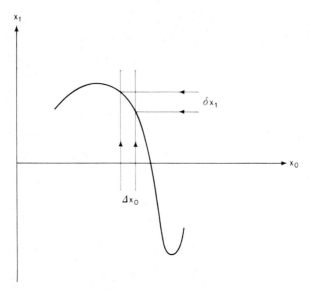

Fig. 17. The x_0, x_1 curve (schematic) showing the image width δx_1 in the recording plane of an entrance slit of width Δx_0.

the same order of magnitude when $\Delta\phi \sim 10^{-4}$ rad. and $\Delta x_0 \sim 1$ μm. In practice slit widths ~ 5 μm are employed with this type of instrument, and when used as an analyser in the energy analysing electron microscope, the entrance aperture of the Möllenstedt system lies in the final image plane of the electron microscope. The use of an objective aperture to form the image of a specimen means that only those electrons scattered within the solid angle subtended by the objective aperture at a point on the bottom surface of the specimen contribute to the intensity arriving at the corresponding point in the image plane. The objective aperture semi-angle θ usually employed in 100 kV microscopes is in the range 10^{-3} to 10^{-2} rad. The electrons contributing to the intensity at a given point in the image plane therefore lie within a cone of semi-angle $\Delta\phi = \theta/M$, where M is the magnification. The value of M is typically $20,000\times$, which gives values of $\Delta\phi$ in the range 5×10^{-8} to 5×10^{-7} rad. The resolution limited by angular divergence is therefore negligible in comparison with the resolution limited by slit width when the analyser is used to examine the energy spectra of electrons forming the image of a specimen in the electron microscope. In the discussion that follows the resolution $(\Delta E)_\phi$ given by equation (2) will be neglected and the total resolution will be assumed to be given by $(\Delta E)_s$.

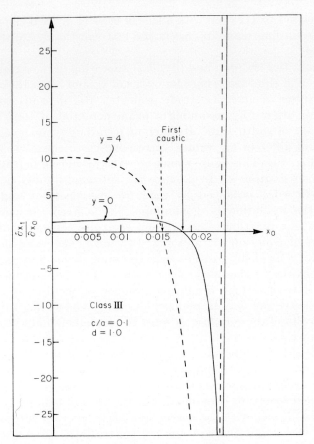

FIG. 18. The gradient $\partial x_1/\partial x_0$ as a function of x_0 for a class III lens. The curve was calculated from the line-charge model with $d = 1$ (Fig. 6).

Equation (1) shows that high resolution $(E/\Delta E)_r$ is obtained when the dispersion $(\partial x_1/\partial E)$ is large and $(\partial x_1/\partial x_0)$ is small. The operating position of the analyser entrance slit is that value of x_0 corresponding to the first extremum in the x_0, x_1 curve. With a class II or IV lens, the recording plane does not in general intersect the innermost caustic envelope and since the dispersion is zero at the second and successive caustic edges, a lens of class II or IV is unsuitable for use as a velocity analyser. The position of the recording plane can of course be adjusted so that intersection with the first caustic edge produced by the system occurs. It is, however, desirable to have high dispersion as well as good resolution, and Fig. 15 shows that higher dispersion can be obtained with a lens of class I or III than with a lens of class II or IV if the condition that the recording plane intersects the first caustic edge of the system is satisfied.

The quantities $(\partial E/\partial x_1)$ and $(\partial x_1/\partial x_0)$ appearing in the expression for $(\varDelta E)_s$ given in equation (1) are functions of x_0, and hence the resolution of an analyser of given electrode geometry also depends on x_0. The energy resolution $\varDelta E_r$, assuming a beam potential of 100 kV and an entrance slit of width 1 μm, at different recording plane levels for two analysers of different c/a values, corresponding to classes I and III, is given in Fig. 19a. This figure shows how the resolution depends on the entrance slit position x_0. In practice the slit is positioned close to the first caustic edge produced by the zero loss electrons and for this position the resolution is a minimum. For a fixed slit position the resolution as a function of energy loss δE is not constant however, as inspection of equation (1) and Fig. 16 shows. The variation of $\varDelta E_r$ with δE for a fixed slit position, corresponding to the first caustic edge produced by the zero loss electrons, is shown in Fig. 19b, and it is apparent that the extremely high resolution obtained at the first extremum of the x_0, x_1 curve can only be realized for energy losses lying within a few electron volts of the zero loss. It must be remembered however that at the present time the resolution for a class III analyser is limited by the energy spread of the beam incident on the specimen, this being about 1 eV and not by the electron optical properties of the analyser system.

In most energy loss experiments calibration of the energy loss axis of the recorded spectrum is required. The problem of calibration is eased if x_1 varies linearly with the energy loss δE for a given slit position x_0. This however is not the case for the electrostatic system, as Fig. 20 shows. The calibration curves of this figure, which refer to analysers of classes I and III, were calculated by assuming the slit to be positioned at the first caustic edge produced by each system. In practice the calibration curve is obtained by altering the potential of the inner electrodes of the analyser by known increments. Each voltage step simulates an energy loss and by recording these simulated losses successively on the same photographic plate the dependence of x_1 on δE can be determined experimentally.

The final factor influencing the choice of analyser design geometry is the range of energy losses that are required to be examined. The energy loss δE is a multi-valued function of x_1 (Fig. 20) and it is therefore important that all the prominent losses in the spectrum should lie within the range $0 - \delta E_R$ where δE_R is the energy loss at which the first extremum in the δE, x_1 curve occurs. The range of energy losses of interest here (section I) is typically not more than about 100 eV and

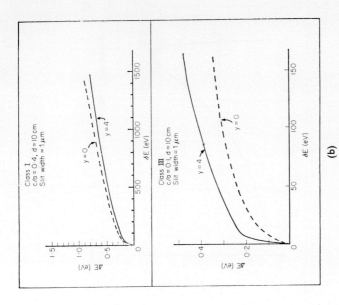

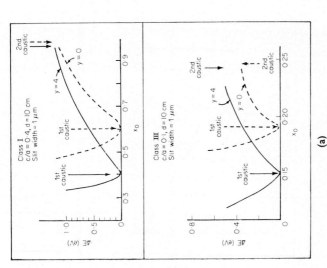

FIG. 19. The resolution ΔE_r plotted as a function (a) of slit position x_0 and (b) energy loss δE for projection distances $y = 0$ and 4. These curves were calculated from the line-charge model by assuming $d = 10$ cm and an incident beam potential of 100 kV.

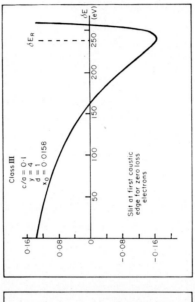

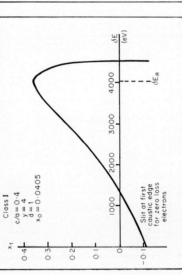

Fig. 20. Calibration curves for class I and III analyser calculated from the line-charge model. The incident beam potential is assumed to be 100 kV and the unit of x_1 refers to $d = 1\cdot0$ (Fig. 6).

Fig. 20 shows that a class III analyser operating at a beam potential of 100 kV can accommodate a range of this order of magnitude. A class III lens is therefore preferable as an analyser since the resolution as a function of energy loss (Fig. 19) is better than that obtained with a class I lens.

B. *The Cylindrical Magnetic Analyser*

At high accelerating voltages the problems of electrical insulation make the use of an electrostatic analyser undesirable. It is obviously preferable to seek a purely magnetic system and so avoid these problems altogether. A suitable magnetic device, of compact form and high energy resolution, is a cylindrical system first described by Ichinokawa (1965), (see also Ichinokawa and Kamiya, 1966; Ichinokawa, 1968; Considine and Smith, 1968; and Considine, 1970). This analyser

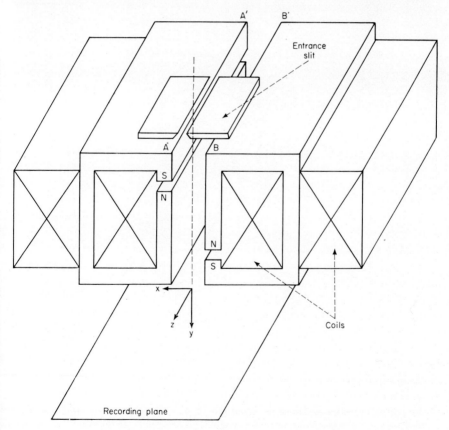

FIG. 21. A view of the cylindrical magnetic analyser.

is the magnetic analogue of the electrostatic system discussed in Section IIA and consists of two pairs of pole pieces displaced relative to each other in the manner indicated in Fig. 21. The coils are energized so that the magnetic polarities across the pole-piece gaps are in opposite senses. The reason for this reverse polarity and also the asymmetric arrangement of the gaps between the pole pieces is considered below. An entrance slit placed above the pole pieces apertures the incoming electron beam. The field distribution in the space between the pole pieces is shown schematically in Fig. 22 together with the dimensions of the system incorporated with the Cavendish 750 kV electron micro-scope (Considine and Smith, 1968, 1971; Considine, 1970). The field distribution deflects the incoming electron beam in both the xy and yz planes (Fig. 21) and if the entrance slit is placed in a suitable off-axis position, electrons of different energies are dispersed in the highly

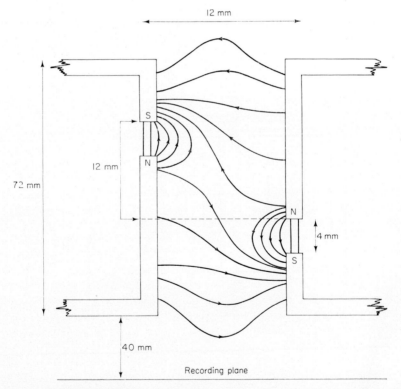

FIG. 22. The field distribution in the region of the pole pieces of the analyser. The dimensions given in this figure are those assumed by Considine (1970) in calculating the curves of Figs. 35, 36 and 38 of this article.

inhomogeneous fields near the pole pieces of the system. The trajectories of electrons of slightly different energies projected on the xy and yz planes (Fig. 21) are given in Figs. 23a and b. The xz projection of the system is given in Fig. 24. In this diagram the broken lines AA', BB' correspond to the pole-piece faces AA' and BB' shown in Fig. 21 and

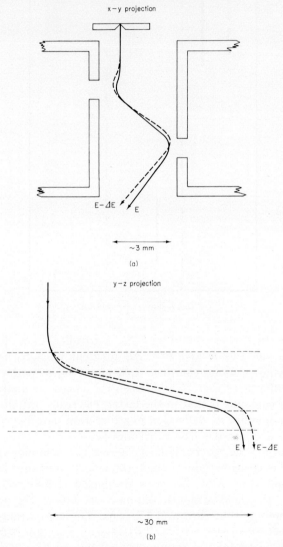

FIG. 23. (a) The xy projections and (b) the yz projections of electron trajectories in the system.

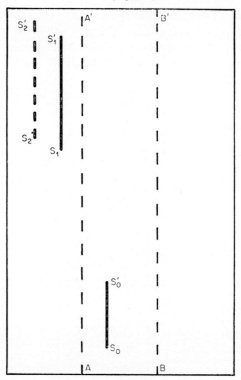

FIG. 24. The *xz* projection.

the entrance slit is represented by $S_0 S_0'$. A parallel beam of electrons of two discrete energies E and $E - \Delta E$ incident on $S_0 S_0'$ arrive at the recording plane xz along the lines $S_1 S_1'$ and $S_2 S_2'$ respectively. One of the major differences therefore between the magnetic analyser and its electrostatic analogue is that in the latter, electrons initially parallel to the y direction suffer no deflection in a direction parallel to the axes of the cylindrical electrodes (the z direction)

The reason for the reverse polarity across the pole-piece gaps is that this mode of excitation reduces the total deflection of the electron beam in the yz plane (Fig. 23b). The gaps are placed at different levels since this reduces the field strength required to deflect the beam into a trajectory such as that shown in Fig. 25a. The field strength required for the trajectory in the system illustrated in Fig. 25b is much higher than that required for the asymmetric system of Fig. 25a. Schematic trajectories of electrons of the same energy entering the system at

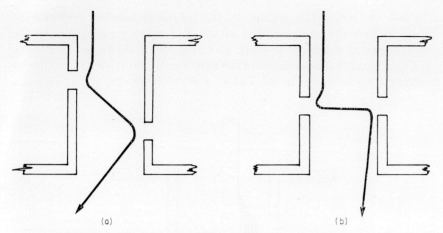

FIG. 25. Electron trajectories in systems where the pole pieces are placed (a) asymmetrically and (b) symmetrically with respect to each other.

different points along the x axis are given in Figs. 26a and b. The excitation of the lens is assumed low in Fig. 26a and the effect of increased excitation is shown in Fig. 26b. If the excitation is further increased electrons incident at 3 will be deflected into the pole-piece face and will be lost from the system. Still further excitation causes electrons incident at 2 also to be lost from the system. In practice the lens is operated at a sufficiently high excitation so that only those

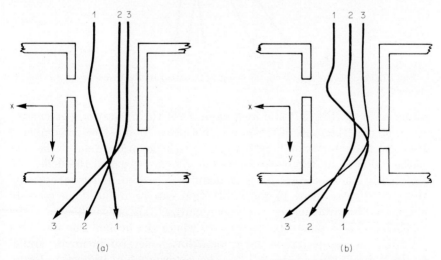

FIG. 26. Electron trajectories in the system when (a) low energizing currents are used and (b) when the energizing current is high.

electrons entering the system in the neighbourhood of trajectory 1 pass through the analyser. The effect of increasing excitation on the trajectories of electrons incident at 1 (Fig. 26) is shown schematically in Fig. 27 and the dependence of the exit coordinate x_1 on the energizing current I is shown in Fig. 28. The x_1, I curve intercepts the x_1 axis at a

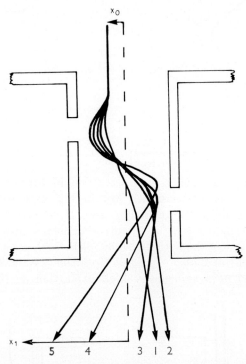

Fig. 27. The effect of increasing the energizing current on the trajectory of an electron incident at x_0.

value $x_1 = x_0$ (Fig. 27) and at a value I_c of the excitation current the beam strikes the edge of the lower pole-piece. It must be remembered that the trajectories of Fig. 27 represent the xy projection of the true paths followed by the electrons and as I is increased so the deflection suffered by the electrons in the yz plane is increased. It is worth noting that the displacement in the z direction can be reduced, or entirely eliminated by rotating the analyser about an axis perpendicular to the yz plane. This is illustrated in Fig. 29 where the broken line indicates the trajectory projected on the yz plane for an electron incident parallel to the y direction and the full line the corresponding projection for an electron incident at an appreciable angle to the y direction.

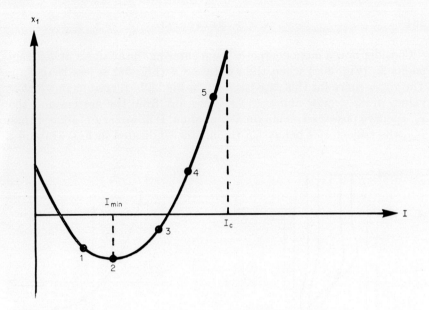

FIG. 28. The variation of x_1 with energizing current I for an electron incident at x_0 (Fig. 27).

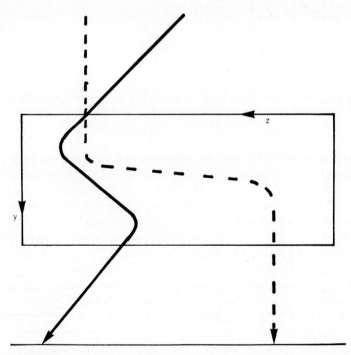

FIG. 29. The yz projections of electron trajectories. The comparatively large deflection of the electron in the z direction can be reduced by allowing the electron beam to be incident at an angle to the y axis.

Consider now a monoenergetic beam entering the analyser at different points x_0 (Fig. 30a) when the excitation I (Fig. 28) is less than I_{min}. The x_0, x_1 curve for this case is shown in Fig. 30b. Electrons incident at values of $x_0 < (x_0)_2$ and $x_0 > (x_0)_1$ are lost from the system and the x_0, x_1 curve possesses no maxima or minima. If however I is greater than I_{min} the trajectories behave in the manner indicated in Fig. 31a and a

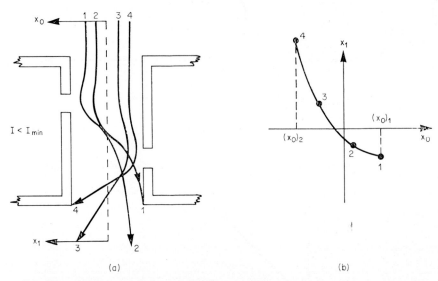

(a)   (b)

FIG. 30. (a) Trajectories of electrons incident at different points x_0 when the energizing current $I < I_{min}$ (Fig. 28). (b) The x_0, x_1 curve obtained when $I < I_{min}$.

minimum appears in the x_0, x_1 curve (Fig. 31b). The system therefore possesses a caustic edge and the behaviour of the x_0, x_1 curve is similar to that of the electrostatic analyser (Section IIA). The existence of this caustic edge is the reason for the high resolving power $(E/\Delta E_r \sim 10^5)$ reported in the literature for this system. The behaviour of the x_0, x_1 curves for different excitation currents is illustrated in Fig. 32 where $I_4 > I_3 > I_2 > I_1 > I_{min}$. The minima of these curves become increasingly sharp as the excitation current is increased and it is apparent that the image width δx_1 (defined in Section IIA, Fig. 17) of a parallel monoenergetic beam entering the system through an entrance slit of width Δx_0, positioned at the appropriate caustic edge, increases as the excitation increases. The resolution is given by $\Delta E_r = (|\partial E/\partial x_1|)$ δx_1 and the behaviour of δx_1 with excitation suggests therefore that the optimum excitation for energy analysis should be close to I_{min}. The dispersion $(|\partial x_1/\partial E|)$, however, is found to increase more rapidly than

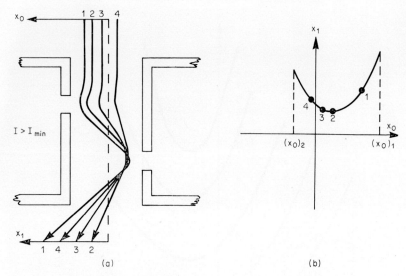

FIG. 31. (a) Trajectories of electrons incident at different points x_0 when $I > I_{\min}$ (Fig. 28). (b) The x_0, x_1 curve when $I > I_{\min}$.

δx_1 with increasing excitation with the result that the smallest value for the resolution ΔE_r occurs at I_c. The effect of excitation on the dispersion is considered in more detail below.

The x_0, x_1 curves of Fig. 32 do not indicate how the displacement of the beam in the z direction is influenced by the pole-piece excitation. The effect of lens excitation on this displacement can best be illustrated by considering the trace that the point of intersection of an electron beam of infinitesimal cross-section makes with the recording plane xz as the entrance position x_0 is varied. The traces obtained for excitations greater than $I_{\min}$ are given schematically in Fig. 33. The arrow on any one of the traces indicates the direction of movement of the point of intersection as x_0 is increased.

The effect of excitation on the dispersion of the analyser can also be illustrated by considering the x_0, x_1 curve for electrons of two discrete energies E and $E - \Delta E$. The full lines of Fig. 34 refer to electrons of energy E and the broken curves to energy $E - \Delta E$. The main points of interest are that the dispersion $(\Delta x_1/\Delta E)$ increases with increasing excitation and at the caustic edge $(\Delta x_1/\Delta E)$ is non-zero. The dispersion obtained at the caustic edge of an analyser with pole-piece dimensions given in Fig. 22, and with the recording plane at a distance 40 mm below the base of the system, is shown in Fig. 35. The dispersion was calculated (Considine, 1970) by assuming a fractional energy change

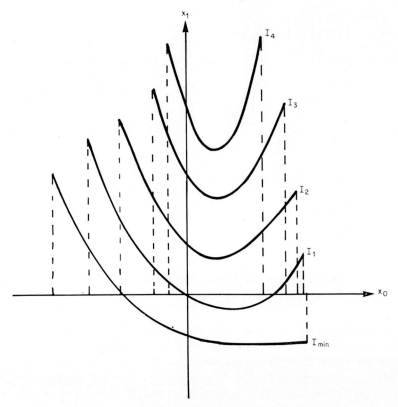

FIG. 32. The x_0, x_1 curves obtained for different energizing currents
$(I_4 > I_3 > I_2 > I_1 > I_{\min})$.

$\Delta E/E = 10^{-3}$. The separation Δx_1 (Fig. 34) of electrons of energies
E and $E - \Delta E$ was then found by using numerical trajectory tracing
techniques. The horizontal axis of Fig. 35 refers to an excitation parameter k defined by the relation

$$NI = k\,V_r^{\frac{1}{2}} \tag{3}$$

where NI is the number of ampere turns used to excite the lens and

$$V_r = V\left(1 + \frac{eV}{2m_0c^2}\right)$$

where V is the beam potential, m_0 the rest mass of the electron, e its
charge and c is the velocity of light. Equation (3) gives the excitation
required to produce a trajectory of a given shape as a function of the
beam potential. Also plotted as a function of k in Fig. 35 is the width

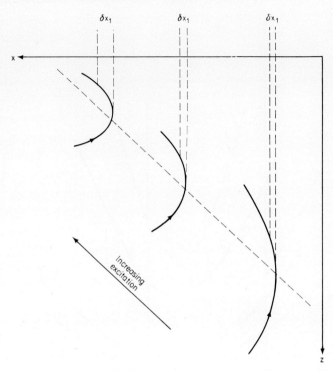

FIG. 33. A schematic diagram showing the traces that would be obtained on the recording plane xz by illuminating an entrance slit of infinitesimal width aligned parallel to the x axis (Fig. 21). The image width δx_1 of an entrance slit of width Δx_0 increases as the excitation is increased.

δx_1 (Fig. 17) measured at the caustic edge for an entrance slit of width 15 μm. Using the expression $\Delta E_r = (|\Delta E/\Delta x_1|)\ \delta x_1$, the resolution as a function of k can be calculated from the information given in Fig. 35 and it is found that ΔE is very nearly independent of k. The specific resolution $(\Delta E/E)_r$ falls from $1\cdot3 \times 10^{-5}$ at $k_{\min}$ to $0\cdot9 \times 10^{-5}$ at k_c where $k_{\min}$ and k_c are the excitation parameters for currents $I_{\min}$ and I_c of Fig. 28. The dispersion for the system of Fig. 22 at excitation $k = 0\cdot52$ amp. turns (volts)$^{-\frac{1}{2}}$ as a function of the fractional energy loss $\Delta E/E$ is given in Fig. 36, which shows that the dispersion is linear for losses up to $\Delta E = 3 \times 10^{-3}E$.

It is to be noticed that the dispersion of this system is extremely small; for example at k_c, the value for a beam potential of 100 kV is only $\sim 5\ \mu$m/volt compared with $\sim 50\ \mu$m/volt for an electrostatic analyser of similar geometry. Obviously the dispersion can be increased by increasing the dimensions of the analyser by some constant factor

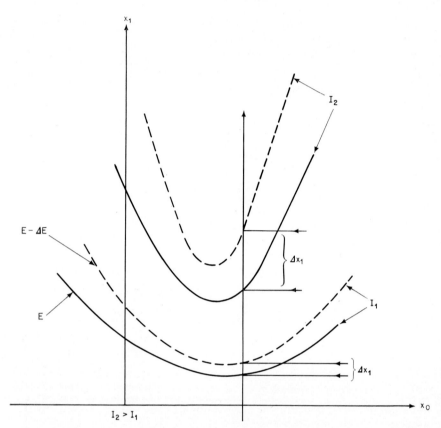

FIG. 34. The x_0, x_1 curves for electrons of different energies E and $E - \Delta E$. The dispersion is the quantity $|\Delta x_1 / \Delta E|$ taken in the limit $\Delta E \to 0$.

and also by increasing the projection distance. If photographic techniques are to be employed for recording the energy loss spectra a dispersion of about 100 μm/volt is necessary if problems associated with the limited spatial resolution of photographic emulsions are to be avoided. To achieve a dispersion of 100 μm/volt with the design dimensions of Fig. 22, a projection distance $\sim$ 100 cm is required which is obviously impractical. To overcome this problem Ichinokawa (1965) mounted the analyser between the intermediate and projector lenses of an electron microscope and used the magnification produced by the projector lens to achieve a reasonable dispersion. It is however highly undesirable to position an analyser between these two lenses, since it can interfere with the normal operation of the microscope and more importantly reduces the *spatial* resolution available for microanalysis

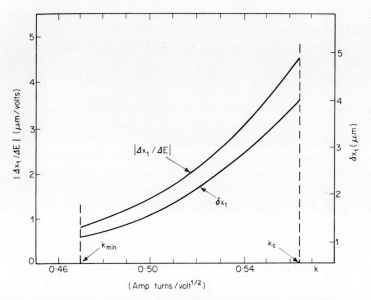

FIG. 35. Variation of the dispersion $|\Delta x_1/\Delta E|$ and image width δx_1 (defined in Fig. 17) with the excitation parameter k. The quantities plotted in this figure refer to an analyser with design dimensions given in Fig. 22. The width of the entrance aperture assumed in calculating δx_1 is 15 μm and the beam potential assumed in calculating $|\Delta x_1/\Delta E|$ is 100 kV.

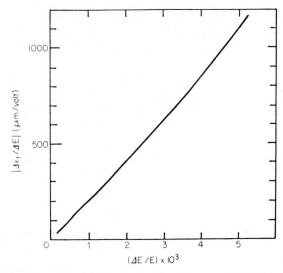

FIG. 36. The dispersion $|\Delta x_1/\Delta E|$ plotted as a function of the fractional energy loss $\Delta E/E$ (after Considine, 1970).

(Section IIIA). If photographic techniques are to be used to record the spectra it is preferable to mount the analyser in the final image plane of the microscope and employ an auxiliary lens placed below the analyser to achieve the required magnification. Alternatively the spectra can be recorded electronically (Section IIIA) in the manner employed by Considine and Smith (1968, 1971) and Considine (1970).

In view of the small dispersion associated with the magnetic system it would seem desirable to excite the pole pieces to a value of k as close to k_c (Fig. 35) as possible. There are, however, several practical considerations to be made before the optimum excitation can be decided on. Consideration must first be given to the range of energy losses that is to be examined. Caustic trajectories for electrons of the same energy E are given in Fig. 37 for excitation parameters k, where $k \geqslant k_{\min}$, and

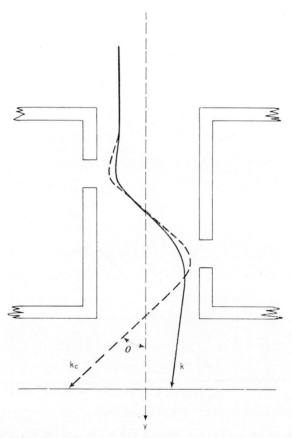

FIG. 37. Electron trajectories for two excitation parameter $k = k_c$ and $k \geqslant k_{\min}$.

k_c. These trajectories, however, are also those for electrons of energy E (full curve) and $E - (\Delta E)_R$ (broken curve) where $(\Delta E)_R$ is the range of losses accepted by the analyser when the excitation is k. Provided $(\Delta E)_R \ll E$, equation (3) shows that the largest excitation parameter (k_{max}) that can be used if the range of losses is $(\Delta E)_R$ is given by

$$k_{max} = k_c \left(1 - \left((\Delta E)_R / E\right)\right)$$

The fractional range $(\Delta E)_R / E$ of losses contributing to the microscope image is usually in the region of 10^{-3}. The energy loss range is therefore an unimportant factor in deciding the optimum value of k to be used, since the equation for k_{max} given above shows that values of k close to k_c can be used if so desired. Consideration must also be given to the angle θ (Fig. 37) that the emerging beam makes with the y axis. If spectra are to be recorded photographically, it is desirable that the beam entering the auxiliary lens should make as small an angle as possible with the y axis. Similarly, if the spectra are to be recorded electronically, alignment coils placed below the analyser (Section IIIA) are required to deflect the beam onto a detector, and the design of the deflection coils is simplified if θ is small. The angle made by the trajectory projected on the yz plane and the y axis is very small (less than $\sim 10^{-5}$ rad. for $(\Delta E)_R / E = 10^{-3}$) whereas the angle made by the trajectory in the xy projection at $k = k_c$ is $\sim 0 \cdot 2$ rad. The variation of θ with k is given in Fig. 38 and it is seen that θ is very small when $k = k_{min}$. High

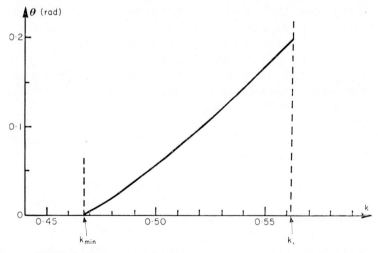

FIG. 38. The angle θ (Fig. 37) plotted as a function of the excitation parameter k (after Considine, 1970).

dispersion and small θ are therefore conflicting requirements. If the spectra are recorded photographically, then values of k close to $k_{\min}$ can be used, provided that an auxiliary lens of sufficient magnification is employed. If, however, electronic detection is used, then it is found that yet another factor, namely the width of the energy selecting aperture placed over the detecting system, influences the choice of excitation parameter used. A full discussion of this point is given in Section IIIA.

C. *Mirror Prism Device*

The possibility of devising a filter lens which could be placed immediately after the objective lens of an electron microscope and allow normal operation of the intermediate and projector lenses to produce highly magnified energy selected images was the subject of theoretical studies by Hennequin (1960) and Paras (1961). Hennequin studied the dispersive properties of magnetic prisms whereas Paras studied a number of devices among which was a system consisting of a combined electrostatic mirror and magnetic prism. This analyser was subsequently developed by Castaing and Henry (1962, 1963, 1964) as an integral part of an energy selecting electron microscope which is described in Section IVA of this article.

The filter lens consists of a double magnetic prism, employing a uniform magnetic field, and a concave electrostatic mirror biased at cathode potential (Fig. 39). Under certain conditions, which are considered below, the system possesses two sets of stigmatic points; one set $R_1 R_2 R_3$ (Fig. 39) is real, the other set $V_1 V_2 V_3$ is virtual. The reason for the existence of these points is illustrated in Figs. 40a and b. A convergent beam of electrons (Fig. 40a) initially focused at V_1, so that V_1 is a virtual stigmatic object point, is deviated by the prism so that it appears to diverge from the point V_2, which is a virtual stigmatic image point; similarly a convergent beam focused initially at V_2 appears to diverge from the point V_3. A beam of electrons diverging from R_1 (Fig. 40b) which is a real stigmatic object point is however brought to a real focus at R_2, and similarly a beam diverging from R_2 is brought to a real focus at R_3. The positions of these stigmatic points, and indeed their very existence and character, are determined by the angle of incidence which the incoming electrons make with the entrance face of the prism.

Before considering the formation of stigmatic object and image points in this system, a general analysis of the electron optical properties of the magnetic prism will be given. Consider a point object A

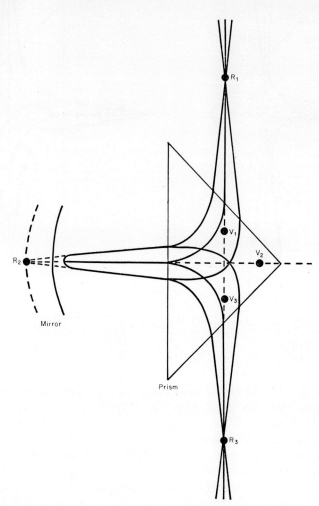

FIG. 39. The double magnetic prism and electrostatic mirror. R_1, R_2 and R_3 are real stigmatic points and V_1, V_2 and V_3 are virtual stigmatic points.

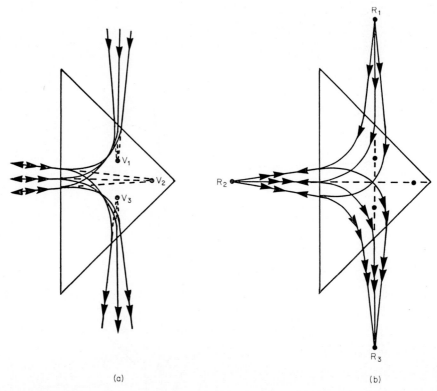

(a)   (b)

FIG. 40. (a) The virtual stigmatic points V_1, V_2 and V_3. (b) The real stigmatic points R_1, R_2 and R_3.

(Figs. 41a and b) emitting a fine cone of electrons in the direction AP, and suppose that the mean trajectory APP' A'_1 A'_2 makes an angle ϵ_1 (Fig. 41b) with the normal to the face DE of the prism. In the region of the uniform magnetic field enclosed by the faces CD and DE of the prism, the mean trajectory has a radius of curvature a and the electron emerges from the second face CD of the prism at an angle ϵ_2, suffering a total angular deviation θ. In the general case two focal lines, A'_1 and A'_2, are formed in image space corresponding to the object point A. The radial focal line A'_1 is parallel to the magnetic induction B and the axial focal line A'_2 is perpendicular to B. The magnetic prism is therefore astigmatic and possesses two sets of cardinal points. The first set corresponds to electrons whose trajectories lie in a plane perpendicular to the magnetic induction B (the plane CDE of Fig. 41a). This plane is called the *first principal section* of the system. The second set of cardinal points corresponds to trajectories which lie in the cylindrical surface whose generators are perpendicular to the first principal section and pass through the mean trajectory $APP'\ A_1'\ A_2'$ (Fig. 41a). This cylindrical surface is called the *second principal section*. Expressions for the positions of the cardinal points have been given by Cotte (1938). In his system of coordinates, the cardinal points lying in object space are measured from the origin P (Fig. 41b) in a direction $\mathbf{T}_1$ tangential to the mean trajectory at P and those in image space from the origin P' and in a direction $\mathbf{T}_2$ tangential to the mean trajectory at P'. These distances are positive if they are parallel to the forward sense of the mean trajectory and negative if antiparallel. The sign convention for ϵ_1 and ϵ_2 is as follows. The angles ϵ_1 and ϵ_2 are positive if $\mathbf{n}_1 . \mathbf{T}_1$ and $\mathbf{n}_2 . \mathbf{T}_2$, where $\mathbf{n}_1$ and $\mathbf{n}_2$ are the outward normals of the prism faces (Fig. 41b), are also positive. In Fig. 41b therefore, ϵ_1 is negative and ϵ_2 is positive. Cotte's equations are given in Table II, where f and h are the distances of the focal and principal points in object space and f' and h' are the distances of the focal and principal points in image space.

For the first half of the double prism ($\theta = \pi/2$, $\epsilon_1 = -\epsilon$, and $\epsilon_2 = 0$), the cardinal points of the first principal section are given by

$$f_1 = 0; \qquad h_1 = a; \qquad f_1' = a \tan \epsilon; \qquad h_1' = a(\tan \epsilon - 1)$$

and those of the second principal section by

$$f_2 = -a \cot \epsilon; \qquad h_2 = 0; \qquad f_2' = a(\cot \epsilon - \pi/2); \qquad h_2' = -a\pi/2$$

For the second half of the prism ($\theta = \pi/2$, $\epsilon_1 = 0$, $\epsilon_2 = \epsilon$) the cardinal points of the first principal section are given by

$$f_1 = -a \tan \epsilon; \qquad h_1 = -a(\tan \epsilon - 1); \qquad f_1' = 0; \qquad h_1' = -a$$

and for the second principal section by

$$f_2 = -a(\cot \epsilon - \pi/2); \qquad h_2 = a\pi/2; \qquad f_2' = a \cot \epsilon; \qquad h_2' = 0$$

These equations show that the cardinal points of the first principal section in the first and second halves of the prism are symmetric about the mirror axis $R_2 V_2$ (Fig. 42a). A similar symmetry holds for the cardinal points of the second principal section (Fig. 42b). Corresponding cardinal points of the two principal sections do not in general coincide. The radial symmetry of a conical beam of electrons incident on the entrance face of the double prism is therefore not preserved in general as it passes through the system.

For *certain* positions of the object point A, however, the radial and axial focal lines coincide at some point A' (Fig. 41b) and for these positions of A, A' is a stigmatic image point. The conditions which

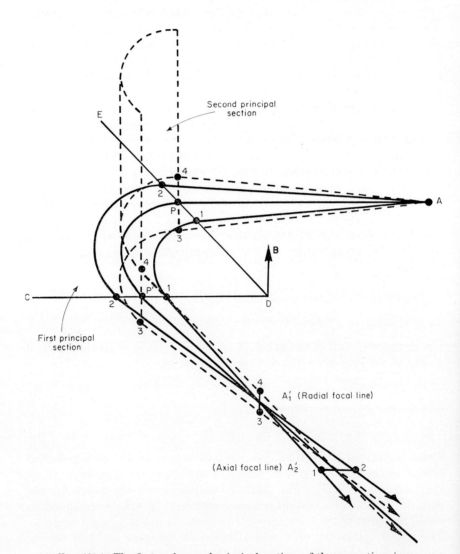

FIG. 41(a). The first and second principal sections of the magnetic prism.

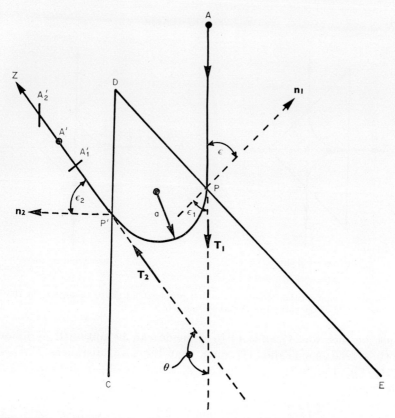

Fig. 41(b). A diagram of the first principal section showing the various angles of incidence, etc.

Table II

Equations Given by Cotte (1938) for the Distances of the Focal and Principal Points of the Magnetic Prism

First Principal Section	Second Principal Section
$f_1 = -\dfrac{a \cos(\theta - \epsilon_2) \cos \epsilon_1}{\sin(\theta + \epsilon_1 - \epsilon_2)}$	$f_2 = \dfrac{a(\theta \tan \epsilon_2 - 1)}{\theta \tan \epsilon_1 \tan \epsilon_2 - \tan \epsilon_1 + \tan \epsilon_2}$
$h_1 = -\dfrac{a\{\cos(\theta - \epsilon_2) \cos \epsilon_1 + \cos \epsilon_1 \cos \epsilon_2\}}{\sin(\theta + \epsilon_1 - \epsilon_2)}$	$h_2 = \dfrac{a\,\theta \tan \epsilon_2}{\theta \tan \epsilon_1 \tan \epsilon_2 - \tan \epsilon_1 + \tan \epsilon_2}$
$f_1' = \dfrac{a \cos \epsilon_2 \cos(\theta + \epsilon_1)}{\sin(\theta + \epsilon_1 - \epsilon_2)}$	$f_2' = \dfrac{a(\theta \tan \epsilon_1 + 1)}{\theta \tan \epsilon_1 \tan \epsilon_2 - \tan \epsilon_1 + \tan \epsilon_2}$
$h_1' = \dfrac{a\{\cos(\theta + \epsilon_1) \cos \epsilon_2 - \cos \epsilon_1 \cos \epsilon_2\}}{\sin(\theta + \epsilon_1 - \epsilon_2)}$	$h_2' = \dfrac{a\,\theta \tan \epsilon_1}{\theta \tan \epsilon_1 \tan \epsilon_2 - \tan \epsilon_1 + \tan \epsilon_2}$

f and h refer to object space whereas f' and h' refer to image space (see Figs. 42 a and b).

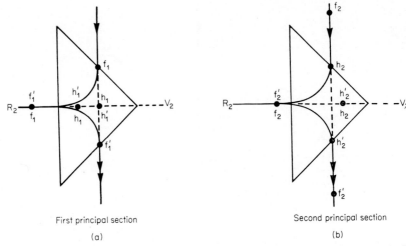

First principal section   Second principal section

(a)   (b)

FIG. 42. (a) The cardinal points of the first principal section ($0 < \tan \epsilon < 1$). (b) The cardinal points of the second principal section ($\cot \epsilon > \pi/2$).

must be met for A'_1 and A'_2 to coincide can be obtained as follows. Cotte's relations (Table II) show that the focal distances $P'A'_1$ and $P'A'_2$ (Fig. 41b) are related to a, ϵ_1, ϵ_2, θ and the object distance PA, by the equations

$$\xi'_1/a = \frac{\dfrac{(\xi/a)\cos(\theta + \epsilon_1)}{\cos \epsilon_1} - \sin \theta}{(\xi/a)\dfrac{\sin(\theta + \epsilon_1 - \epsilon_2)}{\cos \epsilon_1 \cos \epsilon_2} + \dfrac{\cos(\theta - \epsilon_2)}{\cos \epsilon_2}} \tag{4}$$

$$\xi'_2/a = \frac{\theta - (\xi/a)(1 + \theta \tan \epsilon_1)}{(\xi/a)(\tan \epsilon_1 - \tan \epsilon_2 - \theta \tan \epsilon_1 \tan \epsilon_2) - (1 - \theta \tan \epsilon_2)} \tag{5}$$

where $\xi = PA$, $\xi'_1 = P'A'_1$ and $\xi'_2 = P'A'_2$. The condition required for coincidence of the radial and axial focal lines is just

$$\xi'_1 = \xi'_2 = \xi'$$

where $\xi' = P'A'$. The sign convention for ξ and ξ' is the same as that for f, f', h and h' (see above) and therefore in Fig. 41b ξ is negative and ξ' is positive. For the first half of the double prism $\theta = \pi/2$, $\epsilon_1 = -\epsilon$ and $\epsilon_2 = 0$ with the result that equations (4) and (5) reduce to

$$\frac{\xi'_1}{a} = \tan \epsilon - \frac{1}{(\xi/a)} \tag{6}$$

$$\frac{\xi'_2}{a} = \frac{(\xi/a)}{1 + (\xi/a) \tan \epsilon} - \frac{\pi}{2} \tag{7}$$

Using the condition $\xi'_1 = \xi'_2 = \xi'$ for the coincidence of the axial and radial focal lines, and eliminating ξ' from equations (6) and (7), we obtain,

$$\left(\tan^2 \epsilon + \frac{\pi}{2} \tan \epsilon - 1\right)\left(\frac{\xi}{a}\right)^2 + \frac{\pi}{2}\left(\frac{\xi}{a}\right) - 1 = 0 \qquad (8)$$

The solutions for ξ obtained from this equation give the positions of the stigmatic object points as a function of the angle of incidence ϵ. The variation of ξ as a function of $\tan \epsilon$ obtained by solving equation (8) is shown in Fig. 43. On eliminating ξ from equations (6) and (7) we obtain

$$\left(\frac{\xi'}{a}\right)^2 + \left(\frac{\pi}{2} - 2 \tan \epsilon\right)\left(\frac{\xi'}{a}\right) + 1 - \pi \tan \epsilon = 0 \qquad (9)$$

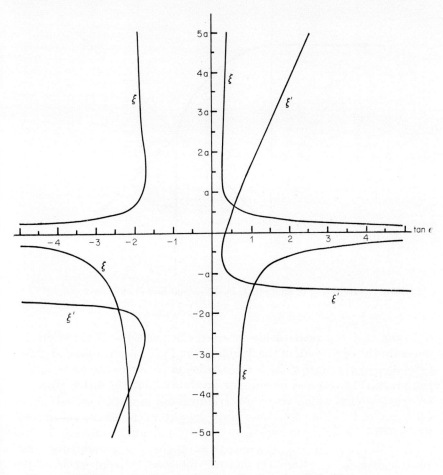

Fig. 43. Variation of the image and object distances ξ and ξ' with $\tan \epsilon$.

and the solutions of ξ' obtained from this equation give the positions of
the stigmatic image points as a function of ϵ. The variation of ξ' as a
function of $\tan \epsilon$ obtained from equation (9) is also given in Fig. 43.
Equations (8) and (9) also show that ξ and ξ' have imaginary roots only,
for values of $\tan \epsilon$ lying between $-1 - (\pi/4)$ and $+1 - (\pi/4)$, and
within this range of values of $\tan \epsilon$ the prism does not possess stigmatic
points. In addition, for values of tan greater than $\frac{1}{4}\{(\pi^2 + 16)^{\frac{1}{2}} - \pi\}$
the prism always possesses two pairs of stigmatic points, one pair being
real and the other virtual.

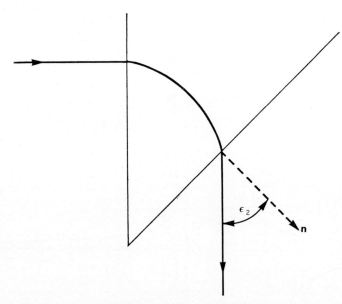

FIG. 44. The trajectory of an electron in the second half of the double prism.

So far only the relationship between the positions of the stigmatic
points in the upper half of the double prism have been considered. After
reflection by the mirror, however, the beam traverses the lower half of
the prism (Fig. 44) and it will now be shown that the latter possesses
stigmatic points which are symmetric to those in the upper half about
the axis $R_2 V_2$ (Fig. 39). The mean trajectory re-enters the double
prism with an angle of incidence $\epsilon_1 = 0$ and after suffering a total
angular deviation of $\pi/2$, emerges at an angle $\epsilon_2 = \epsilon$. Substitution of
these values in equations (4) and (5), followed by application of the
stigmatic condition $\xi_1' = \xi_2' = \xi'$ and successive elimination of ξ and ξ',

yields the equations,

$$\left(\frac{\xi}{a}\right)^2 - \left(\frac{\pi}{2} - 2\tan\epsilon\right)\left(\frac{\xi}{a}\right) + 1 - \pi\tan\epsilon = 0 \qquad (10)$$

$$\left(\tan^2\epsilon + \frac{\pi}{2}\tan\epsilon - 1\right)\left(\frac{\xi'}{a}\right)^2 - \frac{\pi}{2}\left(\frac{\xi'}{a}\right) - 1 = 0 \qquad (11)$$

A comparison of these equations with equations (8) and (9) shows that the only differences are that the object and image points in the first half of the prism correspond to image and object points in the second half, and that ξ and ξ' in equations (10) and (11) are of opposite sign to those in equations (8) and (9). The stigmatic points of the double prism are therefore situated symmetrically about the axis $R_2 V_2$ (Fig. 39) and furthermore the stigmatic image points of the first prism are coincident with the stigmatic object points of the second prism. In practice this coincidence can only be brought about if the apex of the mirror and its centre of curvature are at R_2 and V_2 (Fig. 39). If the apex is at R_2 and the centre at V_2 the system is dispersive, whereas if the apex is at V_2 and the centre at R_2 it can be shown that the device is completely achromatic, and hence the latter configuration is of no interest if the system is intended for use as an energy analyser.

From the preceding analysis it is clear that if an incident beam has its cross-over situated at R_1 or V_1 (Fig. 39), the radial symmetry of the beam is not destroyed by its passage through the mirror-prism system. In practice the filter lens is placed in a position such that the stigmatic object and image points R_1 and V_1 lie in the back focal plane and image plane respectively of the objective lens (Fig. 45) of an electron microscope. In this case a stigmatic image is produced in the plane I_2 passing through the point V_3 and the cross-over of the exit beam is produced at R_3. It is clear from the symmetry of the stigmatic points R_1, V_1, R_3 and V_3 about the mirror axis that the magnification produced by the mirror-prism system is unity, and this can be proved rigorously from Cotte's equations (Table II). The filter lens, therefore, simply inverts the positions of the cross-over and image produced by the objective lens.

So far in this discussion it has been assumed that the electrons passing through the system are monoenergetic. The effect of the filter lens on electrons of different energies will now be considered and the dispersive properties of the system examined. If an electron of energy E follows a trajectory PP' (Fig. 46) of radius a, then an electron of energy $E + \Delta E$ will follow a trajectory PP'_1 of radius $a + \Delta a$. It will be assumed that Δa is small so that to the first order $P'P'_1 = \Delta a$. The calculation of the trajectory of an electron of energy $E + \Delta E$ through the mirror-prism system is largely a matter of geometry. The coordinates of the various

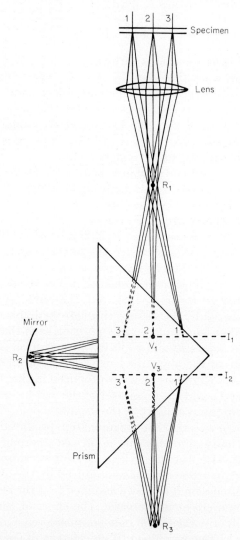

FIG. 45. If the image plane of an objective (or intermediate) lens is made coincident with the virtual stigmatic object plane I_1, a real stigmatic and achromatic image is formed in the plane I_2 after reflection by the electrostatic mirror.

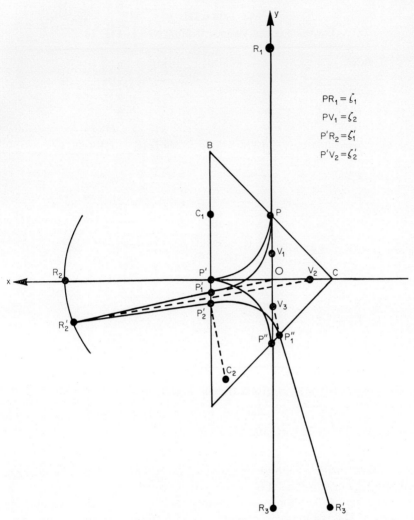

$PR_1 = \zeta_1$

$PV_1 = \zeta_2$

$P'R_2 = \zeta_1'$

$P'V_2 = \zeta_2'$

FIG. 46. The various points R_1, P, C_1, etc. given in Table III.

points P, P', etc., indicated in Fig. 46 are given in Table III and these have been obtained by neglecting second order quantities $(\Delta a)^2$ etc., and by assuming $\tan \epsilon > \frac{1}{4}\{(\pi^2 + 16)^{\frac{1}{2}} - \pi\}$. It is also assumed that the signs of the stigmatic object and image distances ξ and ξ' correspond to those given in Fig. 43, these being the values for the first half of the double prism. The positive value of ξ in Fig. 43 is denoted by ζ_1, the negative value by ζ_2; the positive value of ξ' by ζ_1' and the negative by ζ_2'.

TABLE III

Positions of the Various Points given in Fig. 46

Point	x	y
P	0	a
P'	a	0
P'_1	a	$-\Delta a$
P'_2	a	$-\dfrac{\Delta a}{a}\left\{\dfrac{a(\zeta'_1 + \zeta'_2) + 2\zeta'_1\zeta'_2}{(\zeta'_2 - \zeta'_1)}\right\}$
P''	0	$-a$
P''_1	$2\Delta a\,\dfrac{(\zeta'_1 + a)}{(\zeta'_2 - \zeta'_1)}$	$-a - 2\Delta a\,\dfrac{(\zeta'_1 + a)}{(\zeta'_2 - \zeta'_1)}\tan \epsilon$
C_1	a	a
C_2	$a + \Delta a\,\dfrac{(\zeta'_1 + \zeta'_2 + 2a)}{(\zeta'_2 - \zeta'_1)}$	$-a - \dfrac{2\Delta a}{a}\,\zeta'_2\,\dfrac{(a + \zeta'_1)}{(\zeta'_2 - \zeta'_1)}$
R_1	0	$a - \zeta_1$
R_2	$\zeta'_1 + a$	0
R_3	0	$\zeta_1 - a$
V_1	0	$a - \zeta_2$
V_2	$\zeta'_2 - a$	0
V_3	0	$\zeta_2 - a$

The trajectory PP' is given by the equation

$$\{x - (a + \Delta a)\}^2 + (y - a)^2 = (a + \Delta a)^2$$

and the equation of the tangent $P'_1 R'_2$ is

$$y = - (\Delta a/a)x$$

to the first order of small quantities. Electrons of different energies emerging from the face AB of the prism therefore appear to originate from the point 0, which is an achromatic point of the system. An analysis of the trajectories of electrons of different energies originating from different points of a specimen (Fig. 47) shows that an achromatic (and also stigmatic) image of the specimen is formed on the plane defined by the equation $x = 0$ (Fig. 47), and this plane is therefore an achromatic plane of the system. After reflection by the mirror, electrons of energy $E + \Delta E$ follow the circular path $P'_2 P''_1$ (Fig. 46), whose centre is at C_2, and emerge from the prism at P''_1. The equation of the exit trajectory $P''_1 R'_3$ is, to the first order of small quantities,

$$y = \frac{- a\,(\zeta'_1 - \zeta'_2)\,x}{2\Delta a\,(\zeta'_1 + a)\,\{(\zeta'_2/a) - \tan \epsilon\}} - a\left\{\frac{1}{(\zeta'_2/a) - \tan \epsilon} + 1\right\}$$

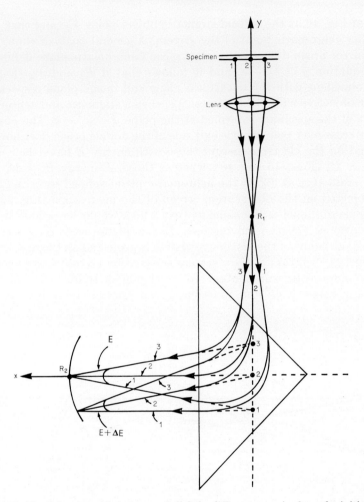

FIG. 47. The behaviour of electrons of slightly different energies brought initially to a cross-over at the point R_1.

Equation (6) shows however, that under the stigmatic conditions assumed,

$$\frac{1}{(\zeta_2'/a) - \tan \epsilon} = \frac{\zeta_2}{a}$$

and therefore that the equation of the trajectory $P_1''R_3'$ can be written as

$$y = \frac{(\zeta_1' - \zeta_2')\,\zeta_2}{2\varDelta a\,(\zeta_1' + a)}\,x + \zeta_2 - a \tag{12}$$

It is clear from this equation that the exit trajectory intersects the y

axis of Fig. 46 at the virtual stigmatic object point V_3, and that V_3 is another achromatic point of the system. A general analysis shows that the mirror prism device possesses a second achromatic plane defined by the equation $y = \zeta_2 - a$, and it follows that if R_1 and V_1 (Fig. 46) are coincident with the back focal plane and image plane respectively of the objective lens of the microscope, then a stigmatic and achromatic image of the specimen is formed at the plane $y = \zeta_2 - a$. The position of the cross-over point on the exit side of the double prism does however depend on the electron energy; those with energy E have their cross-over at R_3 (Figs. 46 and 48), whereas those of energy $E + \Delta E$ have their cross-over at R_3'. If the achromatic plane defined by $y = \zeta_2 - a$ is projected on the observation screen of the microscope (Fig. 49) the image so obtained is the same as that which would be produced by a conventional electron microscope. If the plane containing the exit cross-over points of the dispersive system is projected on the final screen, the energy spectrum of the electrons selected by the objective aperture of the microscope will be observed. In this mode of operation however the apparatus does not function as an energy analysing electron

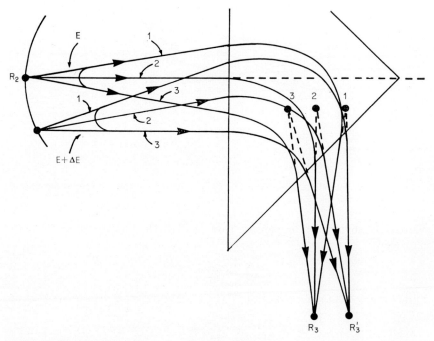

FIG. 48. The behaviour of electrons of energies E and $E + \Delta E$ in traversing the second half of the magnetic prism.

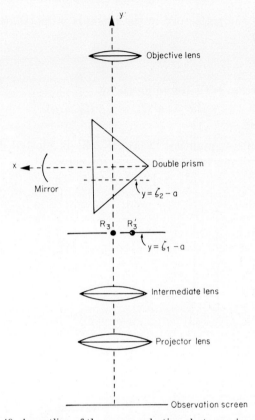

FIG. 49. An outline of the energy selecting electron microscope.

microscope since the observed loss spectrum is produced by electrons which originate over the entire area of irradiation of the specimen. Finally if a selecting aperture is placed at the exit cross-over plane of the filter lens (Fig. 49) and the achromatic plane $y = \zeta_2 - a$ is projected on the final screen, the image so produced is formed by electrons with energies lying between E and $E + \delta E$, where δE depends on the width of the selecting aperture and the dispersion of the system at the exit cross-over plane. An expression for the dispersion can be obtained by determining the point of intersection of the exit trajectory $P_1'' R_3'$ (Fig. 46) with the line $y = \zeta_1 - a$. Equation (12) shows that the x coordinate of R_3' is given by

$$\Delta x = \frac{2 \Delta a \, (\zeta_1 - \zeta_2) \, (\zeta_1' + a)}{\zeta_2 \, (\zeta_1' - \zeta_2')}$$

If the beam potential of an electron of energy E is V then the radius of curvature a of the trajectory in the prism is proportional to $\sqrt{V}$, with the result that

$$\frac{\Delta a}{a} = \frac{1}{2}\frac{\Delta V}{V}$$

and the dispersion is therefore given by

$$\frac{\Delta x}{\Delta V} = \frac{a}{V}\frac{(\zeta_1 - \zeta_2)(\zeta_1' + a)}{\zeta_2(\zeta_1' - \zeta_2')}$$

The dimensionless quantity $(V/a)\,|\Delta x/\Delta V|$ is plotted in Fig. 50 as a function of the tangent of the angle ϵ for values of

$$\tan \epsilon > \tfrac{1}{4}\{(\pi^2 + 16)^{\frac{1}{2}} - \pi\}.$$

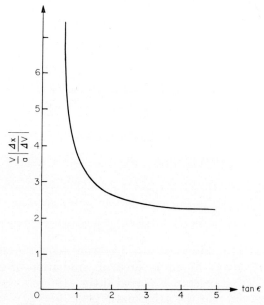

FIG. 50. The dimensionless quantity $(V/a)|\Delta x/\Delta V|$ plotted as a function of $\tan \epsilon$. The quantity $|\Delta x/\Delta V|$ is the dispersion of the filter lens.

III. Energy Analysing Electron Microscopes

A. *Instrumentation*

The idea of incorporating an energy analyser with an electron microscope is not new. The first attempts at combining these instruments to produce a microanalyser of high spatial resolving power were reported by Hillier (1943) and Marton (1944). Both authors used homogeneous field magnetic analysers of poor resolving power ($E/\Delta E_r \sim 10^3$), with the result that much of the fine detail of the energy loss spectra was lost. Furthermore the state of electron microscopy at this period was such that only shadow images of the specimen could be obtained, and it was therefore impossible to relate changes in the loss spectra to changes in the substructure of the specimens examined.

The work of Hillier and Marton was followed by that of Kleinn (1954), Leonhard (1954), Marton and Leder (1954), Gauthe (1954) and Watanabe (1954, 1955, 1956). These authors employed the high resolution electrostatic analyser described in Section IIA of this article. In each case the analyser was placed between the intermediate and projector lenses of the microscope. This position is highly undesirable if full use of the spatial resolution and magnification of modern electron microscopes is to be made. Consider an analyser employing an entrance slit of width 5 μm situated in the image plane of the intermediate lens of a microscope producing an overall magnification of 20,000$\times$ at the final image plane. The objective lens forms an image in the object plane of the intermediate lens and this first stage of magnification is typically 25$\times$. The intermediate lens further magnifies the image by a factor of about 6·4 so that at the analyser entrance aperture the magnification is 160$\times$. The apparent width of the entrance slit, which for the moment will be taken as the spatial resolution, is therefore (5/160) μm or 310 Å. If, however, use is made of the magnification of the projector lens and the analyser is mounted in the final image plane where the magnification is 20,000$\times$, the apparent width of the slit is 2·5 Å. An analyser placed in the latter position therefore takes full advantage of the spatial resolution ($\sim$ 3 Å) afforded by commercial electron microscopes, and selects, in effect, a line in the final image from which the energy spectrum is recorded. One further advantage associated with placing the analyser in this position is that by simply cutting a narrow slot in the final screen of the microscope, simultaneous observation of both image and spectrum is achieved. It is only in recent years that sufficiently sophisticated instruments have been developed to allow realization of Hillier's and Marton's original aim of high resolution microanalysis. The rest of this section is devoted to a discussion of energy analysing electron microscopes which have been developed in the Cavendish Laboratory.

The first system to be described here consists of a cylindrical electrostatic analyser (Section IIA) mounted below the final image screen of a Siemen's Elmiskop I microscope operating at 100 kV (Metherell, 1965; Metherell et al., 1965; Cundy et al., 1966). A schematic outline of the instrument is shown in Fig. 51. The analyser section extends from the base of the camera chamber C_2 of the microscope to the floor. This extension of the microscope column means that the control console

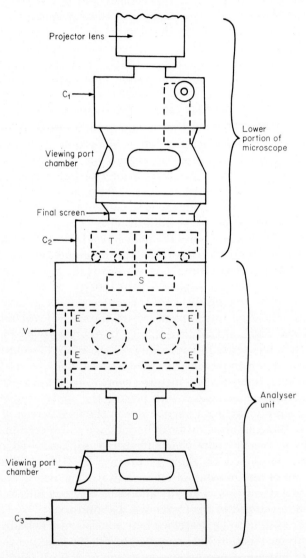

FIG. 51. An outline of the 100 kV energy analysing electron microscope which utilizes the cylindrical electrostatic analyser described in Section IIA.

upon which C_2 usually rests must be removed and mounted on a separate trolley. The entrance slit unit S of the analyser is suspended from a trolley T which allows translation and rotation of the slit assembly for alignment purposes. The trolley is housed in the camera chamber C_2 which means that the image cannot be recorded in the usual manner on photographic plates stored in C_2. For this reason a projector tube camera C_1, mounted on the viewing port chamber, is employed to record the images on 35 mm film. The camera unit in C_1 can be swung into a horizontal position for recording purposes and into a vertical position to allow visual observation of the image on the final screen. The length and width of the aperturing slit of the analyser can be altered by using controls mounted out of vacuum and coupled to the slit assembly by sliding connector rods and universal couplings. The electrode system of the analyser (C and E) is housed in a vacuum chamber V which is attached to a tube D. The function of this tube is to increase the projection distance and hence obtain increased dispersion in the spectrum. The dispersion tube is in turn mounted on a viewing port chamber which allows visual observation of the spectra, which are recorded photographically on plates stored in the camera chamber C_3.

One practical point worthy of note here is the method used to prepare the edges of the analyser entrance aperture. Slits used in optical spectrometers are usually unsuitable for electron energy analysers, since the slit widths employed in the latter are of the order of a few microns and small scale roughness of the edges can lead to undesirable streaking effects in the recorded spectra (Fig. 52). The following slit edge preparation technique, suggested by F. Fujimoto (private communication), has been found to be highly successful in practice (Cundy, 1968; Considine, 1970). Each edge piece of a spectrometer slit mechanism is milled down so that the cross-section has the dimensions indicated in Fig. 53. Fine glass fibres of diameter $\sim 0\cdot2$ mm are glued to the blunted edges using a suitable adhesive such as "Araldite", care being taken to ensure that no adhesive flows on to the outside edges of the glass fibres. After the adhesive has dried the slit edges must be thoroughly cleaned with suitable solvents and then coated with a layer of evaporated gold about 1000 Å thick. To provide a tenacious layer of gold the evaporation must be carried out as slowly as possible.

The H.T. cable, which normally carries the filament and cathode lines directly to the gun of the microscope, is fed into an auxiliary H.T. tank which is illustrated schematically in Fig. 54. This tank acts primarily as a junction box so that the H.T. applied to the electron gun can also be applied to the centre electrodes of the analyser. The two lines F (Fig. 54) carry the filament heating current and since the beam poten-

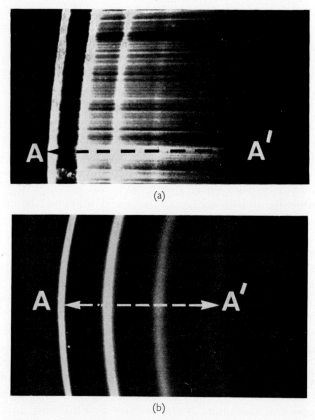

(a)

(b)

Fig. 52. (a) An example of the spectrum obtained when a spectrographic slit is used as the entrance aperture of the analyser. (b) An example of the spectrum obtained when the edge pieces of the slit mechanism have been prepared in the manner discussed in the text. The lines AA' are the energy loss axes of the spectra.

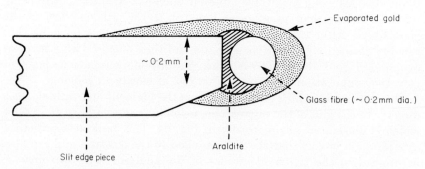

Fig. 53. A cross-section of a slit edge piece modified for electron optical use.

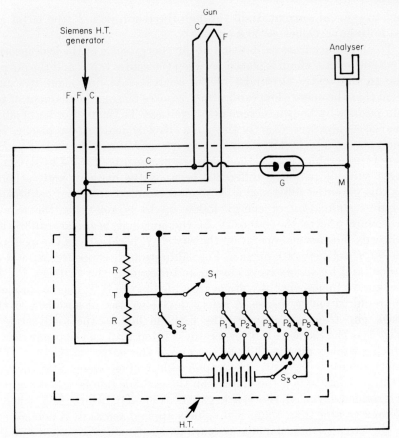

Fig. 54. An outline of the auxiliary H.T. tank and calibration unit (enclosed by the broken lines). The switches S and P are activated by phototransistors triggered by light beams, thus avoiding electrical insulation problems.

tial is the voltage difference between the tip of the filament and earth, it is desirable that the analyser should be biased at the same voltage; the reason being that any small voltage fluctuations produced by the H.T. generator during the time required to record an energy loss spectrum affect both the beam potential and the analyser bias equally. If the analyser is biased in this manner no spatial displacement of the loss spectrum occurs at the recording plane, to the first order of small quantities, with the result that these voltage fluctuations do not affect the energy resolution of the instrument. The lead to the analyser is therefore taken from the centre tap T of two equal resistances R (Fig. 54). The resistance of the electron gun filament line is ~ 10 ohms and to prevent current drain through the resistances R, the latter are made to have values ~ 50 k ohms.

To obtain accurate calibration of the energy scale of the loss spectra, a potentiometer chain is placed between the centre tap T and the output line to the centre electrodes of the analyser. At first sight it would seem that the energy loss calibration requires biasing the filament of the gun *positive* by known increments of voltage. To the first order of small quantities however, exactly the same effect is achieved by biasing the analyser *negative* by the same increments of voltage. This latter procedure is preferable because the beam brightness is controlled by developing a suitable potential difference between the filament and cathode. As this potential difference is $\sim$ several hundred volts and calibration requires simulation of energy losses up to ~ 100 volts, the former procedure affects the intensity of the incident beam as calibration proceeds. For normal operation the switch S_1 is closed and the switches S_2, S_3, P_1, P_2 etc. are opened. For calibration S_1 is opened, S_2 and S_3 closed, and by successively closing and reopening the switches P_1, P_2 etc. known increments of voltage are added to the H.T. bias to simulate the required energy losses. A safety device worthy of note here is the spark gap G which connects the cathode line to the analyser lead (Fig. 54). The Siemens H.T. generator is fitted with an electronic device which switches off the H.T. to the gun if the latter sparks over. The safety device is insensitive to appreciable voltage changes of the filament and only acts if the voltage on the cathode line falls below a predetermined level. If the spark gap G (Fig. 54), which fires when a voltage greater than about 800 volts is applied across it, is not present in the circuit, then damage to various electronic components in the Siemens' generating unit will occur.

The second system (Considine and Smith, 1968, 1971; Considine, 1970), which will now be described, is a cylindrical magnetic analyser unit attached to the Cavendish 750 kV electron microscope (Smith *et al.*, 1966). A discussion of the energy resolution and dispersion of this system can be found in Section IIB. The main factors which influenced the design of this energy analysing attachment are:

(a) the very small dispersion obtained with the magnetic system (~ 5 μm/volt at 100 kV);

(b) the restricted space between the base of the camera chamber and floor of the Cavendish high voltage microscope. Employment of photographic methods for recording loss spectra requires the incorporation of an alignment coil and auxiliary lens with the analyser unit. If the spatial resolution of photographic emulsion is taken as 100 μm and the dispersion as 5 μm/volt at a beam potential of 100 kV, then to achieve an

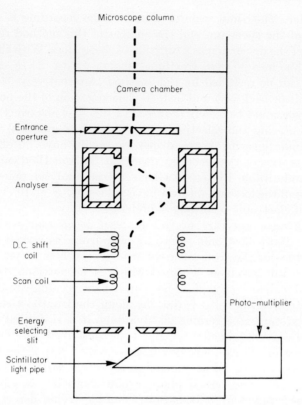

FIG. 55. An outline of the analyser unit attached to the Cambridge 750 kV electron microscope.

energy resolution of 1 eV an auxiliary lens of magnification $20\times$ is required. The size of a lens producing a magnification of this order, with incident beam potentials up to 750 kV, is such as to prohibit the use of photographic techniques with the Cavendish microscope. This problem was overcome by using electronic methods for recording the loss spectra.

A schematic outline of the unit designed by Considine and Smith (1968) is given in Fig. 55. The incoming electron beam passes through the entrance aperture and is dispersed by the analyser into its various energy loss components. One of the loss components is deflected in the x direction (Fig. 21), by a D.C. shift coil, on to an energy selecting slit situated immediately above the scintillator of a photomultiplier unit. The loss spectrum is scanned across the selecting slit by means of a second coil, the output of the photomultiplier is amplified and the resulting signal fed into the y amplifier of either an oscilloscope or an

xy recorder. The time scale or x axis of the recording is the energy loss axis of the spectrum and the success of this method relies on the linearity of the dispersion as a function of energy loss. With the magnetic system this means that fractional losses up to $\Delta E/E \approx 3 \times 10^{-3}$ (Fig. 36) can be recorded by this method. Alternatively, the scan coil could be dispensed with and instead the excitation of the pole pieces of the analyser could be slowly reduced. In this case electrons of different energies entering the selecting slit would follow identical trajectories and any non-linearity of the dispersion would have no effect on the energy loss axis of the recorded spectrum. This method can, however, lead to undesirable hysteresis effects in the analyser pole pieces and since most of the losses of interest have values $\Delta E/E \leqslant 10^{-3}$, the former method of electronic recording is preferable.

The entrance aperture of the analyser is circular with diameter $\sim 30~\mu\mathrm{m}$. With electronic recording no purpose is served by making this aperture in the form of a slit; the reason being that the photo-multiplier unit produces a signal which is proportional to the total intensity incident on the phosphor and if a slit is employed the detector is insensitive to intensity variations along the length of the slit. This puts the system at a serious disadvantage if it is desired to perform microanalysis experiments of the type described in Section IIIB. These experiments can only be carried out with ease if the loss spectra from points lying on a line selected in the image can be recorded simultaneously, as is the case if photographic methods are employed. To obtain the equivalent information when electronic detection is used involves displacing the image repeatedly by small amounts in a given direction and recording the spectrum obtained at each displacement, a process which is at best tedious.

The system can, however, be readily converted into a high voltage energy selecting electron microscope. To achieve this, the scan coil of the unit illustrated in Fig. 55 is dispensed with and the D.C. shift coil is used to deflect a given loss component of the electron beam on to the selecting slit. A set of xy deflecting coils placed between the objective and intermediate lenses of the microscope allows the final image to be scanned in a raster across the entrance aperture of the analyser. The output of the photomultiplier is used to modulate the brightness of an oscilloscope whose raster is in synchronism with the scan of the microscope image. These modifications, if carried out, would result in the display of an energy selected image on the cathode ray tube of a T.V. display unit.

The two coils of the analyser each consist of 500 turns of copper wire encapsulated in vacuum tight aluminium cans. These cans have small flexible pipes which communicate through a vacuum port to the external atmosphere. This alleviates heating, contamination, and vacuum pumping problems which would result if the coils were run in vacuum. The D.C. shift coil, consisting of 750 turns, and the scan coil, consisting of 375 turns are both wound on the *same* pair of pole pieces, the latter being square in cross-section. The use of square pole pieces, rather than circular, eliminates to a first order approximation astigmatism in the x-direction (Fig. 21). This design does not, of course, eliminate astigmatism in the z-direction, but a focusing action in this direction is not a disadvantage since it is perpendicular to the direction of dispersion. The energy selecting aperture is in the form of a slit, rather than a circular aperture, to alleviate the problem of alignment. It is important that the width w of this slit should be nearly the same as the image width δx_1 of the entrance aperture of the analyser (Fig. 33). If w is greater than δx_1 then the energy resolution is determined by w rather than by the electron optical properties of the analyser, and also there is little point in making w less than δx_1 since no increase in the energy resolution occurs, and the only result is a reduction in the intensity arriving at the scintillator of the photomultiplier unit. When electronic methods are used to record the loss spectrum, it is desirable to use values of the excitation parameter k (Section IIB) close to k_{min} (Fig. 35) so that only a weak excitation of the D.C. shift coil is needed to deflect the beam on to the selecting slit (Fig. 38). The variation of the image width δx_1 with excitation parameter k (Fig. 35) shows that at $k = k_{min}$ the slit width which must be employed is ~ 2 μm. This width is difficult to achieve in practice and a more reasonable value for the lower limit of the slit width is ~ 4 μm. A value of $\delta x_1 = 4$ μm is obtained when $k = 0.52$ amp. turns (volt)$^{-\frac{1}{2}}$ and this represents the minimum value of k that can be used with the analyser when electronic methods are used to record the energy spectrum.

Energy loss calibration is carried out by introducing known changes in the accelerating potential of the gun of the electron microscope. The H.T. generator of the Cavendish 750 kV microscope is air insulated and this allows easy access to the electron gun and injector electronics. Insertion of auxiliary equipment inside the torus dome of the accelerator unit allows the beam potential to be changed by known increments and hence calibration of loss spectra is readily achieved.

B. *Applications*

An energy analysing electron microscope can be operated to record either the loss spectrum of electrons forming a selected area diffraction pattern (Figs. 56a and b) or the spectrum of electrons which contribute

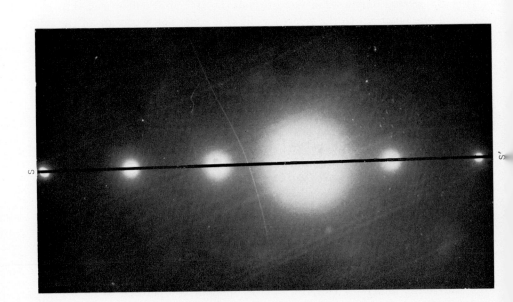

FIG. 56. (a) A diffraction pattern of an aluminium specimen and (b) the energy loss spectrum of electrons

to the image of a specimen (Figs. 58 and 59). The diffraction pattern of a specimen of aluminium at a beam potential of 100 kV is given in Fig. 56a and the loss spectrum of electrons arriving along the line SS' in this pattern is shown in Fig. 56b. The lines $S_0 S_0'$, $S_1 S_1'$, and $S_2 S_2'$ indicated in Fig. 56b are the zero loss, first plasmon loss (15 eV) and second plasmon loss (30 eV) lines, respectively, and the familiar parabolic dependence of the plasmon energy loss on the angle of scatter is clearly evident in this spectrum. It is noticeable that the loss lines $S_0 S_0'$ are curved (see also Figs. 58b and 59b). This curvature is due partly to end caps fitted to the inner electrodes of the analyser and partly due to the

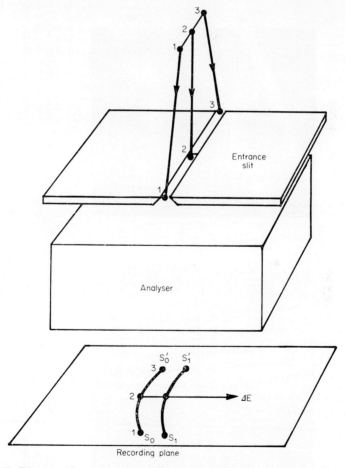

Fig. 57. Diagram illustrating one of the reasons why the energy loss lines appear curved (see text for discussion).

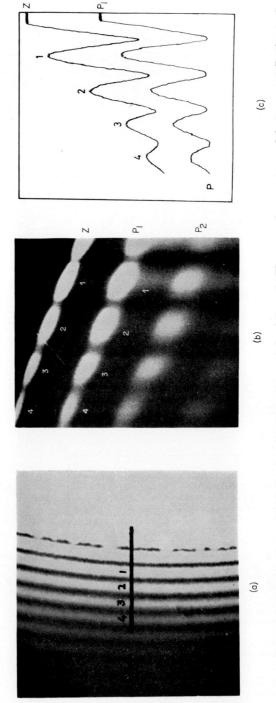

FIG. 58. (a) An electron micrograph of a wedge shaped specimen of aluminium. (b) The energy loss spectrum of electrons contributing to the intensity variation along the line indicated in the micrograph. Z, P_1 and P_2 are the zero loss, first and second plasmon losses respectively. (c) Microdensitometer traces taken from the lines Z and P_1 of the spectrum (after Spalding, 1970).

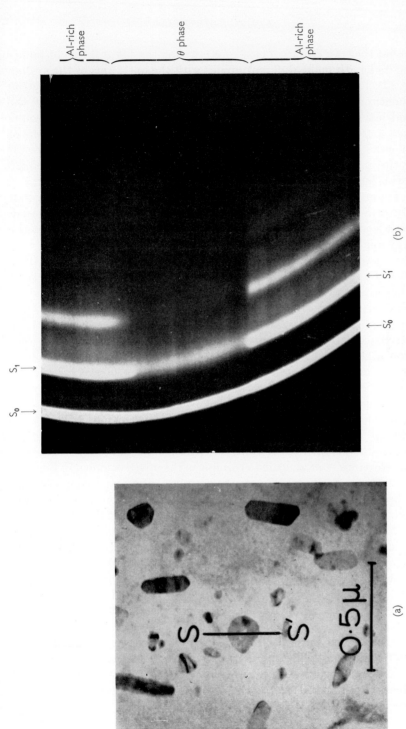

Fig. 59. (a) Micrograph of θ phase precipitates in solid solution Al-4% Cu alloy. (b) The loss spectrum of electrons contributing to the intensity along the line indicated in the micrograph (after Cundy *et al.*, 1968, courtesy of *The Philosophical Magazine*).

fact that the trajectories of electrons entering the slit at different points along its length make different angles with the optic axis of the system (Fig. 57). The analyser is only sensitive to the component of the electron momentum perpendicular to the plane containing the axes of the cylindrical electrodes. Electrons of the same energy arriving at points 1 and 3 (Fig. 57) on the entrance slit have normal components of momenta less than that of an electron arriving at point 2, with the result that the analyser registers an apparent energy loss for electrons 1 and 3 and this leads to the curvature of the loss lines shown in Figs. 56b and 59b. For an energy *selecting* electron microscope utilizing the Möllenstedt analyser (Section IVA) it is important that this curvature be eliminated, since it leads to a first order aberration of the filtered image. For the energy *analysing* electron microscope however, this curvature is unimportant and needs no correction; the reason being that a loss spectrum is always recorded together with a calibration plate and any distortions or aberrations produced by the system affect both equally.

Most of the experiments involving the use of the energy analysing electron microscope have relied on measurements of the loss spectra of electrons contributing to images rather than diffraction patterns. The reason is that a selected area diffraction pattern contains information averaged out over an area of specimen of the order of several microns across, whereas with an image the spatial resolution, as limited by the width of the analyser entrance slit, is of the order of several Ångstroms. Bright field thickness fringes obtained with a specimen of aluminium is shown in Fig. 58a. The horizontal line drawn on this micrograph indicates the position of the analyser entrance slit and the loss spectrum of the electrons contributing to this line in the image is shown in Fig. 58b. In this figure, Z, P_1 and P_2 are the zero loss, first and second plasmon loss lines respectively. In Fig. 58c are shown microdensitometer traces of the zero loss and first plasmon loss components of the spectrum of Fig. 58b, and it is clear from both the loss spectrum itself and the intensity traces of Fig. 58c that the plasmon loss electrons preserve the contrast observed in the image. This preservation of contrast by the inelastically scattered electrons, first observed by Kamiya and Uyeda (1961), has been the subject of intensive experimental studies involving the use of both energy analysing (Cundy *et al.*, 1966, 1967a; Cundy *et al.*, 1969) and energy selecting electron microscopes (see Section IVB for references). The preservation of image contrast by the inelastically scattered electrons allows details appearing in the image to be related to details appearing in the loss spectrum and means that the position of the analyser slit relative to the image need not be known to any high degree of accuracy.

An example of the loss spectrum obtained in an alloy system in which the composition of the specimen varies across the field of view is given in Fig. 59b. An electron micrograph of θ phase precipitates of composition $CuAl_2$ imbedded in a matrix of nearly pure aluminium is shown in Fig. 59a and the line SS' marked on this micrograph indicates the analyser entrance slit position. The lines S_0S_0' and S_1S_1' in Fig. 59b indicate the zero loss and first plasmon loss lines of the spectrum corresponding to the line SS' in Fig. 59a and a marked difference between the loss spectrum of the θ phase precipitate and the matrix material is clearly visible. The rest of this section is devoted to a discussion of the principles involved in the microanalysis of binary alloy systems.

The binary alloys most amenable to microanalysis studies are those with constituents which possess sharply defined and well separated energy losses. One of the best examples is the Al–Mg system which has been studied in some detail by Spalding and Metherell (1968). For aluminium the mean plasmon loss, corresponding to the peak P of Fig. 60 is $15\cdot3 \pm 0\cdot1$ eV, and for magnesium it is $10\cdot4 \pm 0\cdot1$ eV. The value of the ratio of the loss half-widths w/w_0 is $1\cdot7$ for both elements, and since w_0 is typically ~ 2 eV at 100 kV this gives a halfwidth

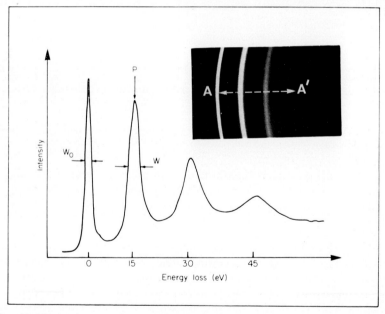

Fig. 60. Energy loss spectrum of aluminium. The line AA' is the loss axis and P is the mean loss (after Spalding and Metherell, 1968, courtesy of *The Philosophical Magazine*).

$w \sim 3\cdot5$ eV. The loss lines of these elements are therefore both well separated and sharply defined. The variation of the mean plasmon loss with alloy composition is given in Fig. 61 and the broken vertical lines of this figure indicate the various phase boundaries of the system (see, for example, Hansen, 1958). The main point of interest here is that for the solid solution a, γ and δ phases the mean plasmon loss varies linearly

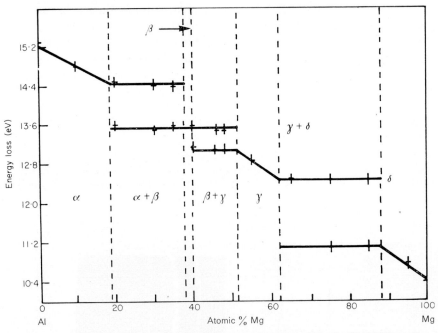

FIG. 61. Variation of the mean loss (P, Fig. 60) with composition in the Al-Mg system (after Spalding and Metherell, 1968, courtesy of *The Philosophical Magazine*).

with composition, within the limits of experimental error. According to the simple free electron theory of plasmon excitation (see, for example, Raimes, 1967) the energy loss E_p is given by

$$E_p = \hbar \, (4\pi n e^2/m)^{\frac{1}{2}}$$

where n is the number of free electrons per unit volume and the other symbols have their usual meaning. The variation of E_p with alloy composition is therefore expected since the addition of magnesium, with two free electrons per atom, to aluminium, with three free electrons per atom, simply dilutes the free electron concentration in the solid solution alloy. Similar measurements of the mean loss in the Al-rich phase of solid solution Al-Cu alloys have been reported by Spalding *et al.* (1969), and their results are shown in Fig. 62. The non-linear dependence of the loss on copper concentration is due to band structure effects (see for example Raether, 1965) which are not taken into account in the simple theory of plasma oscillations.

The loss measurements discussed above were made on alloy samples of known composition. It is obvious that if the dependence of energy loss on composition is known then a measurement of the loss can be used to determine the composition of an unknown specimen, and further-more, if advantage is taken of the spatial resolution afforded by an

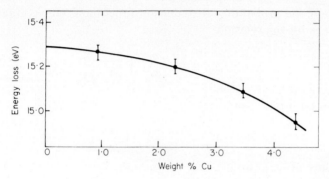

Fig. 62. Variation of the mean loss (*P*, Fig. 60) with composition in the Al-rich solid solution phase of the Al-Cu system (after Spalding *et al.*, 1969, courtesy of *The Philosophical Magazine*).

energy analysing electron microscope, it must be possible to perform a microanalysis of an inhomogeneous alloy. Other factors unconnected with the alloy composition which could influence the energy loss, such as the effect of vacancy concentration (Cundy, 1968; Cundy *et al.*, 1968) or the effect of elastic constraints in an inhomogeneous alloy (Cook and Howie, 1969), have been considered but are thought to be unimportant in any cases of practical interest that have arisen so far.

Before discussing any specific examples of microanalysis experiments it is pertinent to consider the factors limiting the spatial resolution available with this technique. The spatial resolution as limited by the analyser entrance slit can be reduced to a value below that of the microscope simply by increasing the magnification of the image sufficiently. A far more important factor limiting the spatial resolution involves the physics of the scattering process. A plasma oscillation in a metal is simply a quantized compression wave of the free electron gas. The plasmon energy is determined by the conduction electron density averaged over some domain in the crystal, the size Δx of this domain

being the extent of the plasmon wave packet initially excited by the
incident electron. For an inhomogeneous alloy, Δx is therefore the
spatial resolution of the microanalysis.

An approximate value for Δx can be obtained from the uncertainty
relation

$$\Delta x \Delta q \simeq 1$$

where Δq is the spectral width of the plasmon wave packet in k space.
The probability $P(q)$ that the incident electron excites a plasmon into a
plane wave state of wave vector q, where q is restricted to have values
between q_{min} and q_c, is proportional to $1/q^2$ (Ferrell, 1956). Defining the
width Δq of the probability distribution function as the average value
of q, that is by the relation

$$\Delta q = \int_{q_{min}}^{q_c} \frac{q\,dq}{q^2} \Big/ \int_{q_{min}}^{q_c} \frac{dq}{q^2}$$

we obtain on integration

$$\Delta q = \left(\frac{q_c q_{min}}{q_c - q_{min}} \right) \ln \left(\frac{q_c}{q_{min}} \right)$$

The values of q_{min} and q_c for aluminium are $1 \cdot 9 \times 10^{-3}$ Å^{-1} and
$2 \cdot 2 \times 10^{-1}$ Å^{-1} respectively, giving $\Delta q \approx 10^{-2}$ Å^{-1} and hence that
$\Delta x \approx 100$ Å. The spatial resolution as limited by the physics of the
inelastic scattering process is therefore ~ 100 Å and this is considerably
larger than that limited by the width of the analyser entrance aperture.

The principles outlined above have been applied to a study of
segregation and initial stages of precipitation in Al-7% Mg alloy (Cundy
et al., 1968b); a study of the dislocation loop growth mechanism in Al-3%
Mg alloy (Spalding et al., 1969a); and a study of the copper distribution
in lamellar Al-CuAl$_2$ eutectics (Spalding et al., 1968, 1969a). The micro-
structure of a eutectic Al-CuAl$_2$ aged for several months is shown in the
electron micrograph of Fig. 63. By carefully tilting the specimen it is
possible to orientate one of the lamellar interfaces parallel to the
direction of incidence of the electron beam. This procedure is necessary
to avoid the possible superposition of energy losses from both phases
(Cundy et al., 1968a). Measurements of the energy losses of electrons
contributing to the image at points lying in a line, selected by the
analyser entrance slit, perpendicular to the interface boundary are given
in Fig. 64. In region I, which lies in the Al-rich phase and extends over
almost all of this phase, the energy loss is constant and has a value of
$15 \cdot 25 \pm 0 \cdot 03$ eV, the error quoted being the standard deviation of a
number of measurements of the mean plasmon loss. In region II, which

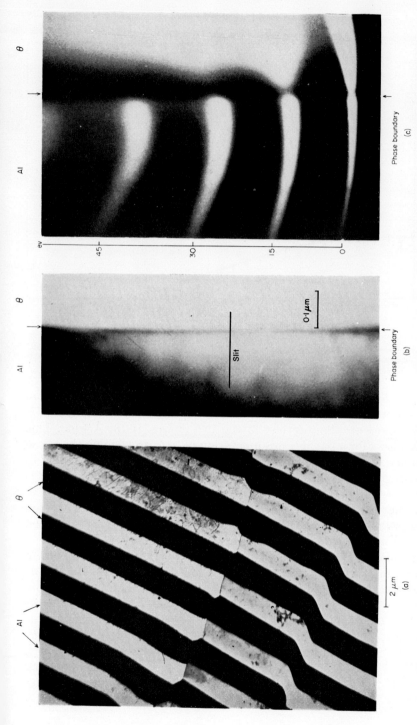

FIG. 63. (a) Micrograph of Al-CuAl₂ eutectic alloy showing the lamellar substructure of the specimen. The light regions are the Al-rich lamellae (courtesy of F. D. Lemkey and W. Tice). (b) Region of an Al-CuAl₂ specimen near the phase boundary (after Spalding et al., 1969b). (c) Loss spectrum taken from the line indicated in (b) (after Spalding, 1970).

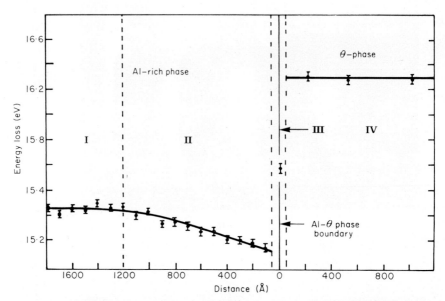

Fig. 64. Variation of the mean loss (P, Fig. 60) as a function of distance from the phase boundary in an aged specimen of Al-CuAl$_2$ eutectic alloy (after Spalding *et al.*, 1969b, courtesy of *The Philosophical Magazine*).

extends along the phase boundary and has a width ~ 1000 Å, the loss varies from $15\cdot25 \pm 0\cdot03$ eV down to $14\cdot97 \pm 0\cdot03$ eV. Region III, which represents the interface of the two phases, is 100 Å wide. This width allows for a possible misorientation of $\pm 3°$ in aligning the phase boundary parallel to the incident direction of the electron beam, and in this region the loss is $15\cdot58 \pm 0\cdot09$ eV. The loss in region IV, which is the θ phase region of the alloy, is constant at $16\cdot34 \pm 0\cdot09$ eV. The smaller error associated with the loss measurements in the Al-rich phase is due to the fact that the energy loss is more sharply defined (Fig. 65) in this region than in the θ phase region.

The solute concentration profile in the Al-rich phase, obtained from the loss measurements and the calibration curve of Figs. 62 and 64 respectively, is given in Fig. 66. At distances greater than ~ 100 Å from the phase boundary, the solute concentration ($\sim 1\cdot5\%$) is much lower than that expected from eutectic growth theory ($\sim 6\%$, see for example Chadwick, 1963). Similarly the increasing concentration in the vicinity of the interphase boundary is also contrary to that predicted by theory. Suggestions for these discrepancies are given in the paper by Spalding *et al.*, (1969b) to which the interested reader is referred. At the phase boundary the incident electrons are tangential to the interface and a

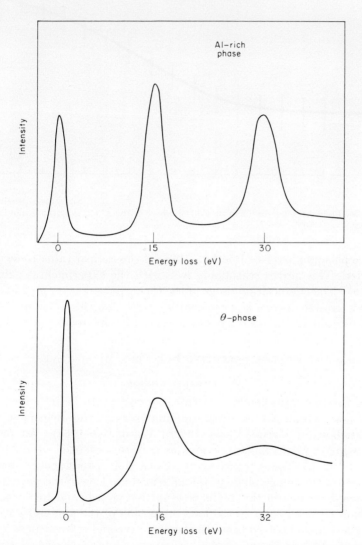

FIG. 65. Energy loss profiles in the Al-rich and θ phase (regions I and IV of Fig. 64) of the Al-CuAl$_2$ eutectic alloy (after Spalding *et al.*, 1969b, courtesy of *The Philosophical Magazine*).

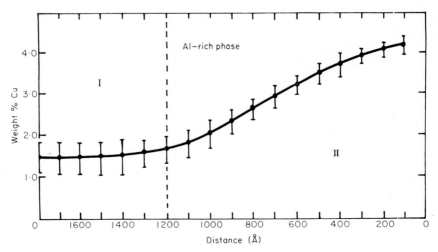

FIG. 66. Concentration profile in the Al-rich phase of the Al-CuAl$_2$ eutectic alloy (after Spalding *et al.*, 1969b, courtesy of *The Philosophical Magazine*).

surface plasmon loss (see Raether, 1965) of theoretical value 15·68 eV is expected. This agrees reasonably well with the experimental value of 15·58 eV. The composition range of the CuAl$_2$ phase region predicted by eutectic growth theory is very small ($\sim 1\%$) and this is confirmed by the constancy of the energy losses observed in this region.

IV. Energy Selecting Electron Microscopes

A. *Instrumentation*

The first energy selecting electron microscope to be described here utilizes a cylindrical electrostatic analyser as the dispersive unit (Watanabe and Uyeda, 1962a, 1962b; Watanabe, 1964). An outline of this instrument, which operates at a beam potential of 100 kV, is given in Fig. 67. The entrance aperture of a Möllenstedt energy analyser is placed in the image plane of the intermediate lens of the microscope. The image formed in this plane is swept backwards and forwards in a direction perpendicular to the long dimension of the analyser entrance slit by a scan coil placed between the objective and intermediate lenses. A second slit selects one of the loss components of the beam passing through the dispersive system, and this component is scanned by a second coil worked in synchronism with the sweep of the image formed by the intermediate lens. After magnification by the projector lens an energy selected image is therefore swept out on the final image plane of the microscope.

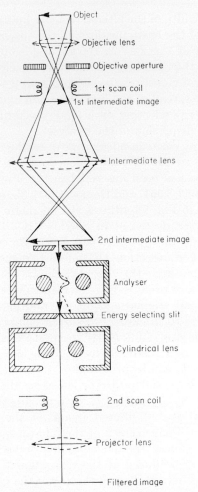

FIG. 67. An outline of the energy selecting electron microscope which employs the cylindrical electrostatic analyser (Section IIA) as the dispersive unit.

A cylindrical electrostatic lens is placed between the analyser and the projector lens to eliminate the curvature of the loss line selected to form the filtered image. The reason for this curvature is discussed in Section IIIB, and although there is no necessity for correcting this curvature in the energy analysing electron microscope, it will, if left uncorrected in the energy selecting microscope, produce a first order distortion in the filtered image. Any curvature of the selected loss line can be eliminated by either translating the cylindrical lens into a suitable off-axis position, or by rotating the lens about an axis parallel to the axes of the cylindrical electrodes.

The energy bandwidth of the filtered image is limited primarily by the energy spread ΔE of the beam incident on the specimen. This spread is usually of the order of 1 or 2 eV at 100 kV and is much larger than the ultimate resolution that can be obtained with the electrostatic analyser (Section IIA). It is obviously desirable to obtain a bandwidth which is equal to the energy spread of the incident beam and this is achieved if the width of the energy selecting slit is made equal to the product of the spread ΔE and the dispersion of the analyser.

An energy selecting electron microscope operating at beam potentials in the range 50 kV to 100 kV and which does not involve the use of scanning techniques has been described in the literature by Castaing and Henry (1962, 1963, 1964) and Henry (1964). This instrument, which is shown in outline in Fig. 68, utilizes the magnetic prism and electrostatic mirror analyser described in Section IIC. The cross-over and image plane of a first intermediate lens are made coincident with the real stigmatic point R_1 (Section IIC) of the magnetic prism and the achromatic plane containing the virtual stigmatic point V_1 respectively. An achromatic image is produced by the dispersive system in the plane containing the virtual stigmatic point V_3 and electrons of different energies are brought to different cross-over points in the plane containing the real stigmatic point R_3. If an aperture is used to select one of the cross-over points produced by a given loss component of the beam, a filtered image can be produced on the final screen by focusing the second intermediate and projector lenses on the achromatic plane containing the point V_3. Alternatively, by removing the energy selecting aperture, a conventional unfiltered image of the specimen can be produced on the final image screen of the microscope.

The energy resolution of the device is determined primarily by the size of the cross-over produced at the real stigmatic point R_1. If the diameter of the cross-over at R_1 is d, then since the magnification of the dispersive system is unity, the energy resolution is given by $\Delta E_r = d/D$, where D is the dispersion of the analyser. The design dimensions of the prism used by Castaing and Henry are (Section IIC) $\epsilon = 45°$, $a = 4$ cm, $\zeta_1 = -8$ cm, $\zeta_1' = 5$ cm, and for this geometry the dispersion at the plane containing the point R_3 is 2 μm/volt for a beam potential of 100 kV. The focal length of the microscope objective lens is $f_1 = 3\cdot2$ mm and for an objective aperture semi-angle $\theta = 10^{-3}$ rad. the cross-over diameter produced by the objective is $d_1 = 2f_1\theta = 6\cdot4$ μm. If the analyser is placed immediately after the objective lens the resolution obtained is ~ 3 eV at 100 kV. To obtain a smaller resolution requires a reduction in the cross-over diameter and this can be achieved by placing the dispersive system after the first intermediate lens. In the

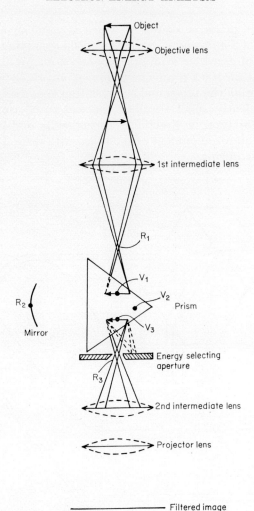

Object
Objective lens
1st intermediate lens
R_1
V_1
V_2
Prism
R_2
V_3
Mirror
Energy selecting
aperture
R_3
2nd intermediate lens
Projector lens

Filtered image

FIG. 68. An outline of the energy selecting electron microscope which utilizes the mirror-prism device (Section IIC).

instrument described by Castaing and Henry this lens is situated at a distance $l = 160$ mm from the objective and has a focal length f_2 which is variable between 15 and 25 mm. The diameter of the cross-over produced by the first intermediate lens is $d_2 = 2f_1\theta f_2/l$, and for the values of f_2 given above, this gives 0.6 μm $< d_2 < 1$ μm. The resolution limited by the cross-over size is therefore ~ 0.5 eV and in practice this means that the resolution is limited mainly by the spread in energy of the incident beam which is typically of the order of 1 or 2 eV at 100 kV.

The magnetic prism, a sketch of which is given in Fig. 69, has a pole-piece gap of 2 mm. and is excited by a coil consisting of 5000 turns of copper wire of diameter 0·13 mm. wound on an Araldite core. At a beam potential of 100 kV an excitation of about 43 amp. turns is required to operate the prism and the resulting heat dissipation ($\sim$ 0·5 watts) is sufficiently small to allow the coil to be run in vacuum. Mechanical translation in directions perpendicular and parallel to the axis of the coil allows alignment of the prism with respect to the axis of the electrostatic mirror.

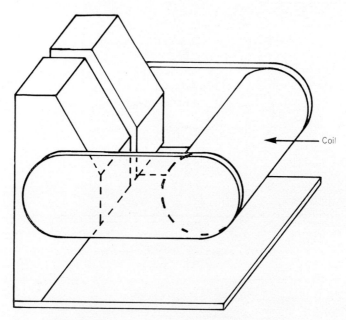

FIG. 69. The magnetic prism (after Henry, 1964).

The mirror consists of a modified electron gun, in which the filament has been replaced by a cylindrical cathode C (Fig. 70). The cathode is biased slightly negative with respect to the beam potential and the Wehnelt cylinder is biased negative with respect to the cathode. The electrode system therefore has a field distribution similar to that of an immersion lens. Adjustment of the position of the pole of the mirror is effected by mechanical movement of the gun and the focal length is adjusted by altering the bias of the cathode and Wehnelt cylinders.

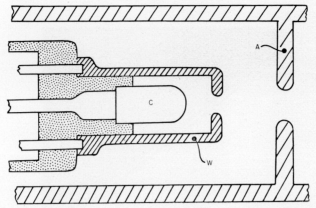

FIG. 70. The electrostatic mirror. A: anode, C: cathode, W: Wehnelt cylinder (after Henry, 1964).

B. *Applications*

Most of the experimental work involving the use of energy selecting electron microscopes has been directed at investigating the preservation of image contrast by inelastically scattered electrons. By displacing the objective aperture of a conventional electron microscope so that no Bragg scattered electrons contribute to the image, Kamiya and Uyeda (1961) demonstrated (Fig. 71) that inelastic scattering does not destroy image contrast. In this simple experiment the energy losses of the electrons are unknown and it is therefore impossible to discriminate between different types of inelastic scattering process. Using the energy selecting electron microscope described in Section IVA, Watanabe and Uyeda (1962a, b) were able to show that electrons which lose energy by plasmon excitation can be used to form filtered images which exhibit contrast effects similar to those observed in the ordinary image (Fig. 72). This result was given theoretical justification by Fujimoto and Kainuma (1961, 1963), Howie (1962, 1963) and Fukuhara (1963). A further series of experiments (Castaing *et al.*, 1966a,b; Cundy *et al.*, 1966; Cundy *et al.*, 1969) have shown that all electrons with losses lying between about 1 eV and 100 eV also appear to preserve image contrast.

The question remains as to whether or not thermal diffuse scattering also preserves contrast. It is not possible with present instrumentation to resolve the energy losses of electrons which have excited phonons, since these lie within the energy spread of the incident beam. An indirect experiment, in which the objective aperture is displaced so that no Bragg beams contribute to the image, can however determine whether or not contrast is preserved by inelastic scattering processes producing energy losses in the range 0–1 eV. Most of the electrons contributing to

(a)

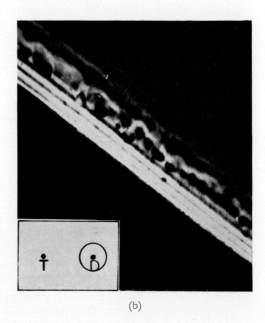

(b)

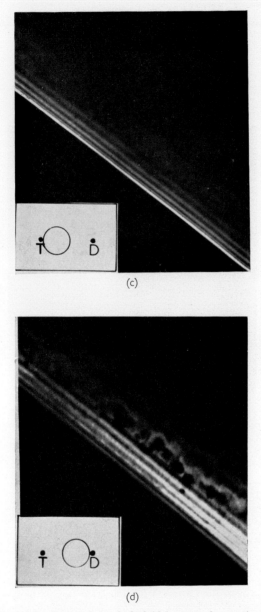

(c)

(d)

FIG. 71. (a) The bright field and (b) dark field images of a specimen of aluminium. (c) The image obtained by displacing the objective aperture so as to exclude the Bragg scattered electrons. The position of the aperture is indicated in the inset where T represents the central spot and D the diffracted spot of the diffraction pattern. (d) The image obtained when the objective aperture is displaced in the vicinity of the diffracted spot D. In (a) and (b) both elastically and inelastically scattered electrons contribute to the contrast observed, whereas in (c) and (d) only the inelastically scattered electrons contribute to contrast.

Ordinary image

(a)

FIG. 72. Examples of energy selected images of Al at 100 kV. (a) The ordinary image. (b) The zero-loss image and (c) the 15 eV loss image showing the preservation of contrast by plasmon scattered electrons (courtesy of Dr. H. Watanabe).

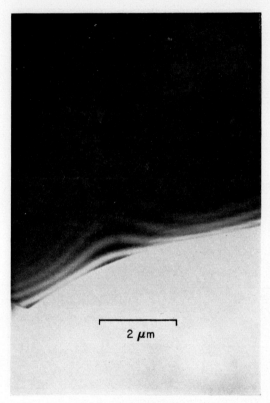

Zero loss image

Fig. 72 (b).

15 eV loss image

FIG. 72 (c).

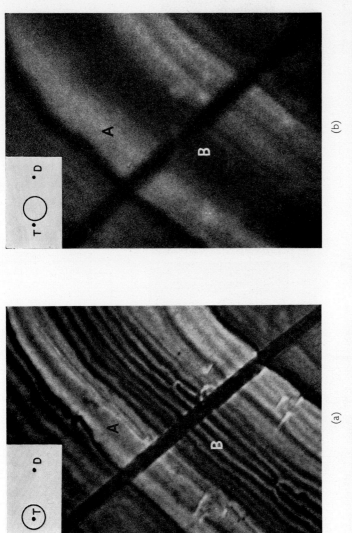

(a)   (b)

Fig. 73. Filtered images of a specimen of Au obtained by selecting electrons with energy losses less than 1 eV. The positions of the objective aperture used to form these images are shown in the insets. In (a) the strong fringe contrast at A and B is due almost entirely to Bragg scattered electrons. In (b) which is the image formed by the inelastic component with losses less than 1 eV, the fringe contrast at A and B has almost entirely disappeared (after Hili, 1967, courtesy of Professor R. Castaing, Dr. L. Henry and *Journal de Microscopie*).

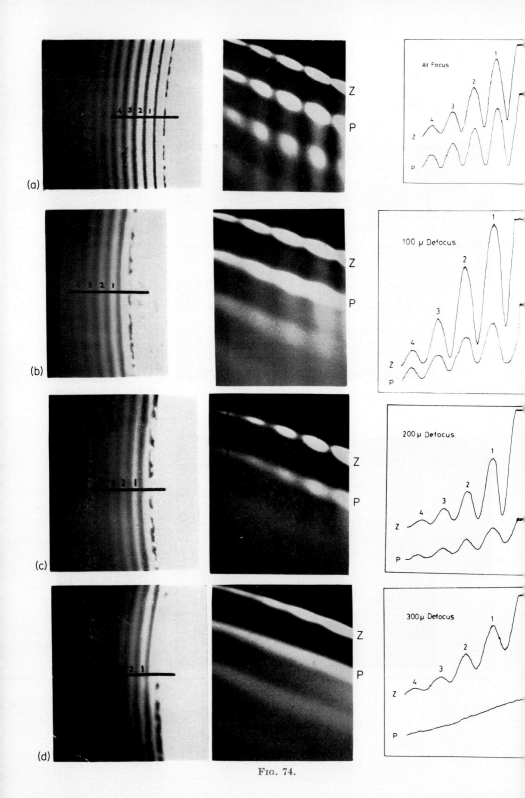

FIG. 74.

this loss range can be expected to have been inelastically scattered by the excitation of phonons, since the displaced aperture selects only those electrons in the diffuse background surrounding the Bragg spots of the diffraction pattern. Filtered images of aluminium specimens obtained with a displaced objective aperture have been published by Castaing et al., 1966a, b, and these show weak contrast effects in the vicinity of thickness fringes and bend contours. Reduced contrast was also observed by Cundy et al., 1967a, at thickness fringes in aluminium, but these authors pointed out that since the contrast was so weak there was some uncertainty as to whether it was due to phonon scattering or due to electrons entering the objective aperture after elastic scattering in oxide or contamination layers present on the surface of the aluminium specimens. In a further series of experiments (Castaing et al., 1967; Castaing et al., 1968) great care was taken to ensure that the specimens examined were free of surface films. This was achieved by using specimens of gold, to eliminate oxide films, mounted in an anti-contamination stage, to reduce the deposition rate of carbon on the specimen surfaces to a negligible level. Even under these stringent conditions weak contrast effects (Fig. 73) are still observed, provided that sufficiently small objective apertures are used, indicating that a proportion of the phonon scattered electrons preserve image contrast. In an experiment reported by Cundy et al., (1969), however, no discernible contrast at stacking fault images could be observed in the inelastic component of the 0–1 eV loss. These authors used an objective aperture somewhat larger than those employed by Castaing and his colleagues and it is possible that the discrepancy between the two sets of results lies in a destruction of contrast by diffuse scattering, as discussed by Metherell (1967b) and Spalding and Metherell (1970) for the case of plasmon scattering.

A study of the preservation of contrast by plasmon scattered electrons in out of focus images has been made by Spalding (1970) and Spalding and Howie (1971) using an energy analysing electron microscope. A through focal series of electron micrographs of a wedge shaped specimen of aluminium is given in Fig. 74a, b, c and d. The loss spectra of electrons arriving at the lines indicated on each image, together with microdensitometer traces which show the thickness fringe intensity variations along the zero loss (Z) and plasmon loss lines (P), are also given in this figure. At a defocus of 300 μm (Fig. 74d) the contrast due to the plasmon scattered electrons

FIG. 74. A through focal series of images of a specimen of Al, together with the loss spectra from the lines indicated in the micrographs. Microdensitometer traces of the zero loss (Z) and plasmon loss (P) lines are also given. (a) at focus, (b) 100 μm defocus, (c) 200 μm defocus, (d) 300 μm defocus. The contrast in the plasmon loss line disappears at a defocus of 300 μm (after Spalding, 1970).

has disappeared completely, whereas some contrast is still observable in the zero loss electrons. The reason for this difference in behaviour of the elastically and inelastically scattered electrons is almost certainly due to the greater divergence angle associated with the latter on emergence from the specimen. The angle of collimation a_0 (Fig. 75) of the incident beam is typically $\sim 10^{-4}$ rad., whereas the mean angle of scatter a_p for $100\,kV$ electrons which have lost energy by plasmon excitation is $\sim 10^{-3}$ rad. The image formed by the plasmon scattered electrons therefore goes out of focus more rapidly than the zero loss electrons as defocus proceeds.

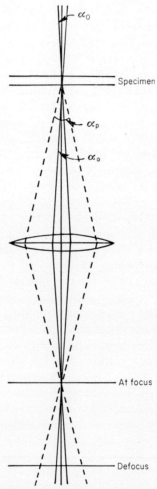

FIG. 75. Diagram illustrating the mechanism responsible for the loss of contrast observed in the through focal series of micrographs of Fig. 74.

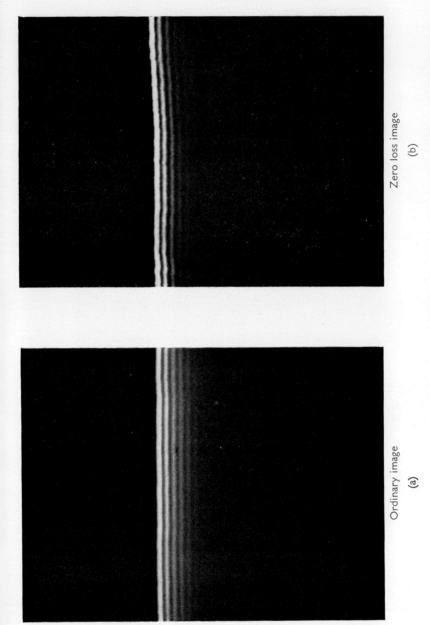

Ordinary image

(a)

Zero loss image

(b)

FIG. 76. (a) The ordinary image and (b) the zero loss image of an MgO specimen at 75 kV. Exclusion of the inelastically scattered component of the electron beam changes the mean absorption coefficient for the ordinary image from μ_0 to $\mu_0 + 1/\lambda$ for the zero loss image, where λ is the mean free path for inelastic scattering (courtesy of Dr. H. Watanabe).

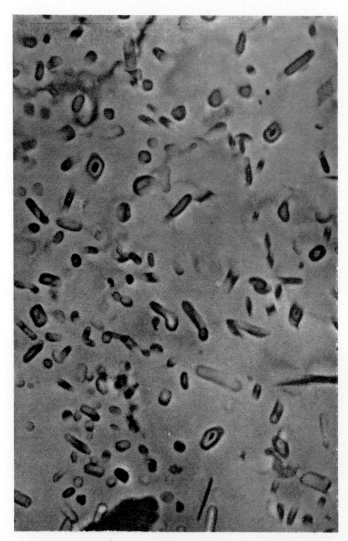

FIG. 77 (a).

FIG. 77. Filtered images obtained with a specimen of Al-7·6% Zn-2·6% Mg alloy. (a) The 14·6 ± 1 eV image showing the η phase precipitates, which appear dark against a light background. (b) The 22·5 ± 2·5 eV image, in which the η phase precipitates appear light against the darker background (after Hili, 1966, courtesy of Professor R. Castaing, Dr. L. Henry and *Journal de Microscopie*).

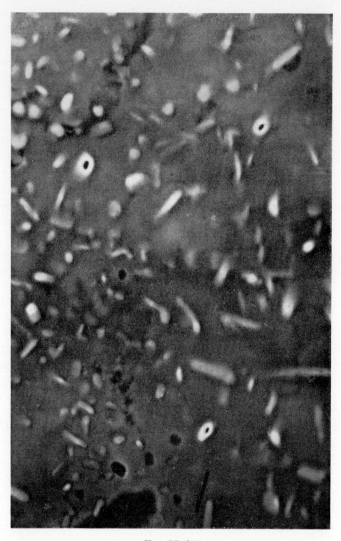

FIG. 77 (b).

The effect of inelastic scattering on absorption contrast has been studied by Watanabe (1966). In his experiments the mean absorption coefficients of the ordinary image (Fig. 76a) and the zero-loss image (Fig. 76b) of an MgO specimen were obtained from measurements of the intensity profiles of thickness fringes. A discussion of his results can be found in papers by Cundy et al. (1967b) and Metherell (1967c) to which the interested reader is referred.

As a microanalyser the energy selecting electron microscope appears at present to be of somewhat more restricted application than the energy analysing electron microscope; the reason being that the band-width of filtered images is limited by the energy spread of the incident beam (~ 1 eV) whereas the microanalysis experiments, described in Section IIIB, involving the use of an energy analysing electron micro-scope, rely on the measurement of the mean plasmon energy and with care random errors associated with these measurements can be reduced $\pm\ 0.03$ eV (see for example Spalding et al. (1969a) for a discussion of errors). It is therefore possible, for example, to detect differences in the solute concentration in the a phase of the Al-Mg alloy amounting to about 1%. Using filtered images with a bandwidth of 1 eV, it would be impossible to detect solute concentration differences in the same alloy of much less than about 10%. The energy selecting microscope can, however, be used to identify different phases of a precipitated alloy system, provided that the energy losses associated with the different phases are sharply defined and separated by more than about 1 eV. In this situation it is possible to form a filtered image using the energy loss characteristic of one of the phases. An example of an alloy which meets these requirements is Al-7·6% Zn-2·6% Mg which, when suitably heat-treated, forms η phase precipitates (composition $MgZn_2$) imbedded in the solid solution phase of the system. The loss peak of the η phase occurs at ~ 22 eV whereas that of the matrix material occurs at ~ 14.5 eV, and the filtered images obtained by selecting the 14·5 $\pm$ 1 eV and 22 $\pm$ 2 eV losses are shown in Figs. 77a and b respectively. In Fig. 77a the η phase precipitates appear dark against matrix material and in Fig. 77b the reverse occurs. Using this technique Castaing and his colleagues found that it was possible to identify η phase precipitates as small as 100 Å across.

Acknowledgements

The author is indebted to Drs. K. T. Considine, P. W. Hawkes and K. C. A. Smith for useful discussions and to Professor R. Castaing, Dr. L. Henry, and Dr. H. Watanabe for allowing him to reproduce examples of filtered images obtained with their energy selecting electron microscopes.

REFERENCES

Archard G. D. (1954). *Brit. J. appl. Phys.* **5**, 179, 395.

Castaing, R. and Henry, L. (1962). *C.r. hebd. Séanc. Acad. Sci. (Paris)* **255**, 76.

Castaing, R. and Henry, L. (1963). *J. Microscopie* **2**, 5.

Castaing, R. and Henry, L. (1964). *J. Microscopie* **3**, 133.

Castaing, R., Henoc, P. and Henry, L. (1968). *Proc. 4th European Reg. Conf. Elect. Microsc., Rome* **I**, 285.

Castaing, R., Henoc, P., Henry, L. and Natta, M. (1967). *C.r. hebd. Séanc. Acad. Sci. (Paris)* **265**, 1293.

Castaing, R., Hili, Ali El and Henry, L. (1966a). *C.r. hebd. Séanc. Acad. Sci. (Paris)* **262**, 169.

Castaing, R., Hili, Ali El and Henry, L. (1966b). *C.r. hebd. Séanc. Acad. Sci. (Paris)* **262**, 1051.

Chadwick, G. A. (1963). *Prog. Mater. Sci.* **12**, No. 2.

Considine, K. T. (1970). Ph.D. Thesis, Univ. of Cambridge.

Considine, K. T. and Smith, K. C. A. (1968). *Proc. 4th European Reg. Conf. Elect. Microsc., Rome* **I**, 329.

Considine, K. T. and Smith, K. C. A. (1971). *Brit. J. appl. Phys.* (to be published).

Cook, R. F. and Howie, A. (1969). *Phil. Mag.* **20**, 641.

Cotte, M. (1938). *Ann. Phys. (Paris)* [11] **10**, 333.

Cundy, S. L. (1968), Ph.D. Thesis, Univ. of Cambridge.

Cundy, S. L., Howie, A. and Valdre, U. (1969). *Phil. Mag.* **20**, 147.

Cundy, S. L., Metherell, A. J. F. and Whelan, M. J. (1966). *J. Scient. Instrum.* **43**, 712.

Cundy, S. L., Metherell, A. J. F. and Whelan, M. J. (1967a). *Phil. Mag.* **15**, 623.

Cundy, S. L., Metherell, A. J. F. and Whelan, M. J. (1967b). *Phys. Lett.* **24A**, 120.

Cundy, S. L., Metherell, A. J. F. and Whelan, M. J. (1968a). *Phil. Mag.* **17**, 141.

Cundy, S. L., Metherell, A. J. F., Whelan, M. J., Unwin, P. N. T. and Nicholson, R. B. (1968b). *Proc. R. Soc.* **A307**, 267.

Dietrich, W. (1958). *Z. Physik* **151**, 519.

Ferrell, R. A. (1956). *Phys. Rev.* **101**, 554.

Fujimoto, F. and Kainuma, Y. (1961). *J. Phys. Soc. (Japan) Suppl. BII* **17**, 140.

Fujimoto, F. and Kainuma, Y. (1963). *J. Phys. Soc. (Japan)* **18**, 1792.

Fukuhara, A. (1963). *J. Phys. Soc. (Japan)* **18**, 496.

Gauthe, B. (1954). *C.r. hebd. Séanc. Acad. Sci.* **239**, 399.

Hansen, M. H. (1958). "Constitution of Binary Alloys," 2nd Ed., McGraw-Hill, New York.

Hennequin, J. F. (1960). Diplôme d'Etudes Supérieures, Univ. of Paris.

Henry, L. (1964). Doctorate Thesis, Univ. of Paris.

Hili, Ali El (1966). *J. Microscopie* **5**, 669.

Hili, Ali El (1967). *J. Microscopie* **6**, 725.

Hillier, J. (1943). *Phys. Rev.* **64**, 318.

Howie, A. (1962). *Proc. 5th Int. Conf. Electron Microsc. (Philadelphia)* **1**, AA–10.

Howie, A. (1963). *Proc. R. Soc.* **A271**, 268.

Ichinokawa, T. (1965). *Proc. Int. Conf. Electron Diff. and Crystal Defects (Melbourne)* IN–4.

Ichinokawa, T. (1968). *Jap. J. appl. Phys.* **7**, 799.

Ichinokawa, T. and Kamiya, Y. (1966). *Proc. 6th Int. Conf. Electron Microsc. (Kyoto)* **1**, 89.

Kamiya, Y. and Uyeda, R. (1961). *J. Phys. Soc. (Japan)* **16**, 1361.

Kleinn, W. (1954). *Optik* **11**, 226.

Klemperer, O. (1965). *Rep. Prog. Phys.* **28**, 77.

Laudet, M. (1953). *Cahiers Phys.* **41**, 72.

Leithäuser, E. (1904). *Ann. Phys. Lpz.* **15**, 283.

Lenz, F. (1953). *Optik* **10**, 439.

Leonhard, F. (1954). *Z. Naturf.* 9a, 727.

Lippert, W. (1955). *Optik* **12**, 467.

Marton, L. (1944). *Phys. Rev.* **66**, 159.

Marton, L. and Leder, L. B. (1954). *Phys. Rev.* **94**, 203.

Metherell, A. J. F. (1965). Ph.D. Thesis, Univ. of Cambridge.

Metherell, A. J. F. (1967a). *Optik* **25**, 250.

Metherell, A. J. F. (1967b). *Phil. Mag.* **15**, 763.

Metherell, A. J. F. (1967c). *Phil. Mag.* **16**, 1103.

Metherell, A. J. F. and Cook, R. F. (1970). *Optik* (to be published).

Metherell, A. J. F., Cundy, S. L. and Whelan, M. J. (1965). *Proc. Int. Conf. Electron Diff. and Crystal Defects (Melbourne)* IN-3.

Metherell, A. J. F. and Whelan, M. J. (1965). *Brit. J. appl. Phys.* **16**, 1038.

Metherell, A. J. F. and Whelan, M. J. (1966). *J. appl. Phys.* **37**, 1737.

Möllenstedt, G. (1949). *Optik* **5**, 499.

Möllenstedt, G. (1952). *Optik* **9**, 473.

Paras, N. (1961). Diplôme d'Etudes Supérieures, Univ. of Paris.

Raether, H. (1965). *Springer Tracts Mod. Phys.* **38**, 84.

Raimes, S. (1967). "The Wave Mechanics of Electrons in Metals," North-Holland Publishing Co.

Septier, A. (1954). *C.r. hebd. Séanc. Acad. Sci. (Paris)* **239**, 402.

Smith, K. C. A., Considine, K. T. and Cosslett, V. E. (1966). *6th Int. Conf. Elect. Microsc. (Kyoto)* **I**, 99.

Spalding, D. R. (1970). Ph.D. Thesis, Univ. of Cambridge.

Spalding, D. R., Edington, J. W. and Villagrana, R. E. (1969a). *Phil. Mag.* **20**, 1203.

Spalding, D. R. and Howie, A. (1971). *Phil. Mag.* (to be published).

Spalding, D. R. and Metherell, A. J. F. (1968). *Phil. Mag.* **18**, 41.

Spalding, D. R. and Metherell, A. J. F. (1970). *Phys. Stat. Sol.* (to be published).

Spalding, D. R., Villagrana, R. E. and Chadwick, G. A. (1968). *Proc. 4th Europ. Reg. Conf. Elect. Microsc. (Rome)* **I**, 347.

Spalding, D. R., Villagrana, R. E. and Chadwick, G. A. (1969b). *Phil. Mag.* **20**, 471.

Watanabe, H. (1954). *J. Phys. Soc. (Japan)* **9**, 920.

Watanabe, H. (1955). *J. Phys. Soc. (Japan)* **10**, 321.

Watanabe, H. (1956). *J. Phys. Soc. (Japan)* **11**, 112.

Watanabe, H. (1964). *Jap. J. appl. Phys.* **3**, 480.

Watanabe, H. (1966). *Proc. 6th Int. Conf. Elect. Microsc. (Kyoto)* **I**, 63.

Watanabe, H. and Uyeda, R. (1962a). *Proc. 5th Int. Cong. Elect. Microsc. (Philadelphia)* **I**, A-5.

Watanabe, H. and Uyeda, R. (1962b). *J. Phys. Soc. (Japan)* **17**, 569.

Waters, W. E. (1956). Ph.D. Thesis, University of Maryland.

Wien, W. (1897). *Verh. Dtsch. Phys. Ges.* **16**, 165.

The Quantimet Image Analysing Computer and its Applications

C. BEADLE

Metals Research Ltd.,
Cambridge, England

I. Introduction	362
II. Desirable Design Features	364
A. Inclusion type	365
B. The volume fraction and number of inclusions present	365
C. Distribution through the bulk	365
D. Inclusion size and shape	366
E. Size distribution	366
F. Speed	366
G. Accuracy	366
H. Resolution	367
I. Visual check	367
J. Flexibility and convenience	367
K. Summary of system requirements	367
III. Description of the Quantimet	368
A. Fields of view	370
B. Detection	371
C. Computer	371
D. Monitoring facilities	372
E. Readout	374
F. Speed	375
IV. Accuracy of the Quantimet to other Problems	375
A. Threshold setting errors	375
V. Application of the Quantimet to other Problems	378
A. Secondary measurements	378
B. Other metallurgical applications	379
C. Ceramics and mineralogy	380
D. Soils and petrology	381
E. Particle size analysis	381
F. Fibres	381
G. Biomedical	381
H. Botany	381
I. Photographs and negatives	382
J. Particle physics	382
References	382

I. INTRODUCTION

THE automatic analysis of microscopic or photographic images has been a subject of much interest for the past 20 years and a number of different systems and instruments have appeared (Becker and Franceschini, 1957; Le Bouffant and Soule, 1954; Dell *et al.*, 1959/60; Dudley and Pelc, 1953; Flory and Pyke, 1953; Hallen and Hyden, 1957; Hawksley *et al.*, 1954; Lagercrantz, 1952; Larsson, 1957; Mansberg, 1957; Mansberg *et al.*, 1957; Morgan and Meyer, 1959; Roberts and Young, 1956; Taylor, 1954; Young and Roberts, 1951). Early instruments were applied to biological problems of cell counting and discrimination (Lagercrantz, 1952; Larsson, 1957), but the methods and techniques were soon extended and applied to particle mensuration (Le Bouffant and Soule, 1954; Dell *et al.*, 1959/60; Flory and Pyke, 1953; Hawksley *et al.*, 1954; Mansberg *et al.*, 1957; Morgan and Meyer, 1959; Taylor, 1954). The technology advanced rapidly with the development of the original flying-spot scanner of Roberts and Young (1956) at University College, London, the wide track scanner at the National Coal Board Research Laboratories (Hawksley *et al.*, 1954) and the photograph scanner at the Mullard Research Laboratories (Dell *et al.*, 1959/60). In America a number of other special instruments were developed, notably the cytoanalyser (Bostrom *et al.*, 1959) which was applied with limited success to the pre-screening of cervical smears, and the nuclear track scanner (Becker and Franceschini, 1957).

There are three main performance criteria for image analysis: speed in picture points surveyed per second, resolution expressed in terms of number of picture points measured in the field of view and discrimination, expressed as the number of detectable grey levels. The importance of speed and grey level discrimination is obvious. However, the importance of resolution is not so obvious when speed is expressed in picture points per second (pps/sec) rather than fields of view per second, since a fast (in terms of fields of view per unit time) low resolution system can use a higher magnification and cover the same area with the same accuracy in the same time as a slow system of high resolution. The advantage of high resolution lies in the ability to process an image to a given accuracy with a wider range between the smallest and largest particles and thus high resolution gives greater versatility to the system.

The many systems of image analysis that have been developed fall, for the purpose of comparison, into three main groups, defined by the method of scanning used. These groups are the mechanical scanners (specimen scan), the cathode ray tube/flying spot scanning systems (source plane scan) and television systems (image plane scan). Electron optical systems have been investigated, but are unlikely to become widely used because of their probable high cost and complexity. The relative merits of the three groups can be summarized as follows.

(i) *Specimen Scan Systems*: These rely upon a mechanical stage and are simple and comparatively cheap. They cannot count features, measure their shape or size distribution or make any measurements specific to individual features because of the difficulty in getting sufficiently accurate positional referencing, and are therefore limited by mechanical accuracy to classical linear analysis. Mechanical systems run at comparatively low speed (in the region of 10^4 pps/sec) and have high resolution. The high resolution is not fully utilizable because of the poor positional referencing. The discrimination is good, typically up to 100 grey levels and is generally limited by glare from lenses etc.

(ii) *Source Plane Scan Systems*. These are expensive systems and the type of specimens that can be assessed is restricted. They operate well when looking through transparent samples, but there is seldom sufficient light for looking at opaque objects. They can count and assess the shape of features. The resolution of these systems can be high (in the region of 10^6 to 10^7 picture points per field of view), but their speed is moderate and is limited by signal to noise ratio, and the number of grey levels required, to typically 10^5 pps/sec. They can discriminate up to about 50 grey levels.

(iii) *Image Plane Scan Systems*. These systems are of intermediate cost and have the highest acquisition rate (up to 5×10^6 pps/sec). This allows the most statistically accurate representation of the bulk material in a given time. The television displays are of great assistance to the operator. The systems can be used with a very diverse range of specimens: the criteria being the ability to project an optical image of the specimen into the television camera. The resolution is comparatively low (up to 5×10^5 picture points per field). The discrimination is normally about 6 grey levels but with shading compensation this can theoretically be improved to over 30 levels.

These considerations have led Metals Research Limited to adopt the image plane scan/television system for their Quantimet Image Analysing Computer. The very high speed, the ability to process virtually any image and the television monitor display have proved to be more important in 90% of known applications than the increased resolution and discrimination of source plane scan systems.

The Quantimet was originally designed specifically for metallurgical applications. Metallurgical problems are very similar to other image analysis applications: the metallurgist is interested in the assessment of non-metallic inclusions that look similar to powder particles when viewed down the microscope. Metal grains exhibit a boundary structure similar to the cell matrix found in many biological problems. The Quantimet is widely used in metallurgical laboratories. However, the instrument is, by its very nature, of general purpose and is finding increasing application outside the field of metallurgy—some other applications are described at the end of the chapter. The use of the title "Computer Television Microscope" or similar has been deliberately avoided since the television system leads to ready adaptation for receiving images other than from conventional optical microscopes—the Quantimet may, for example, be used with an epidiascope or coupled directly to an electron microscope. Other special purpose optical systems have been designed: e.g. for measuring the area of inhibition zones in Petri dishes used for micro-biological assays.

II. Desirable Design Features

It is instructive to approach the specification of a suitable instrument through a consideration of a specific problem. The example discussed here is the assessment of non-metallic inclusions in steel. This is an important metallurgical problem but it is also typical of a broad class of problem in various fields of science and technology. The analogy with many biological problems involving the distribution of certain types of cells in a tissue will be obvious, though the relative importance of different parameters may not be the same. The meaning of technical terms used should be obvious to non-metallurgists.

Routine assessments of non-metallic inclusions are carried out because this assessment helps to predict the likely mechanical properties of the steel. There is as yet little agreement over which particular features of inclusions are the best pointers to which mechanical properties, but attempts to establish this are being made in research laboratories all over the world now that a rapid objective and quantitative assessment is possible. The following list indicates what are probably the six most important parameters:

(i) inclusion type;
(ii) the volume fraction of inclusions present;
(iii) the number of inclusions present;
(iv) the distribution of the inclusions throughout the bulk material, with particular attention to the occurrence of clusters or 'stringers';
(v) the size and shape of the typical inclusion, particularly the length and elongation;

(vi) the spread or distribution of sizes in the inclusion population and the probability of there being very large inclusions in the bulk material.

Any reference to the source of the inclusions has been deliberately avoided since its importance has yet to be established. It could well be that both indigenous inclusions, i.e. those precipitated from the melt or solid on cooling, and exogeneous inclusions, i.e. suspended material trapped in the melt on freezing, are best treated together rather than separately, simply as products of the steelmaking practice.

No metallurgist would seriously disagree with this list and since so little is known about the relative importance of each parameter, any instrument must be capable of adequate performance in the measurement of each.

These six parameters and some practical considerations determine the main design requirements of an instrument for automatic inclusion assessment.

A. *Inclusion Type*

Some inclusions are complex in composition and structure but a good short description is undoubtedly that suggested by the JK chart†, i.e. sulphide, alumina and silicate and globular oxide type. The instrument should, ideally, categorize the inclusions in this way. The most difficult problem is how to deal with fields in which there are mixed types of inclusions.

B. *The Volume Fraction and Number of Inclusions Present*

Volume fraction can be determined accurately and quickly by chemical analysis if the composition of the inclusions is known. However, the number of inclusions is also important. The instrument should therefore measure the amount present preferably in terms of number and volume fraction.

C. *Distribution Through the Bulk*

Many inclusion assessments are initially concerned with deducing the likelihood of there being a particular 'bad region' in the bulk material from measurements made in a small sample. It follows that inclusion

---

† A JK chart comprises a number of reference diagrams each of which shows a particular type and density of inclusion distribution in ferrous metal. To use a JK chart, the investigator examines the specimen under a microscope at a suitable magnification. He then compares the image from the specimen with those on the JK charts, and selects the JK image that most clearly and closely resembles the image of his specimen. The JK rating of his specimen is recorded as the reference number of the selected JK image.

content must be assessed 'region' by 'region' within the sample so as to measure the variation between 'regions' and hence deduce the probability of there being a particularly bad 'region' in the bulk. It is not too clear how to determine the best 'region' size, but the likely use of the steel and the need to be able to resolve the smallest important inclusions should probably be taken into account. A reasonable value of approximately 1 mm diameter field of view is suggested by the classical JK/ASTM Chart and until better substantiated values are proposed, there appears to be no reason to depart from this. The instrument should, therefore, operate in many 'fields of view' of approximately 1 mm in diameter.

D. *Inclusion Size and Shape*

One important dimension of a single inclusion is its length, but its aspect ratio (length/breadth ratio) is significant. The morphology also is probably important in some applications. A jagged outline is probably the result of the brittleness of the inclusions during rolling and may be assessable by the steelmaker from knowledge of the inclusion type and the rolling history. The instrument must therefore measure inclusion length and give some indication of the shape in terms of elongation and jaggedness.

E. *Size Distribution*

The instrument must measure the inclusion size distribution so as to help in assessing the likelihood that there may be very many large inclusions in the bulk.

F. *Speed*

The throughput of specimens in a large steelworks suggests that the instrument should be capable of assessing a specimen in two or three minutes.

G. *Accuracy*

The accuracy requirement of an inclusion assessment system is that it should give the best possible indication of the "cleanness", i.e. freedom from inclusions, of the bulk material. Both sampling errors and measurement errors will exist, but the sampling error will usually dominate. In a correctly designed instrument the measurement errors increase as measuring speed increases because less time is spent on each field, but the sampling errors will diminish because more fields are measured in the fixed time available. Ideally these two errors should be matched to give the lowest possible RMS sum, but this cannot be done without knowledge of the distribution of inclusions within the specimen.

Any comparison of overall accuracy must therefore take into account both types of error, i.e. it must be based on the fundamental data acquisition rate. The best criterion is therefore the amount of specimen covered to a certain accuracy in the given time or, in more fundamental terms, the number of picture elements measured per second at some arbitrary level. Application of this concept to the design of an instrument ensures maximization of overall accuracy and guards against the danger of reducing measurement errors excessively at the expense of sampling errors.

H. *Resolution*

Opinion varies widely on the necessary resolving power for inclusion analysis but 2 μm is probably an acceptable interim figure for an inclusion counting instrument based on the optical microscope.

I. *Visual Check*

The operator must look briefly at the surface of the specimen to check for possible artefacts such as dust, scratches, finger marks etc. The instrument should facilitate this or better still provide continuous monitoring displays.

J. *Flexibility and Convenience*

Steels vary very widely and the instrument must therefore be capable of being set perhaps by a works metallurgist to measure the particular parameters required. Once programmed, however, its operation must be a single 'push-button start' and 'print out result' sequence.

K. *Summary of System Requirements*

(i) It should distinguish between different types of inclusions.

(ii) It should give the volume fraction and number of inclusions present.

(iii) It should assess their distribution in the specimen by many localized measurements in fields of view of approximately 1 mm diameter.

(iv) It should measure inclusion length, elongation and jaggedness.

(v) It should measure inclusion size distribution.

(vi) It should treat a typical specimen in 2 or 3 minutes.

(vii) It should rate the bulk material as accurately as possible—the

     criterion for comparing the potential of different methods being
     the number of picture elements scanned per second to a chosen error.
(viii) It should have 2 μm resolution.
 (ix) It should allow easy visual monitoring by the operator.
  (x) It should be partially pre-programmable but very easy to operate
     in routine assessments.

III. Description of the Quantimet

The instrument is illustrated in Fig. 1 and Fig. 2 is a simplified block
diagram. In principle any microscope or image forming system could be
used but the standard microscope supplied is a Vickers instrument.
Incident, transmitted, or mixed illumination can be used and alternative
light sources include a tungsten filament lamp or Xenon arc. A binocular
viewing head is provided on the microscope, so that the image can be
observed directly as well as on the television screen. A wide range of
objectives of all powers up to oil immersion can be used and there are
facilities for phase contrast, polarized light and other special methods.
The microscope projects an image of the field of view on to the screen of
a television camera. The electrical output from this camera passes into
a closed circuit television monitor to provide a television image, and
also into a detector unit where signals from the camera emanating from
the inclusions are discriminated and selected from the rest of the signal.
The output from the detector, consisting of pulses from the detected
inclusions can be fed into the monitor so that the operator can see which
inclusions he has detected, and into the computer which can be pre-
programmed to measure a variety of parameters on any field. The data
from each field of view can be accumulated in registers to provide grand
totals for the specimen or read out separately via conventional computer
peripherals. The whole sequence of operation, including specimen
movement, is controlled automatically. A Sequence Control Module
drives the stage in X and Y directions by means of digital motors. The
total stage movement is 5 cm in the direction parallel to the television
line scan and 2·5 cm in the direction perpendicular to this. The Sequence
Control Module can be programmed to move the stage in a series of
movements variable from 50 μm to 3·2 mm. The total traverse may be
up to 10,000 fields. The stage moves parallel to the direction of the
television line for a selected distance then returns to the starting point,
moves by a selected distance perpendicular to the direction of the line
and then repeats the scan parallel to the original direction. The specimen
is spring loaded upwards on to a reference surface to give automatic
levelling.

Fig. 1. The Quantimet image analysing computer. The microscope and detector unit are shown in the centre. The display monitor and electric typewriter are on the right. Several alternative means of display and data accumulation are available.

Quantimet
System

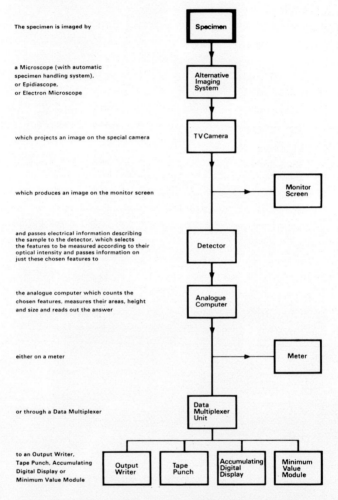

The specimen is imaged by — **Specimen**

a Microscope (with automatic specimen handling system), or Epidiascope, or Electron Microscope — Alternative Imaging System

which projects an image on the special camera — TV Camera

which produces an image on the monitor screen — Monitor Screen

and passes electrical information describing the sample to the detector, which selects the features to be measured according to their optical intensity and passes information on just these chosen features to — Detector

the analogue computer which counts the chosen features, measures their areas, height and size and reads out the answer — Analogue Computer

either on a meter — Meter

or through a Data Multiplexer — Data Multiplexer Unit

to an Output Writer, Tape Punch, Accumulating Digital Display or Minimum Value Module — Output Writer, Tape Punch, Accumulating Digital Display, Minimum Value Module

FIG. 2 Block diagram of the Quantimet system

A. *Fields of View*

The size of field is governed by the magnification used and can be varied by selecting different microscope objectives. Normally, a $\frac{3}{4}$ mm diameter field of view is used with 1·8 μm resolution.

B. *Detection*

Inclusions in steel are detected by their contrast. As the television pick-up tube scans the image of the inclusion, the video signal changes in proportion to the reflectivity of the inclusion. When the video voltage passes a reference threshold the inclusion is detected. Two sets of thresholds are available to the operator so that he can set levels corresponding to oxides in one case and both sulphides and oxides in the other. Subtraction allows the two types to be separated. A characteristic error is found with this method of detection caused by the finite resolution of the system which makes the edge diffuse. Around each black oxide there is, therefore, a diffused grey boundary, and in the process of subtraction part of this is treated as a grey sulphide region. Many methods have been put forward for correcting this effect electronically, including the use of a third 'reference' threshold, but all are open to criticism for introducing further errors. Fortunately, the error is small on most samples and in fact only becomes serious when there are small sulphides in with many large oxides. If great accuracy is required, the final results can readily be corrected by the addition of the known perimeter error. Most inclusion analysis does not justify precision of this order.

C. *Computer*

The signals passing from the detection circuits to the computer consist of a series of television line waveforms quantized to give a series of rectangular pulses. The height of all pulses is fixed and the lengths correspond to the chord lengths intersected by the television lines on the detected inclusions.

To measure the total area of these inclusions, the chord pulses are integrated to give a measurement directly proportional to the time spent by the scanning spot traversing the detected inclusion. This measurement is clearly proportional to its area. The horizontally-projected length of the inclusions is the same as the intercept count, and is measured by counting the total number of intersections of television lines with the detected inclusions. This gives the total length of features in a direction perpendicular to the scan. Counts arising from intercepts shorter than a selected value can be eliminated by means of special size elimination controls that can be set by the operator. The distribution function of the intercepts can be obtained by selecting a range of values. The resulting distribution curve gives information such as form factor variation, depending on the type of specimen under examination.

The counting logic uses a one-line memory system to record the previous scan and to allow the rejection of repetitive intersects from large or re-entrant inclusions by an anti-coincidence circuit. The counting circuit functions only on the outline of the detected inclusions, and since these outlines are in general not vertical, a certain tolerance has to be allowed in the computer logic within which pulses from successive lines intersecting a common inclusion are accepted as coming from one inclusion. This tolerance is called the *acceptance width*, and any two inclusions closer together than this distance will be counted as one. On the other hand, long or re-entrant inclusions lying nearly parallel to the scan direction will demand a fairly large acceptance width to avoid being counted twice.

The acceptance width is therefore made variable by a pre-set control and can be adjusted to suit the type of specimen being examined.

The memory circuit eliminates all miscounts from large inclusions intersecting the television lines more than once and gives the actual number of inclusions viewed. The size elimination controls can be used to eliminate counts from inclusions the length of which in the line scan direction is less than the chosen value. In this way a full size distribution curve can be obtained.

The computer therefore gives all the geometric information needed for inclusion analysis, including volume fraction, length of individual inclusions for a size distribution, local mean shape, number, etc., all referred to individual fields of view. The only departure from the ideal specification is in the use of mean shape per field rather than individual feature shape, but though this would be possible, the minimal gain in information is not worth the added complexity. The further facilities for intercept count and intercept size distribution etc. are found useful in some research applications.

D. *Monitoring Facilities*

By using standard television techniques it has been possible to provide displays for monitoring the correct functioning of the Quantimet. Not only is a bright television display of the specimen available, but a wide variety of clearly recognizable displays of the various computer functions can be superimposed on that of the image. This allows the operator to see at a glance what parameter he is measuring on which inclusions and makes obvious any mis-setting of the controls. Fields of view with scratches, stains or dirt are clearly revealed and can be ignored in the computation of results. Figure 3 shows some typical monitor displays.

The light rectangle which defines the area of measurement can be varied in size, shape and position.

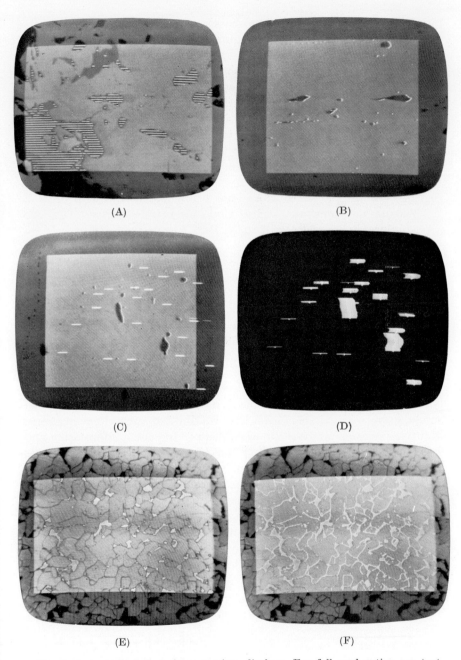

(A)   (B)

(C)   (D)

(E)   (F)

Fig. 3. Some typical Quantimet monitor displays. For full explanation see text. Fisher and Cole (1968), with acknowledgements to "The Microscope".

Fig. 3A. Area. The threshold can be set to allow measurements to be made on one or more of a number of components differing in brightness. Up to six steps between black and white can be distinguished. Here the threshold has been set for very black inclusions. For further examples see Fig. 6.

Fig. 3B. Projection measurement. This gives the sum of the maximum extent of each feature in the horizontal scan direction. In this mode a white spot appears on the trailing edge of each feature. The size elimination control can be set to reject features whose maximum horizontal length is less than a certain value, giving information on chord size distribution.

Fig. 3C. Count. The short white lines below and to the right of each feature indicates only those that are being counted. The size elimination control can be used here to exclude features below a certain size.

Fig. 3D. Acceptance width check. This is the same field as in 3C and shows the distance over which two features close together will be counted as one.

Figs. 3E and 3F. Show two displays of the same specimen (Ferrite and Pearlite in steel) at high (F) and low (E) resolution. The low resolution image may be useful for rejecting unwanted features.

E. *Readout*

A clear presentation of results is important and a wide range of modular units is available to suit both routine and research applications. Routine inclusion analysis can easily lead to too much information and it is imperative that the amount of data referring to any specimen should be condensed to a minimum. Special output devices are therefore used to accumulate information, draw the operator's attention to particularly important fields, and help in the estimation of the distribution through the specimen. For example, it is possible to programme the computer to stop the instrument when a field with more than a preset area, number or projected length of inclusions, is scanned. The operator can either study this field in detail or can reject it if it is atypical. For research the instrument can be used with up to 30 high-speed digital displays and can also produce detailed printed records of the full characteristics of every field of view, using outputwriters or high-speed line printers. For direct on-line computer control of the production process, standard computer peripheral tape punch systems are available. Other arrangements are under development and should further simplify the problem of taking means, standard deviations etc. It is possible that output devices will eventually become available which can make complex judgements comparable to, though more precise than, the JK charts.

F. *Speed*

The instrument is capable of operating faster than 1 field of view/s while taking measurements on 30 different inclusion parameters in each field. Typically, these parameters would be volume fraction, numbers, size distributions and shape information for oxides and sulphides. Each field of view contains approximately 10^5 picture elements and so, when all the parameters are being measured, the system speed is approximately 3×10^6 picture points processed per second.

IV. ACCURACY OF THE QUANTIMET TO OTHER PROBLEMS

A. *Threshold Setting Errors*

The threshold setting error is fundamental to all systems of image analysis. The computing accuracy of the Quantimet of about $\pm 2\%$ is only achieved on fields containing nothing but large features (above $\frac{1}{8}$ of the screen length in diameter). On all fields with smaller features the error is determined solely by the accuracy with which the threshold is set.

The cause of threshold setting errors is illustrated in Fig. 4. This shows the signal response when a finite sized scan spot crosses a feature in the image. Instead of the ideal rectangular electrical signal, the practical electrical signal has sloping edges giving a finite rise time. (Other possible causes include band-width limitations, optical resolution, ill-defined feature boundaries, etc). When this video signal is sliced by a threshold set at some arbitrary level T, it will cause an error δ which is zero when the Quantimet is set at the correct position. This error is a function of the instrument resolution $\varDelta$ and its relative importance can be judged by comparison with the chord length C. The error can always be reduced by increasing the magnification and thus improving the resolution $\varDelta$, i.e. increasing the chord size C for the same resolution. Unfortunately improving the accuracy in this way carries considerable speed penalty since the area examined is reduced as the square of the magnification.

Figure 5 illustrates two examples of what happens when the threshold is set incorrectly. Two curves illustrate the variation in area percentage measurement as the threshold level is changed. The first curve (A) refers to large black features and the all-important slope of the curve is seen to be quite small (5% of the true value per 100 threshold divisions). The second curve (B) shows similar results taken at a much lower magnification so that the features are nearer the resolution limit and the variation

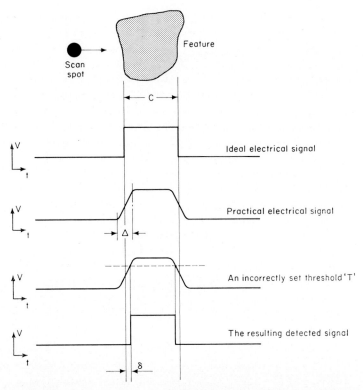

Fig. 4. Threshold setting errors. Instead of the theoretical idea lsquare pulse, the edges of the pulse are sloping. An incorrectly set threshold results in an error in the pulse width.

is seen to be very much greater (80% of the true feature area per 100 threshold divisions). This second example is by no means an extreme case and the instrument is often used in even less favourable regimes so as to improve the sampling speed (e.g. the measurement of non-metallic inclusions in steel of which hundreds of fields must be measured to get statistical accuracies of a few per cent).

Since there are approximately 400 threshold divisions between black and white on the Automatic Quantimet on which these results were obtained, it will be seen that a threshold change as small as 10% will cause errors of some 35% from the true value, even in this relatively favourable case. For this reason great care must be taken in setting the detection threshold.

Recent trials carried out by the BISRA Sub-Committee on Automatic Assessment of Inclusions in Steel (Allmand and Blank, 1968a,b) have confirmed the wide variation in results that is obtained by using different

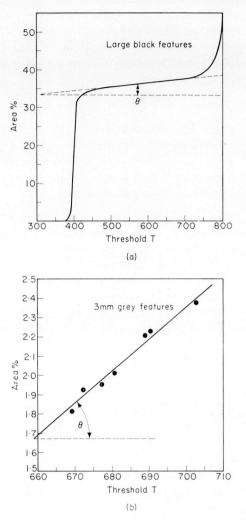

FIG. 5. Examples of the effect of incorrect threshold setting. The error is much greater for grey features especially when examined at a low magnification. For further details see text.

methods of setting the threshold. Various minor procedural alternatives were investigated by this Committee and a method has now been developed capable of giving good and repeatable accuracy on features down to about 1/30 of the scan length.

In practice the threshold of the Quantimet is set by superimposing the detected image on the original video image on the television monitor.

The threshold setting control is adjusted until the two coincide. The BISRA Committee investigated and recommended methods of checking that the two images are in fact coincident. Evans (1970) has also recently emphasized the importance of correct threshold setting.

V. Application of the Quantimet to other Problems

As explained above the Quantimet, though originally designed for metallurgical applications, is essentially a general purpose image analyser. In the field of microscopy it can be used whenever statistical information has to be extracted from microscope images. The instrument essentially measures three basic parameters:

(i) *Area*. The area of detected features as a percentage of the field of view is measured. Depending upon the specimen, up to 6 phases between black and white can be distinguished and measured. Fig. 6 illustrates the measurement of areas in multiphase specimens.

(ii) *Count*. Detected features within the field of view can be directly counted. A minimum chord control will reject those features whose maximum length in the horizontal scan direction is less than a chosen value. By varying this value, and taking different readings, a size distribution can be made.

(iii) *Projection*. The instrument measures the total projection of detected features in the horizontal scan direction, i.e. the maximum vertical extent of the features summed over all the features in the measured field. The minimum chord control also operates in this regime and will provide size distribution information.

A. *Secondary Measurements*

From these basic parameters many secondary or derived parameters can be obtained. These include mean linear intercept, mean form factor, mean feature size, mean features area, number of features per unit area or per unit volume, total perimeter of features, total length of grain boundary, mean fibre diameter, mean fibre length, specific surface area, standard deviations, and parameters of normal, log normal and Rosin-Rammler distributors.

The measurement of these parameters, particularly in metallurgical problems, is described in numerous publications of which some of the more readily available are; Allmand, 1966 a, b; Allmand, 1968; Allmand and Blank, 1968 a, b; Blank 1968 a, b; Cole, 1966; Fisher, 1966; Fisher, 1967; Fisher, 1968; Fisher and Nazareth, 1968; Hofer, 1968; Jesse and Ondracek, 1968; Langhoff and Johnson, 1966; Lindon, 1968; Loveridge

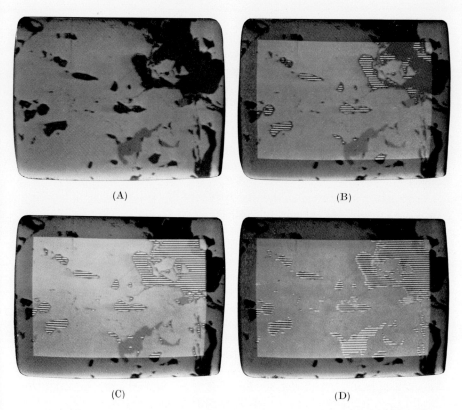

(A)   (B)

(C)   (D)

FIG. 6. Measurement of area in multiphase specimens. A shows the conventional display of the specimen on the monitor. The pale rectangle in B, C and D shows the area of the specimen on which measurements are being made. In B the threshold has been set for measuring the darkest areas only. In C the next gradation has been included and in D all areas from light grey to black are included. In this way the areas occupied by different phases can be measured.

and McInnes, 1968; Martennson, 1965; Melford, 1968; Roche, 1968; Smith, 1968; Strang, 1968.

B. *Other Metallurgical Applications*

The application of the Quantimet for counting and characterizing the non-metallic content in steel has already been described. In the field of metallurgy the instrument has also been used for measuring grain size, measuring the area of ferrite, pearlite, cementite, delta ferrite, retained austenite and resolvable pearlite in ferrous alloys, counting the number and measuring the percentage area of carbide particles in tool steel, measuring the carbon gradient in carburised layers, quantifying the age-hardening reaction in beryllium copper and copper aluminium, and measuring porosity in weld pools.

a. *Iron and Steel.* Non-metallic inclusions—determination of the number, form, factor, average size and size distribution of non-metallic inclusions. Measurement takes about one second per field (Allmand and Blank, 1968a; Jamieson *et al.*, 1968; Schreiber and Radtke, 1966).

Volume fraction measurement—measuring pearlites, graphite, delta ferrite, carbides, and a number of other phases in steel samples. It has also been used to determine the mean size and the mean spacing of carbides, etc.

Segregation and banding and case hardened layers—measuring banding of, for example, carbides in high speed steels by measuring the volume fraction of the banded constituents in a series of thin bands parallel to the banding directions. The Quantimet has been used to determine carbon gradients across case hardened layers by measuring the pearlite volume fraction at successive points across the hardened case.

Grain size determination—grain size determination in plain carbide and carbon and a range of alloy steels; the presence of a second phase, for example pearlite in a ferrite steel, can be allowed for by measuring the area and hence the volume fraction of the phase, and correcting the mean linear intercept figure to allow for it. The instrument gives equal weight to twin and grain boundaries and if many twin boundaries are present and it is required to discount them it can only be done by finding a suitable etchant which does not develop them.

Grain shape and size distribution—measurement of the mean linear intercept in two directions gives a measure of anisotropy; mixed grain size can be expressed quantitatively either by using the minimum chord control to obtain an intercept-size distribution or by using a small measuring rectangle and separately measuring the fine and coarse areas.

b. *Non-Ferrous Metallurgy.* The instrument has been used successfully for measuring grain size of sintered tungsten carbide tools, oxide inclusions in copper and rare earth metals, grain size of a wide range of metals, dislocating density of copper single crystals, cracks in chromium plating.

C. *Ceramics and Mineralogy*

Measurement of porosity and grain size of ceramics, bricks, concrete, high alumina ceramics, etc., volume fraction measurements on several phases in multi-phase minerals, using polarization techniques, and the measurement of volume fraction of different coals of marginally different reflectivity (Jesse and Ondracek, 1968).

D. *Soils and Petrology*

Low power microscopy with a variety of polarization techniques has been used to distinguish and measure a wide range of constituents in soil samples including sands, clays, petrified clays, chalk, iron oxides. Ultra violet fluorescent methods have been used to detect tracer sands in full scale coast erosion and silting experiments.

E. *Particle Size Analysis*

Measurement of varying particle size and particle distribution; measurement of various shapes and factors in powders and particles— the number per unit area and the specific area (Amor and Block, 1968, Fisher, 1966; Strang, 1968).

F. *Fibres*

The mean fibre diameter and length in tangled bundles can easily be found without any combing of the fibres.

G. *Biomedical*

The use of selective staining techniques has opened up a wide range of applications for the Quantimet in biomedicine. The Quantimet has been used to evaluate the volume of colloid in thyroid tissues, to count and measure the areas occupied by mucus-containing cells in the epithelial layer of the bronchus and makes other quantitative measurements on the bronchial tissue of particular relevance to experimental pathologists. In the field of Haematology the Quantimet has been used to count the number of white and red blood cells and platelets in a blood sample. In Pharmacology and Bacteriology it has been used to assess the effectiveness of antibiotics by measuring their actions on bacterial growth and by quantifying measurements taken for microbiological assay. Other biomedical applications include Neurology, in which the area occupied by nerve endings in tissues can be measured and Odontology, in which the extent of decay in teeth can be quantified (Mawdesley-Thomas and Healey, 1969 a, b).

H. *Botany*

The Quantimet has been used to measure the projected area and length of plant roots, the cell size of leaves and tissues, woods, etc. (Natr, 1968). The rate of osmosis in plants can be measured using radioactive tracer techniques.

I. *Photographs and Negatives*

Many X-ray, ultraviolet and electron microscope photographs have been analysed and the use of photographs is also useful when special microscope techniques have been developed—e.g. in radioactive environments.

J. *Particle Physics*

The Quantimet has been used for the determination of electron track density in electron spectrometry (Loveridge and McInnes, 1968) and its application to the measurement of curvature and spacing in bubble chamber photographs is being considered.

REFERENCES

Allmand, T. R., (1966a), *GKN Conf., Rep. no.* 863.
Allmand, T. R., (1966b). *GKN Conf., Rep. no.* 872.
Allmand, T. R. (1968). *The Microscope*, **16,** 163–170.
Allmand, T. R. and Blank, J. R. (1968a) *ISI Publication* **112,** The Iron and Steel Institute, pp. 1–12.
Allmand, T. R. and Blank, J. R. (1968b). *ISI Publication* **112,** The Iron and Steel Institute, pp. 70–71.
Amor, F. A. and Block, M. (1968). *J. Roy. microscop. Soc.*, **88,** 601–605.
Becker, S. and Franceschini, J. B., (1957). *IRE Nat. Conv. Record*, **5,** Part 9, 46–48.
Blank, J. R. (1968a). *The Microscope*, **16,** 189–197.
Blank, J. R. (1968b). *ISI Publication* **112,** The Iron and Steel Institute, 63–69.
Bostrom, R. C., Sawyer, H. S. and Tolles, W. E. (1959). *Proc. Inst. Radio Engineers* **47,** 1895–1900.
Le Bouffant, L. and Soule, J. L. (1954). *Brit. J. appl. Phys.*, Suppl. 3, 143–146.
Cole, M. (1966). *The Microscope*, **15,** 148–156.
Dell, H. A., Hobbs, D. S. and Richards, M. S. (1959/60). *Philips Technical Review*, **21,** No. 9, 253–280.
Dudley, R. A. and Pelc, S. R. (1953). *Nature*, **172,** 992–993.
Evans, D. H. (1970). *The Microscope*. **16,** 85–98.
Fisher, C. (1966). *Proc. Particle Size Analysis Conference*, 77–94.
Fisher, C. (1967). *Biomedical Eng.*, **2,** 351–357.
Fisher, C. (1968). *ISI Publication* **112,** The Iron and Steel Institute, 24–30.
Fisher, C. and Cole, M. (1968). *The Microscope*, **16,** 81–94.
Fisher, C. and Nazareth, L. J. (1968). *The Microscope*, **16,** 95–104.
Flory, L. E. and Pyke, W. E. (1953). *RCA Rev.* **14,** 546–556.
Hallen, O. and Hyden, H. (1957). *Exp. Cell Res.*, **12,** 197–206.
Hawksley, P. G. W., Blackett, J. H., Meyer, E. W. and Fitzsimmons, A. E., (1954). *Brit. J. appl. Physics* Suppl. No. 3, 5165.
Hofer, F. (1968). *The Microscope*, **16,** 171–180.
Jamieson, R. M., Ohennasian, C. E. and Masygan, R. J. (1968), *J. Iron and Steel Institute*, pp. 498–499.
Jesse, A. and Ondracek, G. (1968). *The Microscope*, **16,** 115–121.
Lagercrantz, C. (1952). *Acta Physiol. Scand.*, Suppl. 93, **26,** 1–131.

Langhoff, R. R. and Johnson, A. R. (1966). *Proc. ASTM*, 69th AM., paper No. 228.

Larsson, S. (1957). *Exp. Cell Res.* **12**, 666–669.

Lindon, P. H. (1968). *The Microscope*, **16**, 137–150.

Loveridge, B. A. and McInnes, C. A. J. (1968). *The Microscope*, **16**, 105–114.

Mansberg, H. P. (1957). *Science*, **126**, 823–827.

Mansberg, H. P., Yamagami, Y. and Berkeley, C. (1957). *Electronics*, **30**, Part 12, 142–146.

Martennson, H. (1965). *Fagersta Forum*, 4.

Mawdesley-Thomas, L. E. and Healey, P. (1969a). *New Scientist*, 286–287.

Mawdesley-Thomas, L. E. and Healey, P. (1969b). *Science*, **163**, 1200.

Melford, D. A., (1968). *ISI Publication* **112**, The Iron and Steel Institute, p. 14.

Morgan, B. B. and Meyer, E. W. (1959). *J. Sci. Instr.* **36**, 492.

Natr, L. (1968). *Photosynthetica*, **2**, (1), 39–40.

Roberts, F. and Young, J. Z., (1956). *Proc. Inst. Elec. Engineers*, **99**, Part IIIA, 747.

Roche, R. (1968). *The Microscope*, **16**, 151–161.

Schreiber, D. and Radtke, D. (1966) *Steel Times*, **193**, No. 5118, 246–258.

Smith, M. J. (1968) *The Microscope*, **16**, 123–135.

Strang, A. (1968) *The Microscope*, **16**, 181–187.

Taylor, W. K. (1954). *Brit. J. appl. Phys.*, Suppl. 3, 173–175.

Young, J. Z. and Roberts, F. (1951) *Nature, Lond.* **167**, 231.

Photomicrography and its Automation

F. GABLER and K. KROPP

*Technical University and
C. Reichert Optische Werke AG,
Vienna, Austria*

I. Introduction 385
II. Some Details on Photomicrographic Technique 386
    A. The range of photomicrography 386
    B. Photomicrographic instruments 387
    C. Additional technical details 388
III. The Automation of Photomicrography 390
    A. The measurement of light intensity 390
    B. Processing of the data obtained by measurement of illumination
      intensity. 396
References 413

I. Introduction

During the past 10 years photomicrography has rapidly become an indispensable means of objective recording with the microscope. The number, quality and versatility of photomicrographic instruments has increased tremendously. Operation has been simplified and the achievement of optimum results has become a matter of routine. Essential steps in this development were the efforts to obtain correct exposure time and, subsequently, to make the entire complex comprising exposure time determination, exposure proper, and, if necessary, film transport as completely automated as possible.

In this chapter an attempt will be made to demonstrate the fundamental methods of this automation and to show by typical examples how they can be put into practice. The chapter is not however intended to become a compilation of sales brochures and an attempt to list fully the features of pertinent instruments has been deliberately avoided.

As an introduction to the main topic it is appropriate to go into some relevant details on photomicrographic technique which are essential in order to appreciate the systems and the technical problems involved. The reader is assumed to be familiar with the microscope and with photography. For further details on photomicrography several standard works may be consulted (Allen 1958, Bergner, Gelbke and Mehliss 1966, Claussen 1967, Michel 1967).

II. Some Details on Photomicrographic Technique

A. *The Range of Photomicrography*

Briefly, photomicrography can be defined as photography of real images formed by a compound microscope, which is essentially a combination of two optical systems, i.e. objective and eyepiece. With an objective of magnification $1 \times$ or $2 \times$ and a $5 \times$ eyepiece, the minimum final magnification would be $5 \times$ or $10 \times$. Lower power systems are not often used.

There are, however, other definitions of photomicrography; Michel (1967), for instance, separates photomicrography from photomacrography at a $1 \times$ final magnification. We do not agree with this definition which is after all a matter of opinion. The equipment used for low power magnifications of $2 \times$, for example, would consist of only one magnifying, optical component; it would be a simple, rather than a compound, microscope.

While the minimum final magnification in photomicrography is obviously a matter of definition, the maximum final magnification obtainable depends on the optical system and on the quality expected in the photomicrograph.

Even since Abbe, the opinion has prevailed that in visual microscopy the total magnification should not exceed roughly 1000 times the numerical aperture of the objective. This would mean in practice an upper limit of about $1300 \times$ for the final magnification. (This rule ignores the influence of optical components other than the objective, provided that they do not reduce the aperture of the latter.) This value is based on two facts: first, due to diffraction, an object point is imaged not as a point but as a small disk and secondly, the human retina is unable to distinguish (resolve) disks that are separated by less than a certain minimal distance. A final magnification of 1000 times the numerical aperture of the objective is called the limit of useful magnification. Observation at greater magnification does not show new details but only renders those already observed larger (empty magnification). Furthermore, the exit pupil of the microscope becomes smaller with increasing magnification. This means that the ray bundles converging towards individual image points become narrower, which could introduce defects in image formation. Dirt in the microscope and small inhomogeneities in the observer's ocular media become obtrusive. The latter give rise to the so-called entoptic disturbances, or "muscae volitantes", which may be very troublesome.

Nevertheless, for special cases in visual microscopy, magnifications beyond the useful magnification can be used. Bringdahl (1966) has justified their theoretical value. Lau (1961) has even designed a special microscope which avoids extremely small apertures and quotes many examples of the utility of subjective magnifications up to 5 or 6000×.

Photomicrography obviates several disturbances occurring with visual observation and thus permits direct photography up to say 3000×, using not only high power objectives and eyepieces but additional magnifying intermediate optics or bellows extensions. However, the higher the final magnification, the greater are the demands on cleanliness of all optical elements in the raypath: as in visual microscopy, image formation by narrow bundles of rays emphasizes disturbing shadows in the image arising from out-of-focus dust and dirt.

Of course it is useful to take photomicrographs at much higher total magnification if the pictures, e.g. wall mounts, are to be observed from a greater distance. Their details have to be overmagnified initially in order to subtend a sufficiently large angle at the observer's eye.

B. *Photomicrographic Instruments*

These can be divided into three groups:

a. *Attachment Cameras*. These cameras are equipped with an accessory lens permanently set to infinity. They are mounted directly, or through a supporting arm, on the microscope tube, which carries the photo eyepiece. Apart from certain differences mentioned below, the photo eyepiece is similar to those used in visual microscopy. The attachment cameras are usually designed for medium film sizes such as 6·5 × 9 cm or $3\frac{1}{4} \times 4\frac{1}{4}$ in and especially for the 35 mm size. The system is focused by means of a graticule in a focusing telescope, the graticule indicating simultaneously the picture sizes. Thus the system is focused on an aerial image. A beam-splitting prism placed in the optical path permits continuous observation during the exposure. In order to facilitate focusing of low intensity specimens the beam-splitting prism on some models can be exchanged for a mirror which reflects all the light to the focusing telescope.

The main advantages of 35 mm (strictly 24 × 36 mm) size are the higher illumination intensity at the photographic emulsion (resulting in shorter exposure times), reduced film costs and smaller demands on storage space in the film library. However, meticulous focusing on the object detail is essential in order to avoid the loss in sharpness that may occur after the considerable enlargement usually needed with negatives of this size. Particularly when working with low power objectives, focusing with such precision requires some experience and a high power focusing telescope. Focusing of an aerial image is essential with 35 mm attachment cameras.

b. *Bellows Cameras.* These are mounted separately from the micro-scope, either on a horizontal optical bench or on a rigid vertical stand. Generally, they have no accessory lens; the microscope projects the image directly onto the photographic emulsion. In this case the micro-scope does not work strictly in accordance with the conditions for which it was designed, since the rays forming a real image of a point are no longer parallel to each other when they emerge from the eyepiece. A simple calculation proves that, assuming a given objective-eyepiece distance, to focus an image at a bellows extension of 250 mm, the objective has to be raised so that with a $5 \times$ eyepiece the primary image plane is moved 12·5 mm. The movement required is greater than the correction tolerance of high numerical aperture objectives will allow. (For a $10 \times$ eyepiece the comparable value is 2·5 mm and is tolerable.) Hence, such objectives demand either longer bellows extensions or projection eyepieces (photo eyepieces) adjustable for different bellows lengths.

Bellows cameras are mainly used for large image sizes (plates or sheet films) up to 13×18 cm (5×7 in). The image is focused on a ground glass screen with the aid of a magnifier. Focusing on small object details is facilitated if a small central clear portion is provided on the ground glass so that an aerial image can be focused.

Increased ease in obtaining photographs of maximum sharpness at high final enlargement as compared to 35 mm photography, is the main reason why large size cameras are still used today. A further advantage of the bellows camera is the possibility of varying the enlargement continuously. However, with the development of improved optical Zoom systems to be placed between objective and eyepiece this latter advantage is becoming less important.

c. *Camera Microscopes.* These instruments feature either a built-on or a built-in camera system, presenting an integrated concept in the design and function of a photomicrographic unit. Instruments of this design are available for 35 mm and also for large size photomicrography.

C. *Additional Technical Details*

a. *Imaging Optics.* Photomicrography which accurately records everything presented to the photographic emulsion, requires the best possible illuminating and imaging systems. Planachromatic or plan-apochromatic objectives and eyepieces specially designed for use with them are required for high performance. Modern flat field objectives give an image almost totally flat up to the edge of the field. Since these objectives became available from almost all microscope manufacturers interest in eyepieces with negative power (commercially known as "homals" or "negative eyepieces") has diminished. These eyepieces introduced a field curvature compensating that of an achromatic objective and were very popular not long ago despite several drawbacks (not usable for visual observation, small field of view).

b. *Illumination*. High intensity light sources should be used in photomicrography in order to avoid long exposure times which cause difficulties because of unavoidable vibrations and the reciprocity failure that is discussed in Section Cc.

If, in extreme cases, especially short exposure times are necessary, e.g. when photographing rapidly changing or moving objects, the use of a microflash is suggested. Modern quartz iodine lamps, however, provide a means of illumination which is more than adequate in most cases, including such light-wasting methods as polarization, darkfield, phase contrast and interference contrast. Only rarely will there be need for one of the familiar high-pressure Xenon or metal vapour lamps. Mercury vapour lamps are of course essential for fluorescence photomicrography, because of their high content of short wave light. Another prerequisite is a properly designed illuminating system offering the possibility of adjustment according to the Koehler principle. This point has to be emphasized again and again. Only exact Koehler illumination which includes the use of a well corrected microscope condenser, guarantees a really evenly illuminated photograph and a maximum of image contrast, especially when using high power objectives.

Appropriate colour filters are indispensable to increase contrast in black and white photomicrography. In colour photomicrography they are used as conversion filters for matching the colour temperature of the light source to the requirements of the photographic material as well as compensating filters for the correction of possible colour shifts which can occur with long exposure times. Fluorescence photomicrography requires special excitation and barrier filters which are usually provided by the manufacturers.

c. *Long-exposure Photomicrographs*. It is necessary to make a few remarks about the complications occurring with long exposure times.

The density of an exposed photographic emulsion generally depends only on the quantity of light absorbed. For work within the linear range of the characteristic curve (optical density versus log exposure), differing light intensities result in the same density of the exposed emulsion if the product of light intensity and exposure time remains constant. However, this reciprocity law (Bunsen-Roscoe) is valid only within a certain range of values. It no longer holds for relatively low light intensities and correspondingly long exposure times. Then, for example, in order to double the density of an emulsion more than twice the exposure time is necessary. This phenomenon is called the" Schwarzschild effect" or "reciprocity failure". This effect and the resulting prolongation of an already relatively long exposure time, depends on the kind of photographic emulsion used. It leads to serious problems in the assessment of exposure times, particularly with automation of photomicrography.

Some black and white emulsions show a considerable reciprocity failure at exposure times of only 20 seconds. At exposure times of about 5 minutes the prolongation due to failure of reciprocity may be as much as 16-fold. Colour films, especially those of high speed, often show failures of reciprocity of the same order of magnitude at exposure times of only 10 seconds. A further difficulty is that the failure of reciprocity for each of the three emulsions of which a colour film consists may be different. This causes uncontrollable colour shifts as shown in Fig. 1 (p. 392). It is clear from these facts that long exposures, particularly with colour film material, should be avoided. If this is not possible experience is needed with the film material to be used in order to obtain optimum results. Not only because of reciprocity failure, but for general reasons one should restrict oneself to work with a minimum of different film materials whose properties, advantages and drawbacks are familiar. As in conventional photography the question: "Which are the best films?" can only be answered: "Those you work with all the time."

III. The Automation of Photomicrography

A. *The Measurement of Light Intensity*

Reliable determination of exposure times is the basis of photomicrography and therefore also of its automation. This determination is based on a knowledge of illumination intensity in the image plane. The first problem is where the intensity should be measured. Next comes the selection of suitable light detectors and last but not least the question of how to process the information in order to derive either a reading on a scale calibrated in exposure time or an automatically controlled exposure.

a. *Where Should the Light Measurement be Performed?* There are two obvious solutions to this problem: measurement in the image plane or in an area conjugate to it. Illumination intensity measurement in the image plane proper is quite complicated for mechanical reasons and difficulties are encountered due to the excessive size of the required light detectors if an integrated (average) value of the illumination is to be measured. On the other hand if the light detector is placed in an optically conjugate plane it is necessary to place a beam splitter in the light path and to lead a fraction of the light to the detector via an intermediate optical system. By proper design of the magnification of this system the image of the object can be made to match the size of the light detector. The fraction of the light split off and the image size in the measuring plane have to be adapted to the sensitivity of the light detector, since the interaction of these values determines the sensitivity of the entire measuring system. Certain additional light losses have also to be taken into account because in most cases another beam splitter is necessary for observation (see Section IIBa). Most photomicrographic applications call for integrated light measurements since most microscopic specimens, as opposed to those in conventional photography, display only relatively small differences in brightness, i.e. specimen contrast. Yet even specimens with high contrast yield correct exposure time readings if bright and dark specimen details are more or less uniformly distributed. It is different, however, where small specimen details are randomly distributed within a large field differing considerably from them in brightness. For example, in darkfield or fluorescence images with only a few bright details on a dark background exposures based on integrated readings would be too long by a factor depending on the ratio of dark to bright portions of the field. If the light measurement is carried out at the film plane, light detectors with small measuring area but high sensitivity are best suited for "spot scanning". The illumination intensity then determines the correct exposure time for selected image details. A spot measurement under similar conditions can be performed in a conjugate plane by placing a small diaphragm in front of the light detector. In both cases the disturbing influence of the surrounding field is eliminated and only important image details contribute to the measurement.

Despite the advantages of measurement in the image plane or in a conjugate area the light detector can also be placed elsewhere in the light path, depending on the design of the system. There are, however, dangers in placing the detector in the light path between the objective and the eyepiece diaphragm. Figure 2 demonstrates that in this case parts of the light beam (shaded in the drawing) which do not contribute to image formation are included in the exposure time measurement. This can be compensated by suitable calibrations but only if neither objectives nor eyepieces are changed. Such a condition is unacceptable in automatic photomicrography, however.

On the other hand light measurements behind interchangeable eyepieces, close to the exit pupil, as frequently employed with many exposure meters, are only correct if separate calibration is carried out for each eyepiece. If this is not done the diaphragm of all eyepieces used would have to be imaged within the film size, so that the detector does not receive those beams which fall outside the image. In general this is

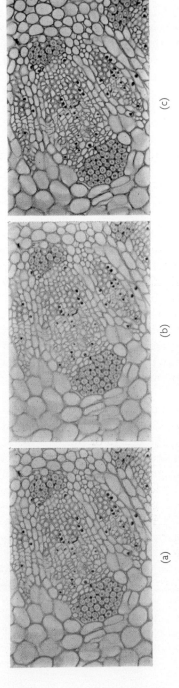

FIG. 1. Influence of reciprocity failure: (a) Image taken with $\frac{1}{8}$ second exposure (measured by exposure meter) on a 100 ASA colour film; (b) Same specimen and same equipment but light reduced by an absolutely neutral grey filter. The calculated exposure time of 15 seconds has been prolonged by a factor of 6 (actual exposure time 90 seconds). Colour shift is obvious. (c) For matching the colour shift a Kodak colour compensating filter CC 30 R was inserted. A further prolongation of exposure time by a factor of 2 up to 180 seconds was necessary. Specimen cross section through stem of *Viscum album* (mistletoe) stained by Fuchsin.

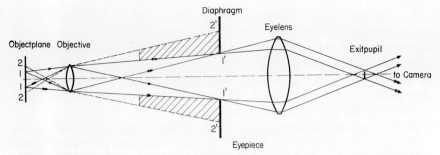

FIG. 2. The objective forms an intermediate image of the object plane on the eyepiece diaphragm. The camera only receives the light coming from an object of diameter 1'—1'. A light detector between objective and eyepiece diaphragm would measure a greater light flux (coming from an object of diameter 2'—2'), a part of which (shaded) is stopped by the diaphragm and does not reach the camera.

not the case; for instance photography with a conventional attachment camera of film size $6\frac{1}{2} \times 9$ cm, an accessory lens f = 200 mm and eyepiece $5\times$, 19 mm field or $10\times$, 14 mm field results in the situation shown in Fig. 3. The image of the $5\times$ eyepiece diaphragm only extends to 15% beyond the film size whereas the corresponding value for the $10\times$ eyepiece is more than 50%! Individual calibration for each eyepiece is therefore essential.

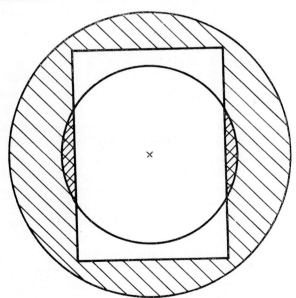

FIG. 3. An image size $6\cdot5 \times 9$ cm is drawn to scale with the images of diaphragms of an eyepiece $5\times$, 19 mm field (inner circle) and an eyepiece $10\times$, 14 mm field (outer circle), respectively. Different proportions of the field are covered.

b. *Accessories for Measurement of Illumination Intensity.* When attempts were first made to determine exposure time it was tempting to measure illumination intensity by means of visual (subjective) photometry. It soon became evident, however, that whatever advantages subjective methods might offer, they would ultimately hinder simple operation and automation. Today photo-electric detectors are used exclusively for exposure time assessment.

It is necessary to distinguish between light detectors utilizing the external photo-electric effect such as photocells and photomultipliers and those utilizing the internal photo-electric effect such as barrier layer cells, photoresistors, photodiodes and phototransistors.

Light detectors based on external photo-electric effect. A photocell consists essentially of a glass bulb containing two electrodes: an anode and a light sensitive photo-cathode. If light of suitable wavelength falls on the cathode, electrons are emitted from its crystal lattice, and are collected by the positively biased anode. The current generated in an external electric circuit (photo-current) depends on the intensity of the incident light and is proportional to it over a certain range. If the glass bulb is evacuated, the unit is called a vacuum photocell. Its sensitivity is not very high; with low light intensities considerable amplification of the photo-current is required. The photo-current can be increased to some extent by filling the bulb with a rare gas. The photoelectrons leaving the cathode collide with gas atoms thereby generating secondary electrons. The amplification obtained in this way should not usually exceed a factor of about 5, or saturation of the system may occur.

The photomultiplier has become increasingly popular due to its very high inherent amplification. The photomultiplier, like the vacuum photocell, contains a photo-cathode as well as several auxiliary anodes or dynodes. These are usually made of antimony-caesium or oxidized silver-magnesium alloys. When the electrons from the photo-cathode strike the first dynode, secondary electrons are emitted. These electrons strike the second dynode and release more secondary electrons. This process is repeated in several steps. By connecting all the dynodes in series amplifications of the order of 10^6—10^9 can be obtained.

Even with the completely unilluminated cathode a movement of electrons towards the anode occurs due to thermionic electron emission. This so-called dark current depends on the material of the cathode, the design of the system etc., and limits the ultimate sensitivity of such a device; for in order to measure minute light intensities at the photo-cathode the final photo-current must be readily distinguishable from the dark current. The minumum detectable light intensity, NEI (noise equivalent input), generating a photo-current equal to the dark current is a useful physical criterion of the sensitivity of a light detecting system.

The spectral sensitivity of the photo-cathode depends on the material used and may have its maximum in the blue or red end of the spectrum. For exposed time determination photomultipliers and photocells with maximum sensitivities in the blue range are used. Their spectral distribution of sensitivity corresponds to that of an unsensitized photoemulsion. In order to use the photo-cathode with normal p\unchro-matic emulsions it is essential to insert suitable light filters in front of the cathode. Naturally this involves a certain loss of sensitivity, but this can generally be tolerated.

Photomultipliers offer the advantages of high sensitivity and considerable linearity. The proportionality between illumination intensity and current extends from dark current to maximum permissible signal current over several decades. For that reason the design of light measuring attachments becomes very simple since no special measures to ensure linearity have to be taken. Because of the high sensitivity of photomultipliers it is as a rule sufficient to deflect a small fraction, say 10%, of the light in the photographic ray path towards the photo-multiplier and still be able to measure specimen details of relatively low intensity.

Apart from the relatively large size of the photomultiplier and its resulting drawbacks the power supply for the unit is complex and expensive. Highly stabilized operating voltages between 1000-2000 V are required. The high amplification factor of the multiplier makes its electrode system vulnerable to destructive overload caused by exposure to high light intensities; protective circuits have to be provided. These requirements and precautions are reflected in the rather high price of photomultiplier equipment.

Light detectors based on internal photo-electric effect. These devices use semiconductors. Photodiodes and phototransistors are rarely used in photomicrography and will not be dealt with here.

The barrier-layer cell, generally a selenium cell, is one of the best known light detectors. Light falling on a selenium layer between a metal plate and a transparent thin metal film causes electrons to move from the selenium to the metal film. This photo-current is reasonably proportional to the illumination intensity. The spectral sensitivity curve has a peak in the green region, similar to that for the human eye, and is easy to match to the sensitivity of panchromatic emulsions. Unfortunately the sensitivity of the selenium cell is low; this restricts its use in photomicrography to simple exposure meters.

Conditions are more favourable with photoresistors. When light falls on certain semiconductors electrons can move from atoms into what is known as the conduction band. This alters the conductivity of the material and if a voltage is applied via two electrodes current can flow through an external circuit. Photoresistors nowadays are constructed from various semiconducting materials of which cadmium sulphide or cadmium selenide are the commonest. The spectral sensitivity curve of a photoresistor can be tailored during manufacture so that it becomes mainly sensitive to visible light and matches well with the sensitivity curve of panchromatic photoemulsions.

Further advantages of the photoresistor are its small size, the ability to withstand excessive illumination for short periods, and to pass fairly high currents with supply voltages of less than 100 V. The nonlinearity of response which has to be considered in the circuit design is a disadvantage. They also have a rather slow response which is particularly noticeable with low light intensities and which limits their application at these levels. This makes it necessary to deflect a larger fraction of the light from the ray path to the photoresistor than is required for a photomultiplier. Many of these problems can be controlled nowadays. Cadmium sulphide photoresistors especially are used to an increasing extent as light detectors for exposure meters, and, less frequently, for automatic exposure control.

B. *Processing of the Data Obtained by Measurement of Illumination Intensity*

One of the main difficulties in measuring illumination intensity is the wide range of intensities occurring in photomicrography. Whereas in conventional photography the usual exposure times range from say 1/1000 sec to 1 sec, exposure times in photomicrography may vary from 1/125 sec to several minutes. This results from variations in light source, microscopic technique, specimen structure, etc. The range of exposure times thus extends over 5 to 6 decades.

The control of such a wide intensity range causes difficulties, especially since the variety of sensitivities of available photographic material, of different film sizes and the special problems inherent in light measurement in the microscope, also have to be considered. Finally the best method of processing the measured information has to be selected. Before considering the problem of automation in detail it is helpful to deal briefly with the methods and instruments used for exposure determination.

a. *Exposure Determination.* For instruments which only serve for exposure time determination, but not for automatic exposure control, simplicity and convenience are of primary importance and photo-multipliers are rarely used.

A desirable feature in all exposure meters is that they should be applicable to any type of microscope or in the form of an integral unit that can be fitted to the instrument.

The simplest version has a selenium barrier-layer cell as light detector, the current from which is directly measured by a sensitive meter. Readings are made on a scale which has to be calibrated in exposure times. Devices of this type are only suitable for relatively intense illumination.

Exposure meters which use photoresistors as light detectors are inherently more sensitive. Many devices of this type are used in

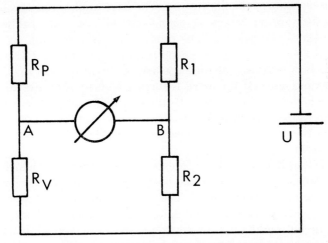

Fig. 4. Photoresistor R_p in a Wheatstone bridge (explanation in text).

conventional photography and have been adapted to the microscope. The non-linear characteristic of the photoresistor has to be taken into account by a corresponding non-linear meter scale. Exposure indication is frequently given by a computing ring which shows the necessary exposure times for different film speeds. The range can be extended by inserting a neutral filter in the front of the photoresistor. The readings have to be adapted to the special conditions in microscopy by means of test exposures.

Another type of photoresistor exposure meter is simple to operate and covers a very wide range of exposure times. This uses the principle of optical compensation in conjunction with a Wheatstone bridge. (Circuit diagram in Fig. 4)

Voltage U is connected to two arms of the bridge, one of which consists of two fixed resistors R_1 and R_2, the other of the photoresistor R_p and a variable resistor R_v. A meter is connected to A and B. For given values of R_1, R_2 and R_v the meter only indicates zero for a well defined illumination intensity at the photoresistor R_p. The basic setting of R_v is chosen so that zero is indicated at the weakest illumination to be measured, which corresponds to a certain point on the characteristic of the photoresistor. In order to read higher illumination intensities, the incident light is attenuated by neutral filters until zero is indicated again on the meter. In practice a series of neutral density filters is provided, the absorption steps of which follow the geometrical sequence 1 : 2 : 4, etc. With this arrangement it is not always possible to get a reading of zero but the meter deflection can be taken as linear with light level over one step of the filter range. The neutral filters can be arranged on a rotating drum which is connected with a scale calibrated in exposure times. To take various film speeds into account, as on conventional exposure meters, a setting ring with ASA and DIN scales is provided. It is rotatable against a mark in the drum and takes up the exposure time scale. If the mark is further split up into several indices, it is possible to take into account other variables such as camera size, camera accessory lenses, position of camera relative to exposure meter, as well as the effects of magnification changes, bellows extensions or beam splitters.

There is another type of photoresistor exposure meter which has no optical compensation and needs no accurate neutral density filters. Electrical compensation is used. The variation of R_p produced by incident light, is compensated by changing the resistance of R_v step by step until the meter reading approaches zero. As in the neutral density filter changer, the variable resistor control is equipped with an exposure time scale, and, if needed, the further refinements described above can be added. The meter can be replaced by other indicator devices such as coloured panel lights to signal zero balance. Modern electronics makes the indication of conditions such as "too much light" or "too little light" easily feasible at little cost.

Exposure meters of these types are commercially available from several microscope manufacturers.

b. *Automation Proper*. The problems of where and how the illumination intensity can be measured in photomicrography and how this measurement can be interpreted in terms of exposure time have now been discussed. The next step is the transmission of this information to a camera shutter, which must be of a type suitable for automation.

Simple attachment cameras almost always have a conventional shutter with exposure times of 1/125 sec to 1 sec, B and T adjustment and usually synchronization for flash photography.

Because of their size these shutters do not work smoothly enough at short exposure times. The inevitable vibrations blur the image at high magnifications under the microscope. Operating such shutters by cable release is very inconvenient and hardly suitable for automation.

There are also cameras with normal central shutters operated by solenoids. The cable opens the shutter end and the magnet holds it open during exposure. An electrical impulse releases the magnet and the shutter closes. To reduce vibration, the whole system must be well sprung but this leads to an undesirable increase in size of the mechanical parts. It was therefore necessary to develop special shutters which could be fully automated and fitted into the ray path of the microscope. Some of these will be discussed briefly.

The shutter is best sited at the exit pupil or its immediate neighbourhood because the cross section of the ray bundles is very narrow there. Shutters so placed can be small and light and as vibration is minimal, short exposure times down to 1/250 sec are possible.

Transmitting the information from the light detector to the shutter and transforming it into exposure times is the next step.

Two principles in automatic exposure control. There are two main procedures: Either the well known principle of capacitor charging in which the exposure time is not actually read, or a newer method in which the exposure time is calculated and can also be read and then fed to a timer, is applied. Most commercial photomicrographic devices use the first principle.

Principle I works as follows:

On operating a switch the shutter opens, light falls onto the photographic emulsion and exposure begins. At the same time part of the light falls on the photocell situated in a plane conjugate to the image and which on being suitable energized, supplies a current proportional to the incident illumination. This current charges a capacitor. Two analogous processes take place: charging of the capacitor and (latent) blackening of the emulsion. Charging continues until there is a certain voltage across the capacitor and the emulsion has undergone the right degree of blackening. The relationship between these two processes is established by calibration. Charging lasts throughout exposure. As soon as a critical voltage has developed across the capacitor, a device such as a cold cathode tube or switching transistor operates a relay circuit which closes the shutter and terminates both exposure and charging of the capacitor.

This method demands strict proportionality between the illumination intensity and photo-current. Photomultipliers are usually preferred because of their high sensitivity. During the exposure two fractions of the available light must be split off, one for specimen observation and one for exposure control. Neither fraction should exceed 20—25%, otherwise the exposure time would be undesirably prolonged. The exposure time can be lengthened or shortened by increasing or reducing the value of the capacitor. Differences in film speed are dealt with in the same way, e.g. by doubling the capacitor value when the emulsion sensitivity decreases by 3 DIN or the ASA value is halved. Factors such as the ratio of the bright to dark area in the image, camera factors, slight over- or underexposure desired, etc., can be taken into account by varying the capacitor.

From this description it can be seen that an important piece of information may be lost—the knowledge of exposure time. Nevertheless for many purposes this is not essential and excellent results can be obtained with routine material. On the other hand there are occasions when a knowledge of actual exposure time is needed in order to obtain the best results.

If specimens move or alter, a short exposure is required. A knowledge of exposure time is also important in very long exposures where reciprocity failure not only extends the basic time but also results in a colour shift (see Fig. 1). In these cases it is essential to make test exposures or to have a timer attached for measurement of the exposure. Only then is it possible to vary lighting and exposure conditions to deal with the situation.

The practical realization of principle II requires a completely different arrangement. One possible system is described here. (Fig. 5)

The required exposure time is calculated at a plane conjugate to the image plane by means of a photoresistor in one arm of a Wheatstone bridge (see Fig. 4). The non-linear characteristic is simulated in a transistorized linearizing component by a series of resistors. These are arranged in such a way as to allow compensation of the photocell current at all levels of illumination. Another transistorized amplifier measures and amplifies the bridge current and translates it into a plus or minus indication on a null indicator (e.g. green panel lights). Independently of this mechanism for measuring and indicating the exposure time, the equipment incorporates a transistorized timer. By means of resistor and capacitor (RC-) combinations, a series of suitably graduated exposure times, mostly in geometrical progression, is "stored" in the timer.

By rotating a control knob the bridge resistors for exposure calculation,

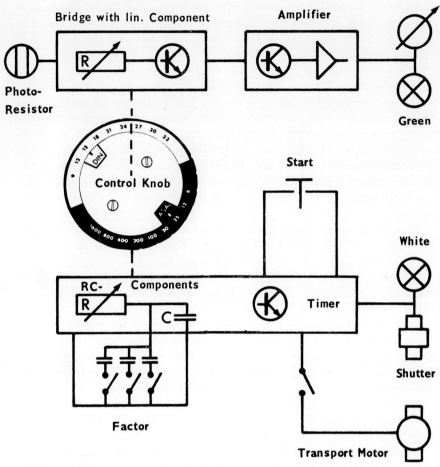

FIG. 5. Principle of exposure time calculation with a photoresistor by means of electrical compensation (explanation in text).

as well as the RC-combinations in the timer can be interconnected by two coupled coaxial rotary switches. This coupling is calibrated in such a way that exposure times which are optimal for a given emulsion correspond to a certain illumination intensity. The speed of the emulsion can be taken into account by changing the coupling setting.

Thus, during the balancing of the signal to zero, two functions are performed; the specimen brightness is determined and the calculated exposure time is indicated. Simultaneously the exposure time, corresponding to this brightness and emulsion speed, is set on the timer. Corrections to the indicated exposure time can still be made after zero balance. It is possible to allow for factors prolonging the exposure time by connecting additional components to the RC-combinations. The starter switch is then operated and the shutter opens for the duration of the set exposure.

A special feature of the system is the photographic shutter located near the exit pupil. For increasing sensitivity, the actual shutter consists of a mirror mounted obliquely in the ray path on an arm connected to the axle of a small motor, which swings the mirror into or out of the ray path. This makes the entire light flux available for intensity measurement as well as for exposure. Only one beam splitter leading to the focusing telescope is necessary. Thus light economy is increased. If the shutter mirror is moved by a magnetic core motor with ironfree armature, the inertia of the moving parts is very low and short exposures are feasible.

Special photomicrographic instruments. These can be divided into automatic and semi-automatic types. Such a classification is of course, arbitrary as the boundaries are hard to define.

In the following a device will be classed as "fully automatic" if exposure time is assessed (and in some cases indicated), the measured information is processed, exposure is controlled and the film is transported after exposure. The latter feature is particularly valuable to the microscopist. In practice, however, this probably means that only devices using 35 mm film can be regarded as truly automatic. These are special instruments of great convenience, but tend to be expensive and do not always have the facility for using other film sizes.

Instruments which offer all the features mentioned above with the exception of automatically controlled film transport will be defined as "semi-automatic".

Fully or partially automatic instruments have been built as attachment cameras, camera microscopes or bellows cameras. Many of the features mentioned above can be found in commercially available instruments. A very extensive survey has been published by Cowen (1969).

Fully automatic photomicrographic instruments as defined above include the "Orthomat" (Leitz GmbH, Wetzlar), the "Photomicroscope" (Zeiss, Oberkochen), the "Photoplan" (Vickers, York), the "Photoautomatic" (Reichert AG, Vienna) and the "Photo Automat" (Wild, Heerbrugg). Film transport in these instruments is achieved by a motor controlled by a contact which closes automatically after exposure.

The "Orthomat" (see Fig. 6) is an attachment camera for 35 mm films and is based on principle I. The beam splitter for observation is not built into the camera but is located in the trinocular body of the microscope.

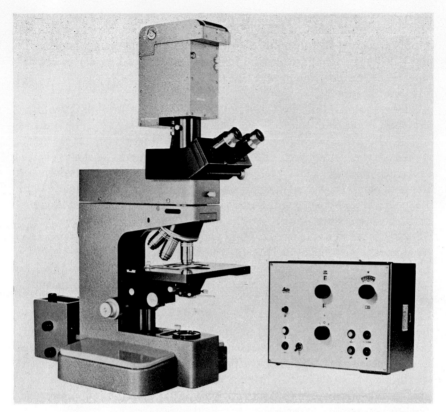

Fig. 6. Automatic 35 mm attachment camera "Orthomat", mounted on a research microscope "Orthoplan", and its control unit. (Courtesy of E. Leitz GmbH, Wetzlar, West Germany.)

This makes binocular observation possible during photography. On the other hand the "Orthomat" cannot be attached to any kind of microscope. All other functional parts, including a $10 \times$ eyepiece, are assembled in the attachment camera which is mounted on the vertical limb of the trinocular tube. The image formed by the microscope is projected onto the film plane by a $0.32 \times$ accessory lens. Between lens and film there is a beam splitter which diverts 10% of the light to the photomultiplier, located on a plane conjugate to the image. In front of the multiplier there is a drawslide containing a rectangular diaphragm to match the picture size, or alternatively a small hole. The rectangular diaphragm serves for integrating, the hole for spot measurements with an area of 1/100 of the normal image field. Above the beam splitter, near the exit pupil of the microscope a special shutter is arranged, the blades of which work as armatures of an electromagnet giving short and vibration free exposures (minimum time 1/200 sec). After exposure an electrical impulse releases the automatic film transport mechanism whereby the film—in its interchangeable film holder—is advanced by the exact distance between one image and the next. The camera is then ready for further exposure.

These various functions are interlocked to ensure an automatic sequence. Signal lamps indicate "shutter open", "film end", etc. Because of the capacitor charging principle, additional functions like "test shot", are necessary, to be able to read the exposure time by a clock. In this case charging takes place with closed shutter and without exposure and film transport. If specimen brightness and film speed are extremely high, the estimated exposure times may be shorter than 1/200 sec. A signal lamp gives warning of this. If the shortest possible exposure time of 1/200 sec is to be used, several blind shots must be made by pushing the "test shot" button and adjusting the illumination in steps until the warning signal no longer lights up. Trial exposures with light intensities varied by neutral density filters are also needed when electronic flash is used.

In the "Photomicroscope II"—also based on principle I (see Fig. 7)—the beam splitter, light detector, electronic shutter and film transport are completely built into the microscope. A photomultiplier serves as light detector and the exposure procedure is similar to that with the "Orthomat". Because of its high sensitivity there is, even at low light levels, sufficient reserve to make not only integrated (average), but also spot measurements (over 5% of the field area).

The new "Photoplan" (Fig. 8) is a particularly versatile instrument built on a modular system which allows a wide variety of attachments to be fitted. These will give any degree of automation required, including automatic transport of 35 mm film. (The automatic 35 mm camera, shutter and photometer can be provided as a separate unit which can be fitted to any microscope.) The 35 mm camera can be replaced by units accepting any format up to 4 in × 5 in and the series of Polaroid backs. A photomultiplier photometer based on principle I (but with a meter indication of exposure in addition) is normally provided but for high sensitivity (e.g. in fluorescence microscopy) or when spot measurements are essential a photomultiplier photometer with a coupled timer (principle II) can be attached. Spot measurements over 1/10, 1/100 or 1/500 of the field area are possible.

All these instruments use photomultipliers but even these sometimes have to work near the limit of their sensitivity so that errors from dark current may arise. Methods of compensating for dark current are known

(a)

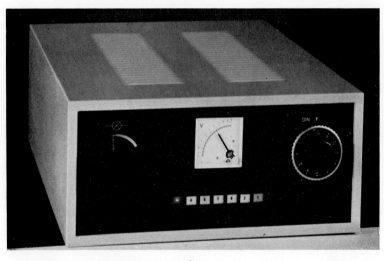

(b)

Fig. 7. Fully automated "Photomicroscope II":
(a) The 35 mm camera is shown on the right of the limb of the microscope;
(b) The control unit. (Courtesy of C. Zeiss, Oberkochen, West Germany.)

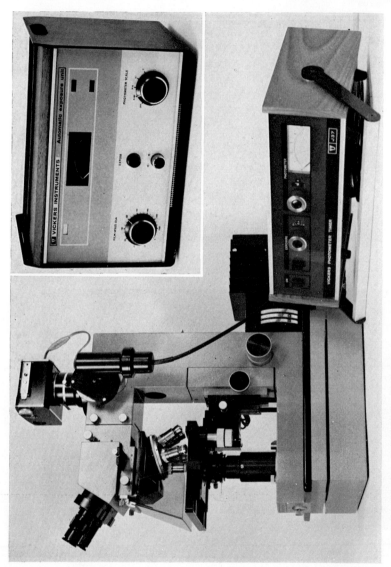

FIG. 8. "Photoplan" microscope showing mechanized 35 mm camera, photomultiplier mounted below camera and high sensitivity partial field measuring photometer timer. The fluorescence version of the microscope is shown. Inset: Control box for alternative automatic exposure type of photometer. (Courtesy of Vickers Instruments, York, England.)

(Frenk, 1960) but for reasons of economy have not been used in commercial instruments.

The "Photoautomatic" (Fig. 9), based on principle II, is another attachment camera which like the "Orthomat" is designed for 35 mm interchangeable film holders. However, a beam-splitter is incorporated,

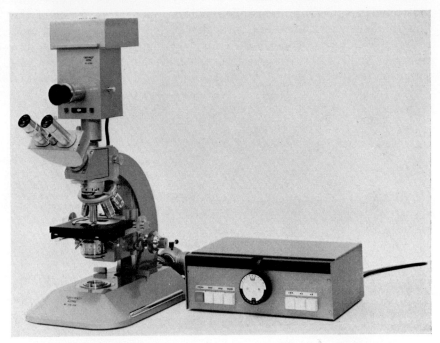

Fig. 9. Automatic 35 mm attachment camera "Photoautomatic", mounted on a research microscope "Zetopan". In front of the camera the viewing telescope and beneath it the signal lights. At the right of the microscope the control unit. (Courtesy of C. Reichert Optische Werke AG, Vienna, Austria.)

leading to a viewing telescope so that it can be used on any microscope. It includes a Zoom eyepiece with magnifications continuously variable between $6\cdot3\times$ and $10\times$. The camera accessory lens has a magnification factor of $0\cdot5\times$. The special shutter which allows a minimum exposure time of 1/250 sec has already been described. Integrating and spot measurements are possible, the latter with 1/64 of the field area. Exposure times can be multiplied by $0\cdot5$, 2 and 4 times, by means of additional keys. Signal lights warn of imperfect attachment or closure of the cassette and also indicate the end of the film. The "Photoautomatic" requires one more manual operation than instruments based on principle

I, namely the setting of the zero signal, however, this is compensated for by the advantages of principle II.

Compensation for reciprocity failure is described in the patents literature (Millendorfer, 1961). As each exposure can be timed, correction can be made either manually or automatically. The RC values that determine exposures in the timer can be increased by switching in capacitors in parallel. If the prolongation factor of the film is known, it can be fed through keys to the control unit and reciprocity failure is compensated for that film by a longer exposure time. The Wild "Photo Automat" (Fig. 10) also achieves partial compensation of reciprocity failure by means of a photoresistor. The curves

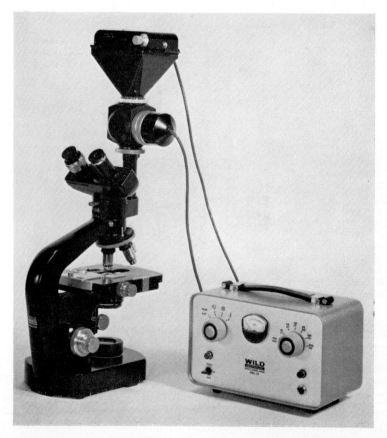

FIG. 10. Attachment camera "M Ka 5", mounted on a laboratory microscope "M 20". Focusing is carried out through the binocular tube. On top of the attachment body the 35 mm camera with motor driven film transport; on the right the light detector (photoresistor) in its housing; at the right of the microscope the control unit. (Courtesy of Wild, Heerbrugg, Switzerland.)

relating light intensity and current for the photoresistor and light intensity and emulsion sensitivity at low illuminations are generally similar. An increase in exposure time required by virtue of reciprocity failure corresponds to a decrease in current resulting from non-linearity of the photoresistor response. The deviations from linearity of both film and photodetector thus tend to compensate. However this compensation is only partial because the exposure prolongation factor varies considerably for different emulsions and the photoresistor characteristic will only be strictly correct for one value of this factor.

This partial compensation also has the disadvantage that if one wants to time the exposure by a clock, one does not read an exactly defined value, but the estimated time multiplied by an unknown factor. Corrections by multiplying factors given by the film manufacturers are hard to apply and test exposures are needed.

The fully automatic versions of all these instruments use motorized 35 mm cameras. Some are available in semi-automatic forms which may be combined with attachment cameras, fully integrated camera microscopes or bellows cameras and are based on either principle I or II. Some manufacturers only produce semi-automatic devices at present (e.g. Zeiss Jena, Nikon, Union).

The simplest version is a camera attachment system with the function of exposure control and shutter on which cameras for various film sizes can be mounted, all using the same accessory lens. The image covers the largest format, while all the other cameras show only sections of this image, depending on their size. The magnification at the film plane remains constant, resulting in both optical and electrical simplification. However some advantages of small film size—higher brightness at the film and shorter exposure time—are lost.

Other semi-automatic instruments, also based on principle I, can be used with various cameras, each equipped with a proper accessory lens matching the film size. Not only different film speeds but also the magnifications which differ for each camera, can be taken into account by varying the capacitor values.

A simple but versatile version of the semi-automatic exposure control based on system II has been put into practice in the camera attachment "Kam ES" (Reichert, Vienna, Fig. 11). Here, too, the beam splitter for observation as well as the special shutter with the light meter are mounted together. Cameras (even cine cameras) with individually chosen accessory lenses can be attached. Exposure timing and shutter setting are carried out as described for the "Photoautomatic" by the same manufacturer. The only difference lies in the control dial that switches the bridge resistors of the exposure meter; it does not have a

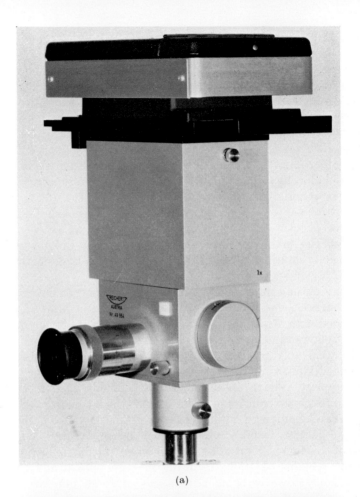

(a)

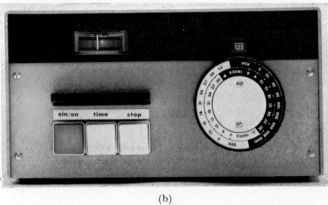

(b)

Fig. 11. (a) Attachment camera "Kam ES" equipped with Polaroid sheet film holder $3\frac{1}{4} \times 4\frac{1}{4}$ in. In front the viewing telescope; on the right the light detector (photoresistor). (b) The control unit; on the right the control knob and above, the window in which the exposure time can be read. On the upper left the zero signal.

(Courtesy of C. Reichert Optische Werke AG, Vienna, Austria.)

single index mark, but is divided into a series of settings to allow for various film sizes (Fig. 11b). By setting the film speed to the corresponding setting, the equipment automatically compensates for the change in magnification. The exposure time is also indicated, and can be adjusted as already described.

This system, unlike those based on principle I, can easily be adapted to micro-cinematography. As is well known, for a standard cine camera the exposure time of each single image is given by the frame frequency and the sector opening, e.g.: 1/50 sec at frame frequency 24/sec. The illumination must therefore be matched to the exposure time and not vice versa as is more usual in photomicrography. Having shutters of their own, cine cameras do not need the special electrically operated shutter provided, which remains open. The light is no longer measured by the fully reflecting shutter mirror but by a partially transmitting cine mirror, provided for the purpose. It is placed in the ray path, just before the shutter mirror. This cine mirror reflects a small fraction of the light to the photoresistor which is adequate for exposure measurement in micro-cinematography.

An index mark is provided on the control dial for this kind of timing. After setting the correct exposure time the illumination must be varied until the zero light flashes and shows the correct brightness. This method fulfils two important conditions for micro-cine work: constant checking of image focus and framing by means of a telescope, as well as correct exposure indication by means of the zero signal lamp.

If changes in image brightness occur during filming, they can be noted and corrected by neutral wedges or polarizing filters. This correction can be made automatically (Kropp and Millendorfer, 1961) by fitting a light controlling device between the lamp and microscope, e.g. a neutral wedge driven by a servomotor. The illumination is first set to optimum brightness, and is kept at this level by the servomotor which is controlled by a servoamplifier that responds to changes in signal from the photoresistor.

One more semi-automatic device, the "Bellows Camera 4 × 5 in" of Leitz, Wetzlar (Fig. 12), will be briefly discussed as an example of an arrangement restricted to large formats. The system is again based on principle I. A sensitive photomultiplier is needed as light detector as illumination is low because of the large size. The photomultiplier is located in the image plane, just out of the field of view, at the side of the focusing screen of a reflex attachment. The light sensitive area of the multiplier is rather small compared to the field size, so that spot measurements are made. Because of the specially arranged ray path of this device which is vertically attached to the microscope

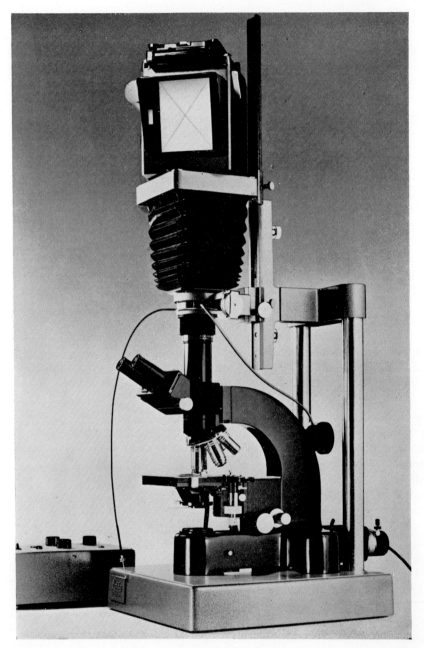

Fig. 12. Large size bellows camera with automatic exposure control mounted on a research microscope "Ortholux". (Courtesy of E. Leitz GmbH, Wetzlar, West Germany.)

(and can also be used for photomacrography) an electro-mechanical shutter with large aperture is used. The shutter is well sprung to reduce vibrations. A wire cable starts the exposure, capacitor charging determines the exposure time during which the shutter opens and a solenoid closes it after exposure. The shortest automatically controlled exposure time is 1/15 sec, which is sufficient for large size exposures.

The semi-automatic exposure control unit of the "Ultraphot 2" (Zeiss, Oberkochen) works in much the same way but is fitted into the microscope.

From the descriptions given, one can see that some so-called semi-automatic photomicrographic equipment is extremely versatile in application but possibly more difficult to operate than fully automated apparatus. It tends to require more concentration and attention on the part of the observer.

REFERENCES

Allen, Roy M., (1958). *In* "Photomicrography", second ed. D. van Nostrand Co., New York.

Bergner, J., Gelbke, E. and Mehliss, W., (1966). *In* "Practical Photomicrography". The Focal Press, London and New York.

Bringdahl, O., (1966). *J. Opt. Soc. Am.* **56,** 811.

Claussen, H. C., (1967). *In* "Mikroskope, Encyclopedia of Physics". (ed. S. Flugge.) Vol. XXIX "Optical Instruments". Springer, Berlin, Heidelberg, New York.

Cowen, B. C., (1969). *Proc. R. microsc. Soc.*, **4,** Pt. 2, 71.

Frenk, H., (1960). *German Patent* 1,167.556.

Kropp, K. and Millendorfer, H., (1961). *Austrian Patent* 223391.

Lau, E., (1961). *Naturwissensch. Rundschau*, **14,** 156.

Lau, E., Schuller, A. and Roose, G., (1960). *Geologie*, **9,** 426.

Michel, K., (1967). "Die Mikrophotographie", 3. Aufl., Springer, Wien, New York.

Millendorfer, H., (1961). *Austrian Patent* 235141.

Author Index

A

Abalmazova, M. G., 226, *260*
Abramowitz, M., 192, *259*
Albert, L., 36, 81, *82*
Allen, Roy M., 385, *413*
Allmand, T. R., 376, 378, 380, *382*
Altman, J. H., 136, 138, *158*
Amor, F. A., 381, *382*
Archard, G. D., 33, *82*, 238, *259*, 267, *359*
Ash, E. A., 256, 257, *261*

B

Barnett, M. E., 164, 176, 249, 251, 253, 254, 256, 257, 258, *259*, *261*
Bartz, G., 164, 212, 237, 241, 244, *259*, *260*
Bates, C. W., 257, *259*
Becker, S., 362, *382*
Berkeley, C., 362, *383*
Berger, J. E., 105, *119*
Bergner, J., 385, *413*
Bethge, H., 164, 205, 237, 238, *260*
Bialas, H., 205, *260*
Blackett, J. H., 362, *382*
Blank, J. R., 376, 378, 380, *382*
Block, M., 381, *382*
Boersch, H., 2, 14, 80, *82*
Bok, A. B., 164, 215, 227, 238, *260*
Born, M., 8, 25, 80, *82*
Borries, B. von, 81, *82*, 163, *261*
Bostanjoglo, O., 248, *260*
Bostrom, R. C., 362, *382*
Bringdahl, O., 386, *413*

C

Castaing, R., 302, 342, 345, 353, *359*
Chadwick, G. A., 336, 337, 338, 339, 340, 358, *359*, *360*
Challis, L. J., 248, *261*
Chao, S. C., 150, *158*
Claussen, H. C., 385, *413*

Coan

Coan, M. G., 81, *84*
Cole, M., 373, 378, *382*
Considine, K. T., 287, 288, 295, 299, 300, 301, 321, 324, 325, *359*, *360*
Cook, R. F., 267, 335, *359*, *360*
Cosslett, V. E., 324, *360*
Cotte, M., 305, 307, *359*
Cowen, B. C., 402, *413*
Cowley, J. M., 14, *82*, 116, *119*
Cundy, S. L., 319, 321, 332, 335, 336, 345, 353, 358, *359*, *360*

D

Deeter, C. R., 226, *260*
Dell, H. A., 362, *382*
De Rosier, D., 105, *119*
Dietrich, W., 267, *359*
Dowell, W. C. T., 67, 81, *82*
Dudley, R. A., 362, *382*
Düker, H., 80, *84*

E

Eberhardt, E. H., 148, *158*
Edington, J. W., 334, 335, *360*
Efron, E., 156, *158*
Eisenhandler, C. B., 81, *82*, *84*
England, L., 257, *259*
Evans, D. H., 378, *382*
Eyer, J. A., 136, *158*

F

Faget, J., 81, *82*
Fagot, M., 81, *82*
Farrant, J. L., 81, *82*
Fernandez-Moran, H., 105, *119*
Ferré, J., 81, *82*
Ferrell, R. A., 336, *359*
Fert, Ch., 81, *82*
Fischer, H., 81, *82*

Fisher, C., 373, 378, 381, *382*
Fitzsimmons, A. E., 362, *382*
Flory, L. E., 362, *382*
Forst, G., 164, *260*
Franceschini, J. B., 362, *382*
Frenk, H., 407, *413*
Friedman, H. D., 140, *158*
Fujimoto, F., 345, *359*
Fukuhara, A., 345, *359*

G

Gabor, D., 60, 62, 81, *82, 83*
Garrood, J. R., 203, 249, *260*
Gauthe, B., 319, *359*
Gelbke, E., 385, *413*
Glaser, W., 4, 9, 68, 81, *83*, 168, *260*

H

Hacking, K., 136, *158*
Hahn, M., 81, *83*
Haine, M. E., 81, *83*, 211, *260*
Hallen, O., 362, *382*
Hamming, R. W., 81, *83*
Hansen, M. H., 334, *359*
Hanszen, K.-J., 4, 6, 18, 21, 25, 35, 36, 37, 39, 42, 43, 45, 46, 47, 48, 49, 51, 52, 56, 59, 60, 61, 62, 63, 64, 66, 67, 71, 72, 74, 75, 76, 78, 80, 81, *83*
Harger, R. O., 150, *158*
Hauser, H., 10, 19, 22, 43, 80, *83*
Hawksley, P. G. W., 362, *382*
Healey, P., 381, *383*
Heidenreich, R. D., 81, *83*
Hellgardt, J., 164, 237, 238, *260*
Henneberg, W., 163, *260*
Hennequin, J. F., 302, *359*
Henoc, P., 353, *359*
Henry, L., 302, 342, 344, 345, 353, *359*
Herd, H. H., 138, 145, *159*
Heydenreich, J., 164, 237, 238, 242, *260*
Heyning, J. M., 136, *158*
Higgins, G. C., 136, *159*
Hili, Ali El, 345, 351, 353, 357, *359*
Hillier, J., 318, *359*
Hobbs, D. S., 362, *382*
Hofer, F., 378, *382*
Hopkins, H. H., 10, *83*

Hopp, H., 164, *260*
Hoppe, W., 43, 44, 59, 81, *83, 84*
Hottenroth, G., 163, 237, *260*
Howie, A., 332, 335, 345, 353, *359, 360*
Huang, T. S., 129, 136, *158*
Hyden, H., 362, *382*

I

Ichinokawa, T., 287, 298, *359*
Igras, E., 226, 247, *260*
Ivanov, R. D., 226, *260*

J

Jamieson, R. M., 380, *382*
Jansen, S. Z., 163, *261*
Jeschke, W., 14, *82, 83*
Jesse, A., 378, 380, *382*
Johnson, A. R., 378, *383*
Jones, R. C., 136, *158*

K

Kainuma, Y., 345, *359*
Kamiya, Y., 287, 332, 345, *359*
Katerbau, K.-H., 81, *84*
Kiselev, D. J., 105, *119*
Kleinn, W., 319, *360*
Klemperer, O., 266, *360*
Klug, A., 105, *119*
Knoll, M., 162, *260*
Komoda, T., 15, 67, 73, 74, 81, *83*
Kramer, J., 164, 227, *260*
Kranz, J., 205, *260*
Krimmel, E., 165, *260*
Kropp, K., 411, *413*
Kunath, W., 36, *83*

L

Labaw, Louis W., 105, *119*
Lagercrantz, C., 362, *382*
Langer, R., 43, 44, 81, *83, 84*
Langhoff, R. R., 378, *383*
Larsson, S., 362, *383*
Lau, E., 387, *413*
Laudet, M., 267, *360*

Lauer, R., 4, 81, *83*
Le Bouffant, L., 362, *382*
Leder, L. B., 319, *360*
Leithäuser, E., 266, *360*
Lenz, F., 25, 34, 59, 60, 80, 81, *82, 83,*
    165, *260,* 267, *360*
Leonhard, F., 319, *360*
Le Poole, J. B., 164, 207, 214, 227, *260*
Le Rütte, W. S., 231, *260*
Levi, L., 136, *158*
Leybourne, J. J., 138, 145, *159*
Lindon, P. H., 378, *383*
Linfoot, E. H., 26, 77, *83*
Lippert, W., 267, *360*
Little, W. A., 248, *261*
Lohmann, A., 12, *83*
Longley, W., 105, *120*
Loveridge, B. A., 379, 382, *383*
Lukjanow, A. E., 249, *260*

M

MacAdam, D. L., 136, *159*
McInnes, C. A. J., 379, 382, *383*
McLean, J. D., 81, *82*
Maffitt, K. N., 226, *260*
Mansberg, H. P., 362, *383*
Martennson, H., 379, *383*
Marton, L., 318, 319, *360*
Masygan, R. J., 380, *382*
Mawdesley-Thomas, L. E., 381, *383*
Mayer, L., 164, 212, 226, 237, 238, 244,
    247, *260*
Mehliss, W., 385, *413*
Melford, D. A., 379, *383*
Menzel, E., 19, 22, 80, *83, 84*
Metherell, A. J. F., 267, 269, 280, 319,
    332, 333, 335, 336, 345, 353, 358, *359,*
    360
Meyer, E. W., 362, *382, 383*
Michel, K., 385, 386, *413*
Millendorfer, H., 408, 411, *413*
Möllenstedt, G., 80, 81, *84,* 267, *360*
Moody, M. F., 116, *120*
Morgan, B. B., 362, *383*
Morgenstern, B., 15, 16, 18, 21, 25,
    39, 42, 46, 59, 60, 61, 66, 67, 71, 72,
    74, 76, 80, 81, *83, 84*
Mulvey, T., 81, *83,* 238, *259*

N

Nagandra Nath, N. S., 19, *84*
Nathan, R., 116, *120*
Natr, L., 381, *383*
Natta, M., 353, *359*
Nazareth, L. J., 378, *382*
Nicholson, R. B., 336, *359*
Niedrig, H., 14, *83*
Nixon, W. C., 164, 176, 203, 249, 251,
    253, 254, 257, 258, *259, 260*

O

Ohennasian, C. E., 380, *382*
O'Neill, E., 32, 35, *84,* 131, 138, 139,
    158
Ondracek, G., 378, 380, *382*
Oppenheim, A. V., 138, *158*
Orthuber, R., 164, 237, 244, *261*
Otsuki, M., 67, *83*

P

Paras, N., 302, *360*
Pelc, S. R., 362, *382*
Perrin, F. H., 130, *158*
Pyke, W. E., 362, *382*

R

Radtke, D., 380, *383*
Raether, H., 335, 340, *360*
Raimes, S., 334, *360*
Raith, H., 14, *82*
Recknagel, A., 163, *261*
Reimer, L., 81, *84*
Richards, M. S., 362, *382*
Rickett, R., 238, *260*
Riecke, W. D., 36, *83*
Roberts, F., 362, *383*
Roche, R., 379, *383*
Röhler, R., 26, 31, 32, 68, 70, 80, *84*
Rosenbruch, K. J., 18, 25, 59, 71, 72,
    76, 81, *83*
Ruska, E., 163, *261*

S

Sadashige, K., *158*
Sawyer, H. S., 362, *382*
Schade, Otto H. Sr., 136, 151, *158*

Schafer, R. W., 138, *158*
Scheffels, W., 81, *83*
Scherzer, O., 68, 81, *84*
Schneider, R., 81, *82*
Schreiber, D., 380, *383*
Schwartz, M., 143, *159*
Schwartze, W., 164, 165, 212, 237, 238, 241, *261*
Septier, A., 267, *360*
Shannon, C. E., *84*
Shelton, C. F., 138, 145, *159*
Siegel, B. M., 81, *82, 84*
Siegel, G., 248, *260*
Smith, K. C. A., 287, 288, 300, 324, 325, 334, *359, 360*
Smith, M. J., 379, *383*
Soule, J. L., 362, *382*
Spalding, D. R., 330, 333, 334, 335, 336, 337, 338, 339, 340, 353, 358, *360*
Speidel, R., 81, *84*
Spivak, G. V., 164, 226, 237, 244, 247, 249, *260, 261*
Stegun, I. A., 192, *259*
Stenemann, H., 238, *260*
Stockham, T. G., 138, *158*
Strang, A., 379, 381, *383*
Stultz, K. F., 136, *159*
Sunder-Plasmann, F. A., 76, 81, *83*
Szentesi, O. I., 249, 256, 257, *261*

T

Taylor, W. K., 362, *383*
Thon, F., 51, 54, 56, 57, 58, 59, 61, 71, 79, 81, *84*
Tolles, W. E., 362, *382*
Trepte, L., 52, 66, 67, 75, 76, *83*

U

Unwin, P. N. T., 336, *359*
Uyeda, R., 81, *84*, 332, 340, 345, *359, 360*

V

Valdre, U., 332, 345, 353, *359*
Van der Ziel, 143, *159*
Villagrana, R. E., 334, 335, 336, 337, 338, 339, 340, 358, *360*
Vorobev, Y. V., 81, *84*
Vyazigin, A. A., 81, *84*

W

Walli, C. R., 140, *159*
Wang, S. T., 248, *261*
Warminski, T., 226, *260*
Watanabe, H., 319, 340, 345, 358, *360*
Waters, W. E., 267, *360*
Wegener, H., 12, *83*
Weimer, P. K., 145, *159*
Weissenberg, G., 237, 241, 244, *259, 260*
Wende, B., 164, *260*
Whelan, M. J., 267, 269, 280, 319, 332, 335, 336, 345, 353, 358, *359, 360*
Wien, W., 266, *360*
Wilska, A. P., 76, *84*
Wiskott, D., 189, 237, 241, *260, 261*
Wolf, E., 8, 25, 80, *82*

Y

Yamagami, Y., 362, *383*
Young, J. Z., 362, *383*

Z

Zernike, F., 80, *84*
Zweig, H. J., 136, *159*

Subject Index

A

Aberration, spherical, 86

Abrikosov-Vortex structure, imaging, 248

Alignment, mirror EM, 215–217

Alloys
  energy selected images, 356, 357, 358,
  loss spectrum of electrons, 331, 333,
    334, 337, 338–340

Aluminium
  energy selected images, 346–350, 353
  loss spectrum of electrons, 265, 328,
    336, 332, 333

Analog video data, conversion to digital,
  88

Aperture contrast, 71–74

Attenuation correction, high frequency, 101–102

B

Biomedical uses, image analysing computer, 381

Botany, use of image analysing computer, 381

Bunsen-Roscoe law, photomicrography
  389

C

Cadmium sulphide photoresistors, 396

Cameras
  electronic, digital image processing,
    145–149
  mirror EM, 214
  photomicrography, 387–389
  television, 88–105

Catalase, image enhancement, 105–112

Ceramics, use of image analysing computer, 380

Cervical smears, pre-screening, 362

Cinematography, micro-, 411

Circuits, solid state, imaging, 248

Cobalt crystal, mirror EM image, 257

Coherent optical system, image processing, 129

Computers
  digital image processing, 153–158
  high resolution by, 113–119
  image analysing, 361–383
  image processing, enhancement, 87–
    102
    geometric correction, 89
    high frequency attenuation cor-
      rection, 101–102
    photometric correction, 89
    random noise removal, 95–97
    scan-line noise removal, 101
    system noise removal, 97–101

Contrast
  aperture, 71–74
  and high resolution, 86–87
  mirror EM, 167–207
  by phase shift, 46–67, 81
    axial illumination, 46–64
    oblique illumination, 65–67
  transfer, 75–79

Crystals, lattice spacings, 81

Cylindrical electrostatic analyser, 267–
  287, 319–324, 340–342

Cylindrical magnetic analyser, 287–302,
  324–327

Cytoanalyser, 362

D

Damage to specimens, and high resolution, 87

Deflection bridge, mirror EM, 212–214

Detector noise, digital image processing,
  141–145

Diffraction
  electron, 113–114
  X-ray, 113, 115, 118

Digital system, image processing, 127–
  159
  computer system, 153–158
  data recording, 149–153

Digital system—*continued*
  electronic cameras, 145–149
  quantization, 135–145
  resolution, 130–135
Digital tapes, conversion to visual presentation, 89

E

Electrical supplies, mirror EM, 214
Electrode, interdigital, image, 255, 266
Electron diffraction, 113–114
Emission electron microscope, 162–163
Energy analysing and selecting microscopes, 263–360
  cylindrical electrostatic analyser, 267–287
  cylindrical magnetic analyser, 287–302
  energy analysing microscopes, 318–340
  energy selecting microscopes, 340–358
  mirror prism device, 302–318
Energy loss spectrum, 264–266, 329–340, 353–358
Exposure, photomicrography
  determination, 396–402
  long, 389–391, 400

F

Ferro-electric domains, use of mirror EM, 226, 244
Fibres, use of image analysing computer, 381
Film
  grain, digital image processing, 138–139
  and long exposure, 389–391
Flying-spot scanner, 362

G

Geometric correction, computer image processing, 89
Germanium, dislocations, mirror EM image, 247
Gold, energy selected images, 351–352, 353

H

Hexamethylbenzene, electron density map, 115, 118
Holograms, ultrasonic, 258

I

Illumination
  contrast transfer, 75–79
  in mirror EM, 207–210
  optical transfer theory, 7–9
  photomicrography, 389
Image
  formation, 1–3
  image analysing computer, 361–383
  processing, digital system, 127–159
    computer system, 153–158
    data recording, 149–153
    electronic cameras, 145–149
    quantization, 135–145
    resolution, 130–135
  processing, enhancement, 85–125
    computer system, 87–102
    high resolution, 86–87
    periodic images, 102–112
    resolution by computer synthesis, 113–119
    sensor noise, 87
Imaging optics, photomicrography, 388–389
Imaging system, mirror EM, 210–211
Inclusions in steel, image analysis, 364–368
Indanthrene, diffraction pattern, 119
Inhomogeneities, use of mirror EM
  electrical surface, 242–246
  geometric, 241–242
  magnetic, 247
Instrument instability, 86
Interdigital electrode, mirror EM image, 255, 266

L

Lens, projector, mirror EM, 214
Light intensity, measurement, 391–396
  data processing, 396–413
Local work function, mirror EM, 226–227

M

Magnesium oxide, energy selected images, 355, 358
Magnetic contrast, mirror EM, 258–259
Magnetic domain patterns, mirror EM, 226

Magnetic flux, superconductors, imaging, 248
Metals, diffusion, mirror EM, 226
Metallurgy, and image analysing computer, 364–368, 379–380
Microcircuits, use of mirror EM, 226
Microfields, semi-conducting surfaces, imaging, 248
Mineralogy, and image analysing computer, 380
Mirror electron microscopy, 161–260
    applications, 226–227
    without beam deflection, 238, 239
    contrast formation, 167–207
    description, 207–217
    with magnetic prism, 238, 240, 241
    practice, 237–249
    results, 218–226
    scanning mirror EM, 227–232
    shadow projection mirror EM, 249–259
    specimen perturbation, 233–7
Mirror prism, for energy analysis, 302–318, 342–345

N

Negatives, use of image analysing computer, 382
Noise
    and computer image processing
        random, 95–97
        scan-line, 101
        system, 97–101
    digital image processing, 135–145
        detector noise, 141–145
        digital step distribution, 136–138
        film grain, 138–139
        quantization noise, 140–141
        system noise, 141
        image sensor, 87
Nuclear track scanner, 362

O

Object, optical transfer theory, 9–23
Optical bench simulation, 115–117
Optical system, coherent, image processing, 129
Optical transfer theory, 1–84
    aperture contrast, 71–74
    contrast by phase shift, 46–67

Optical transfer theory—continued
    history, 80–82
    illumination, 7–9
    image formation, 1–3
    object, 9–33
    optical image, formation, 23–45
    partially coherent or incoherent illumination, 75–79
    point resolution, 67–71
    symbols and definitions, 3–7
    test objects, 79–80
Optics, imaging, photomicrography, 388–389

P

Particle mensuration, 362
Particle physics, and image analysing computer, 381-382
Petrology, and image analysing computer, 381
Photograph scanner, 362
Photographs, use of image analysing computer, 382
Photomicrography, 385–413
    automation, 391–413
        data processing, 396–413
        light intensity, 391–396
        special instruments, 402–413
    instruments, 387–388
    range, 386–387
    technical details, 388–391
Photomultipliers, 394–395
Photomultiplier tube, noise, 141–145
Photoresistors, 396, 397–398, 400–401
p-n junctions, use of mirror EM, 244
Point resolution, 67–71
Projector lens, mirror EM, 214

Q

Quantimet image analysing computer, 361–383
Quantization, digital image processing, 135–145
    detector noise, 141–145
    digital step distribution, 136–138
    film grain, 138–139
    quantization noise, 140–141
    system noise, 141

R

Recording of data, digital image processing, 149–153
  conversion to output, 150–153
  display, 153
  electronic recording, 149–150
Reflection electron microscope, 163
Resolution
  by computer synthesis, 113–119
  digital image processing, 130–135
  high, problems, 86–87
  point, 67–71
Rock salt, mirror EM image, 204, 252, 253

S

Scanning electron microscope, 162, 244
Scanning mirror EM, 249
  with magnetic quadrupoles, 227–232
Scanning, in image analysis, 362–363
Selenium cells, 395, 397
Shadow projection mirror EM, 249–259
Sodium chloride, mirror EM image, 242–243
Soils, use of image analysing computer, 380
Solid state physics, use of mirror EM, 247–248
Sorption phenomena, use of mirror EM, 227
Space mission images
  Mariner, 92, 93, 94, 96
  Ranger, 88, 90, 91, 95, 97, 98, 99
  Surveyor, 104, 105

Specimen stage, mirror EM, 211–212
Specimen damage, and high resolution, 87
Specimen perturbation, and reflected electron beam phase, 233–237
Steel inclusions, and image analysing computer, 364–368
Stroboscopic mirror EM, 249, 256, 257
Surface conductivity, use of mirror EM, 226

T

Television cameras, 145–148
Test objects, for contrast transfer properties, 79–80
Thin film electronic devices, imaging, 248
Thin films, use of mirror EM, 226
Threshold setting errors, image analysing computer, 375–378

U

Ultrasonic holograms, use of mirror EM, 258

V

Vacuum system, mirror EM, 215

W

Wide track scanner, 362

X

X-ray diffraction, 113, 115, 118

Cumulative Index of Authors

Allen, R. D., **1**, 77

Barnett, M. E., **4**, 161

Beadle, C., **4**, 361

Benford, J. R., **3**, 1

Bethge, H., **4**, 161

Billingsley, F. C., **4**, 127

Bok, A. B., **4**, 161

Bostrom, R. C., **2**, 77

Brandon, D. G., **2**, 343

Brault, J. W., **1**, 77

Dupouy, G., **2**, 167

Ferrier, R. P., **3**, 155

Gabler, F., **4**, 385

Hanszen, K. J., **4**, 1

Heydenreich, J., **4**, 161

Holcomb, W, G., **2**, 77

Humphries, D. W., **3**, 33

Krop, K., **4**, 385

de Lang, H., **4**, 161

Marinozzi, V., **2**, 251

Mayall, B. H., **2**, 77

Mendelsohn, M. L., **2**, 77

Metherell, A. J. F., **4**, 263

Millonig, G., **2**, 251

Monro, P. A. G., **1**, 1

Nathan, R., **4**, 85

Pelc, S. R., **2**, 151

le Poole, J. B., **4**, 161

Prewitt, J. M. S., **2**, 77

Reynolds, G. T., **2**, 1

Roos, H., **4**, 161

Rosenberger, H. E., **3**, 1

Ruska, E., **1**, 115

Saylor, C. P., **1**, 41

Septier, A., **1**, 204

Thomas, R. S., **3**, 99

Valentine, R. C., **1**, 180

Welford, W. T., **2**, 41

Welton, M. G. E., **2**, 151

Williams, M. A., **3**, 219

Zeh, R. M., **2**, 77

Cumulative Index of Titles

Accurate microscopical determination of optical properties on one small crystal, **1**, 41

Autoradiography and the photographic process, **2**, 151

Digital transformation and computer analysis of microscopic images, **2**, 77

Electron microscopy at very high voltages, **2**, 167

Electron microscopy, fixation and embedding, **2**, 251

Energy analysing and energy selecting electron microscopes, **4**, 263

Field-ion microscopy, **2**, 343

Fixation and embedding in electron microscopy, **2**, 251

Image contrast and phase-modulated light methods, polarization and interference microscopy, **1**, 77

Image intensification applied to microscope systems, **2**, 1

Image processing for electron microscopy
I. Enhancement procedures, **4**, 85
II. A digital system, **4**, 127

Interference microscopy: image contrast and phase-modulated light methods in polarization, **1**, 77

Mach effect and the microscope, **2**, 41

Mensuration methods in optical microscopy, **3**, 33

Methods for measuring the velocity of moving particles under the microscope, **1**, 1

Microincineration techniques for electron-microscopic localization of biological minerals, **3**, 99

Microscope and the Mach effect, **2**, 41

Microscope systems: image intensification, **2**, 1

Mirror electron microscopy, **4**, 161

One small crystal, accurate microscopical determination of optical properties, **1**, 41

Optical transfer theory of the electron microscope; fundamental principles and applications, **4**, 1

Past and present attempts to attain the resolution limit of the transmission electron microscope, **1**, 115

Phase-modulated light methods and image contrast, in polarization and interference microscopy, **1**, 77

Photographic emulsions and their response to electrons, **1**, 180

Photographic process: autoradiography, **2**, 151

Photomicrography and its automation, **4**, 385

Polarization and interference microscopy, **1**, 77

Quantimet image analysing computer and its applications, **4**, 361

Response of photographic emulsions to electrons, **1**, 180

Resolution limit of the transmission electron microscope, **1**, 115

Small angle electron diffraction in the electron microscope, **3**, 155

Spherical aberration in electron optics, **1**, 204

The assessment of electron microscopic autoradiographs, **3**, 219

Transmission electron microscope, past and present attempts to attain its resolution limit, **1**, 115

Zoom systems in microscopy, **3**, 1